Molecular Aspects of Biotechnology: Computational Models and Theories

NATO ASI Series

Advanced Science Institutes Series

A Series presenting the results of activities sponsored by the NATO Science Committee, which aims at the dissemination of advanced scientific and technological knowledge, with a view to strengthening links between scientific communities.

The Series is published by an international board of publishers in conjunction with the NATO Scientific Affairs Division

A Life Sciences	Plenum Publishing Corporation
B Physics	London and New York
C Mathematical	Kluwer Academic Publishers
and Physical Sciences	Dordrecht, Boston and London
D Behavioural and Social Sciences	
E Applied Sciences	
F Computer and Systems Sciences	Springer-Verlag
G Ecological Sciences	Berlin, Heidelberg, New York, London,
H Cell Biology	Paris and Tokyo
I Global Environmental Change	

NATO-PCO-DATA BASE

The electronic index to the NATO ASI Series provides full bibliographical references (with keywords and/or abstracts) to more than 30000 contributions from international scientists published in all sections of the NATO ASI Series.
Access to the NATO-PCO-DATA BASE is possible in two ways:

– via online FILE 128 (NATO-PCO-DATA BASE) hosted by ESRIN,
Via Galileo Galilei, I-00044 Frascati, Italy.

– via CD-ROM "NATO-PCO-DATA BASE" with user-friendly retrieval software in English, French and German (© WTV GmbH and DATAWARE Technologies Inc. 1989).

The CD-ROM can be ordered through any member of the Board of Publishers or through NATO-PCO, Overijse, Belgium.

Series C: Mathematical and Physical Sciences - Vol. 368

Molecular Aspects of Biotechnology: Computational Models and Theories

edited by

J. Bertrán

Department of Chemistry,
Universitat Autònoma de Barcelona,
Bellaterra (Barcelona), Spain

Springer-Science+Business Media, B.V.

Published in cooperation with NATO Scientific Affairs Division

Proceedings of the NATO Advanced Research Workshop on
The Role of Computational Models and Theories in Biotechnology
Sant Feliu de Guíxols, Spain
June 13–19, 1991

Library of Congress Cataloging-in-Publication Data

NATO Advanced Research Workshop on the Role of Computational Models
  and Theories in Biotechnology (1991 : San Feliu de Guíxols, Spain)
    Molecular aspects of biotechnology : computational models and
  theories : proceedings of the NATO Advanced Research Workshop on the
  Role of Computational Models and Theories in Biotechnology, Sant
  Feliu de Guíxols, Spain, 13-19 June 1991 / edited by J. Bertrán.
      p.   cm. -- (NATO ASI series. Series C, Mathematical and
  physical sciences ; vol. 368)
    Includes index.

    1. Biochemistry--Computer simulation--Congresses.
  2. Biotechnology--Computer simulation--Congresses.   I. Betrán, J.
  (Juan), 1931-   . II. Title. III. Series: NATO ASI series. Series
  C, Mathematical and physical sciences ; no. 368.
  QP517.M3N38   1991
  574.19'2'0113--dc20   92-8531

ISBN 978-94-010-5121-7   **ISBN 978-94-011-2538-3 (eBook)**
DOI 10.1007/978-94-011-2538-3

TABLE OF CONTENTS

FOREWORD

Molecular biotechnology is a new area of scientific development at the interface between molecular biology, molecular genetics, biochemistry, biomolecular crystallography, protein engineering, and modern theoretical chemistry. A group of specialists from different fields gathered at a NATO Advanced Research Workshop at Sant Feliu de Guíxols in order to gain a deeper insight into the problems, the state-of-the-art, and perspectives of this new speciality. The meeting was a unique experience due to the high level of communication among participants, which is partially reflected in the discussions following each chapter of this book.

As mentioned above, molecular biotechnology is a multidisciplinary field of research where experiments, simulations and theories interact. Nowadays, computational simulations have become a leading factor in modern research. When they are applied to problems involving biological structure and functions, they become a pervasive approach within molecular biotechnology, because they provide a clear way to interpret structural information in its relation to biological function. This job has been achieved thanks to the tremendous increase in computer power. However, a fundamental question emerges concerning the sophistication of the model used to capture the main features of biomolecules and biochemical processes. In the present book, leading specialists in molecular biotechnology provide an answer to this question.

Aplications of computer modelling to biotechnology would involve models to predict protein structure from amino acid sequences first. This is the topic of the first four Chapters. In Chapters One and Two, Profs. Scheraga and Maggiora present the latest advances in the characterization of the conformational space of medium-sized molecules. In turn, Profs. Csizmadia and Mezey offer a topological analysis of protein conformations in Chapters Three and Four. Theoretical models must also predict the stability of different protein structures and pathways to folding and unfolding, as well as experimental properties. This is the subject of Chapters Five, Six, and Seven, by Profs. Van Gunsteren, Tapia and Weinstein, respectively. Chapters Six and Seven relate this aspect to another fundamental point in the application of computer modelling to biotechnology, namely the prediction of the catalytic properties of proteins. The remaining chapters of this book deal with enzymatic processes: in Chapters Eight, Nine and Ten Profs. Warshel, Miller and Robb analyse the possibilities and advantages of describing chemical processes occurring in enzymes by means of the Valence Bond approach. Furthermore, in Chapters Eleven through Thirteen Profs. Kollman, Bertrán and Lesyng use the Molecular Orbital approach for the same purpose. The key point turns out to be the coupling between the chemical system, handled through quantum chemical methodology, and the environment, described in a classical way.

This book will hopefully provide the reader with the main clues and guidelines of computer modelling in biotechnology. The participants in this succesfull meeting did indeed receive them.

J. Bertrán

January 7, 1992

INTERNATIONAL ORGANIZING COMMITTEE

Prof. Juan BERTRAN   Departament de Química
                               Universitat Autònoma de Barcelona
                               08193 Bellaterra (Barcelona)
                               SPAIN

Prof. Fernando BERNARDI   Dipartimento di Chimica
                               Università di Bologna
                               Via Selmi, 2
                               40126 Bologna
                               ITALY

Prof. William H. MILLER   Department of Chemistry
                               University of California
                               Berkeley, California 94720
                               USA

Prof. Orlando TAPIA   Department of Physical Chemistry
                               University of Uppsala
                               SWEDEN

Prof. Jose M. LLUCH   Departament de Química
                               Universitat Autònoma de Barcelona
                               08193 Bellaterra (Barcelona)
                               SPAIN

LOCAL ORGANIZING COMMITTEE

Antoni Oliva

Miquel Duran

Agustí Lledós

Vicenç Branchadell

Miquel Moreno

Angels González

Joana Martínez, Workshop Secretary

LIST OF LECTURERS

Prof. Armando ALBERT

Consejo Superior de Inv. Cientif.
Serrano, 117
28006 Madrid
SPAIN

Prof. Juan BERTRAN

Dept. Química
Facultat de Ciències
Universitat Autònoma de Barcelona
08193 Bellaterra (Barcelona)
SPAIN

Prof. Imre G. CSIZMADIA

Department of Chemistry
University of Toronto
80, St. George Street
Toronto, Ontario
CANADA M5S 1A1

Prof. Wilfred Van GUNSTEREN

Lab. für Physikalische Chemie
ETH-ZENTRUM
CH-8092 Zürich
SWITZERLAND

Prof. Peter A. KOLLMAN

Dep. of Pharmaceutical Chem.
School of Pharmacy
University of California
San Francisco, California 94143
USA

Prof. Bogdan LESYNG

Dep. of Biophysics
Inst. of Experimental Physics
University of Warsaw
93 Zwirki and Wigury
02-089 Warsaw
POLAND

Prof. Per-Olov LÖWDIN

Quantum Chemistry Group
University of Uppsala
Box 518
S-75120 Uppsala
SWEDEN

Prof. Gerald M. MAGGIORA

Director of Computational Chemistry
The Upjohn Company
Kalamazoo, Michigan 49001
USA

Prof. Paul G. MEZEY

Department of Chemistry
University of Saskatchewan
Saskatoon, Saskatchewan
CANADA S7N OWO

Prof. William H. MILLER

Department of Chemistry
University of California
Berkeley, California 94720
USA

Prof. Jean-Louis RIVAIL

Laboratoire de Chimie Théorique
Université de Nancy I
B.P. 239
54506 Vandoeuvre-Les-Nancy Cedex
FRANCE

Prof. Michael A. ROBB

Department of Chemistry
King's College London
University of London
Strand
London WC2R 2LS
UNITED KINGDOM

Prof. Harol A. SCHERAGA

Department of Chemistry
Baker Laboratory
Cornell University
Ithaca, New YOrk 14853-1301
USA

Prof. Orlando TAPIA

Department of Physical Chemistry
University of Uppsala
SWEDEN

Prof. Arieh WARSHEL

Department of Chemistry
University of Southern California
Los Angeles, California 90089-0482
USA

Prof. Harel WEINSTEIN

Dept. of Physicology and Biophysics
Mount Sinai School of Medicine
of the City University of New York
New York 10029
USA

LIST OF PARTICIPANTS

Prof. Juan M. ANDRES   Dept. de Química Física
Facultad de Química
Dr. Moliner, s/n
46100 Burjassot (Valencia)
SPAIN

Prof. Xavier AVILES   Dept. de Bioquímica
Universitat Autònoma de Barcelona
08193 Bellaterra (Bacelona)
SPAIN

Prof. Fernando BERNARDI   Dipartimento di Chimica
Universitá di Bologna
Via Selmi, 2
40126 Bologna
ITALY

Prof. A. E. CH. EL-KETTANI   Lab. de Physique Quantique
Université Paul Sabatier
118, Route de Narbone
31062 Toulouse Cedex
FRANCE

Prof. Caterina GHIO   CNR-Inst. de Chimica Quant. and
Energetica Molecolare
Via Risorgimento,35
56126 Pisa
ITALY

Prof. J.A.N.F. GOMES   Dept. de Química
Faculdade de Ciencias
Universidade do Porto
4000 Porto
PORTUGAL

Prof. Juan J. PEREZ   Dept. d'Enginyeria Química
Univ. Politècnica de Catalunya
ETSEIB
Diagonal 647
08028 Barcelona
SPAIN

Prof. Josep Mª LLUCH   Dept. Química
Universitat Autònoma de Barcelona
08193 Bellaterra (Barcelona)
SPAIN

Prof. G. NARAY-SZABO

Chinoin Research Center
POB 110
Budapest 1325
HUNGARY

Prof. Leonardo PARDO

Dept. Bioestadística
Universitat Autònoma de Barcelona
08193 Bellaterra (Barcelona)
SPAIN

Prof. M. PONS

Dept. Química Orgànica
Universitat de Barcelona
Martí i Franquès, 1-11
08028 Barcelona
SPAIN

Prof. J. SMITH

SBPM, Dept. Biol. Cellulaire
et Moléculaire
CEN Saclay
91191 Gif-sur-Yvette
FRANCE

Prof. A. TAYMAZ

Faculty of Sciences and Arts
Akdeniz University, P.K. 750
Anatala
TURKEY

Prof. J. WILKIE

School of Chemistry
University of Bath
Claverton Down
Bath BA2 7AY
ENGLAND

Prof. Ray Y.K. YANG

Department of Chemical Engineering
West Virginia University
Morgantown, WV 26506-6101
USA

CONFORMATIONAL ENERGY CALCULATIONS ON POLYPEPTIDES AND PROTEINS

Harold A. Scheraga
Baker Laboratory of Chemistry
Cornell University
Ithaca, New York 14853-1301, U.S.A.

ABSTRACT. Empirical conformational energy functions are used to try to compute the three-dimensional structures of polypeptides and proteins. The conformational energy surfaces of such molecules have many local minima, and conventional energy minimization procedures reach only a local minimum (near the starting point of the optimization algorithm) instead of the global minimum (the multiple-minima problem). Several procedures have been developed to surmount this problem. A summary is given here of five of these methods, (i) build-up, (ii) Monte Carlo-plus-minimization (MCM), (iii) relaxation of dimensionality, (iv) pattern-recognition-based importance-sampling minimization (PRISM), and (v) the diffusion equation method which smoothes out the potential surface, leaving only the potential well containing the global minimum. These and other procedures have been applied to a variety of polypeptide structural problems. These include the computation of the structures of open-chain and cyclic peptides, fibrous proteins and globular proteins. Present efforts are being devoted to scaling up these procedures from small polypeptides to proteins, to try to compute the three-dimensional structure of a protein from its amino sequence.

INTRODUCTION

Ever since Anfinsen demonstrated that proteins fold spontaneously to achieve their native conformation [1], attempts have been made to try to compute the three-dimensional structure of a native protein as the one for which the free energy of the system (protein plus solvent) is a minimum. Empirical potential functions and procedures for generating a polypeptide chain and minimizing its conformational energy have been developed for this purpose. While reasonably-good potential functions and minimization procedures are available, a difficult problem that had to be surmounted arose from the presence of many local minima in the conformational energy surface (the multiple-minima problem). I have reviewed the developments in this field on several occasions, the most recent being in 1991 [2]. In this paper, I will concentrate on the multiple-minima problem, in particular on five methods that we have recently developed to solve this problem, and refer the reader to the 1991 review [2] which contains references to a variety of other proce-

1

J. Bertrán (ed.), Molecular Aspects of Biotechnology; Computational Models and Theories, 1–15.
© 1992 *Kluwer Academic Publishers.*

dures that were developed for the same purpose. The five methods to be discussed here are (i) build-up [3-5], (ii) Monte Carlo-plus-minimization [6,7], (iii) relaxation of dimensionality [8,9], (iv) pattern-recognition-based importance-sampling minimization [10-12] and (v) the diffusion equation method [13-15].

BUILD-UP PROCEDURE

In the build-up procedure [3-5], the polypeptide chain is broken into small fragments whose energies are minimized. These small fragments are combined into larger ones, whose energies are then minimized. At each stage of the build-up procedure, an ensemble of conformations (not only the global minimum of each fragment) is retained. As the fragments become larger and larger, more and more of the long-range interactions are built into the computations.

The smallest fragment, of course, is the terminally-blocked amino acid residue. A complete search of the conformational space of this molecule identifies all local energy minima. Terminally-blocked dipeptides are then built from all combinations of local minima of the two terminally-blocked amino acid residues, and their energies are then minimized; this minimization introduces the inter-residue interactions that had not yet appeared in the calculations on the single residues.

Continuation of this process until the whole polypeptide chain is generated would lead to an enormous number of conformations to consider. Therefore, at each stage, the ensembles of conformations are reduced in size by two criteria: (1) high-energy conformations are eliminated, and (2) all members of the ensembles of two fragments being connected (that do not have identical conformations of their overlapping portions) are eliminated. In building intermediate fragments, advantage is also taken of statistical correlations among amino acid sequences to determine the optimum sizes of the building blocks [5] (see below).

This technique has been applied to open-chain and cyclic oligopeptides and to fibrous and globular proteins. Examples of these include the pentapeptide enkephalin, as a single chain [16] and in a crystalline array [17], the cyclic decapeptide gramicidin S [18], the collagen-like poly(Gly-Pro-Pro) [19], and the 58-residue bovine pancreatic trypsin inhibitor (BPTI) [3,4]. In the case of enkephalin, intermolecular hydrogen bonds influence the conformation in the crystal [17]. For a cyclic structure, such as gramacidin S, an exact procedure [20] is used to close the ring. The computed structure [18] has been verified by 2D NMR measurements [21]. The coordinates of the calculated structure of poly(Gly-Pro-Pro) [19] agree with those from an X-ray diffraction study [22] of $(Gly-Pro-Pro)_{10}$ within an rms deviation of 0.3 Å. In an interim calculation on BPTI [3,4], a limited number of simulated NMR distance constraints was used to reduce the sizes of the ensembles in the build-up procedure, and the computed structure had an rms deviation from the X-ray structure of as low as 1.1 Å for the α-carbons; however, enough computing power is now available (as indicated by a calculation on the 155-residue leukocyte interferon [23]) so that it is now possible to repeat the calculations on BPTI without introducing the simulated NMR distance constraints.

To avoid the need to introduce such distance constraints, we have recently modified the build-up procedure [5] by taking account of a statistical analysis of known amino acid sequences that indicates that there is non-random pairing of amino acid residues in short segments along the chain [24-28]; i.e. if two residues are separated by j residues in the amino acid sequence, then it is observed from a protein sequence data bank that not every type of residue occurs with the same frequency at a position j+1 residues from a given i[th] one as its frequency in the compositional data base. Instead, there is a non-random correlation between the type of residue at position i and the type that is j+1 residues along the chain, and this non-random correlation fades out when j exceeds 8. These correlations suggest that one should take advantage of such non-random pairing when forming the building blocks in the build-up procedure, and that the low-energy oligopeptides used as building blocks should vary in length from three to eight residues, depending on the amino acid sequence.

Such building blocks were assembled from low-energy ensembles of tripeptides. The sizes of the ensembles of tripeptide conformations were reduced by shifting each successive overlapping tripeptide by one residue, and removing all conformations, in both a forward and backward generation of the whole BPTI chain, that were not compatible with the conformations of the immediately preceding and following overlapping tripeptides. This procedure is based on the assumption that the final structure of the whole protein, as well as properly-chosen overlapping segments thereof, are simultaneously in low-energy (not necessarily the lowest-energy) conformational states. This assumption enabled the tripeptides to be assembled in three- to eight-residue building blocks, whose conformations are determined primarily by short-range interactions; the overlapping building blocks were then combined to form the whole chain. After screening the generated chains by a succession of approximations, and minimizing the energies of the whole protein, two groups of conformations were obtained. Most of the residues in representative structures from each group fell in the same conformational state, defined by the 16 possible conformational letter codes of Zimmerman et al [29], as in the native structure. Even such good agreement of conformational states, however, can still lead to a poor structure [30]; but, further refinements by procedures outlined below can improve the structure, and such calculations are in progress.

The resulting low-energy structures of representatives of the two groups are not yet as good as those of references 3 and 4. However, whereas a limited number of distance constraints, taken from the known structure of BPTI, were introduced in the calculations of references 3 and 4, such information was not introduced in the procedure of reference 5. The latter procedure takes advantage of an assumed natural selection that presumably leads to a limited number of conformations for a polypeptide with selected amino acid sequences in which the overlapping building blocks are simultaneously in low-energy conformations.

MONTE CARLO-PLUS-ENERGY MINIMIZATION (MCM)

Because the usual Metropolis Monte Carlo procedure [31] does not search conformational space efficiently, the MCM procedure (which searches only the space of local energy minima) was introduced [6,7]. A starting conformation is chosen randomly, and its total ECEPP energy (Empirical Conformational Energy Program for Peptides) [32-34] is minimized. Then backbone and side-chain dihedral angles are selected randomly, and random changes (in the range from -180° to +180°) are made in these dihedral angles. The energy of this altered conformation is then minimized, and the Metropolis criterion is used to determine whether or not to accept it. The procedure is then iterated. In 18 randomly selected starting conformations of the pentapeptide Met-enkephalin, the MCM procedure led to the identical (global) minimum in all cases.

The efficiency of the MCM procedure has been compared [35] with that of simulated annealing (SA) [36,37], in an application to the pentapeptide Met-enkephalin. SA explores a continuous space of the independent variables, whereas MCM explores a discrete space consisting of the local energy minima on that space. Starting from random conformations chosen from the whole conformational space in both cases, it was found that, while SA converged to low-energy structures significantly faster than MCM, SA did not converge to a unique minimum whereas MCM did. Furthermore, the root-mean-square deviations with respect to the global minimum showed no correlation with the observed overall energy decrease in the case of SA, whereas such correlation was quite evident with MCM; this implies that, even though the potential energy decreases in the annealing process, the Monte Carlo SA trajectory does not proceed towards the global minimum. It appears that, while SA offers attractive prospects for possibly improving or refining given structures, it seems to be inferior to MCM, at least in problems where little or no structural information is available for the molecule of interest. The reader is referred elsewhere for the details of this comparison [35].

RELAXATION OF DIMENSIONALITY

A procedure that is similar in spirit to that of Crippen [38,39] has been used to carry out optimization in a space of high dimensionality, where the high barriers of three-dimensional space are absent, or at least lower, and then the system is relaxed back to a three-dimensional space [8,9].

In this procedure the energy of interaction E_{ij} between every pair of atoms i and j, which depends only on their separation d_{ij}, is assigned its minimum value, so that the total energy of the molecule E $= \sum E_{ij}$ has its lowest possible energy. Such a structure, however, is not embeddable in a three- dimensional space. Therefore, the system is relaxed back to three dimensions, and the energy rises; i.e. the global minimum is approached from below rather than from above.

The relaxation procedure is based on the following theorem [40]. A set of n points (the n atoms) in (n-1)-dimensional space is

embeddable in three dimensions if the following three necessary and
sufficient conditions hold:

1.   There exist four points (p_1, p_2, p_3, p_4) in the set
      that are exactly three-dimensional (these points can, e.g.,
      be the planar peptide group, or an external regular tetra-
      hedron).
2.   The four-dimensional volume formed by the simplex of points
      $(p_1, p_2, p_3, p_4, p_i)$ is zero for all $i=1,\ldots,n$.
3.   The five-dimensional volume formed by the simplex of points
      $(p_1, p_2, p_3, p_4, p_i, p_j)$ is zero for all $i,j = 1,\ldots,n$.

Based on this theorem, and using distances as the variables, the
following objective function F is minimized.

$$F = w_E F_E + w_{4D} F_{4D} + w_{5D} F_{5D} + W_B F_B \qquad (1)$$

where the w's are weighting factors, F_E is the ECEPP energy, F_{4D}
and F_{5D} are Cayley-Menger determinant constraints on the four- and
five-dimensional volumes, and F_B incorporates information on upper
and lower bounds on the distances. Application of this procedure led
to the same (global) minimum-energy structure that was obtained by the
MCM procedure.

PATTERN-RECOGNITION-BASED IMPORTANCE-SAMPLING MINIMIZATION (PRISM)

Pattern recognition techniques are used to predict a series of
probable backbone structures, whose energies are then minimized to
locate the global minimum [10-12]. The (ϕ,ψ) map of each residue is
divided into four regions (α, ε, $\alpha*$ and $\varepsilon*$), and all possible tripeptides from a properly selected set of X-ray structures from the
Brookhaven Protein Data Bank are collected and grouped according to
conformation (e.g. $\alpha\alpha\alpha$, $\alpha\varepsilon\varepsilon$, $\alpha\varepsilon\varepsilon*$, etc.). The pattern recognition procedure uses amino-acid properties [41] to map peptide sequences into a
multivariate property space. Particular tripeptide conformations tend
to map to particular regions of the property space. These regions are
represented by multivariate Gaussian distributions, where the parameters of the distributions are determined from tripeptides in the
Protein Data Bank. These data are then used to calculate the probability that each tripeptide in a protein under study has a given conformation.

The polypeptide chain is built up from the N-terminus, fitting the
most probable tripeptide conformations together, one tripeptide at a
time, allowing for proper overlap of the tripeptides. As the build-up
proceeds, the probabilities of the growing chain (conformation) are
calculated, and only the 1000 most probable are retained. Thus, when
the C-terminus is reached, there are 1000 different predictions of the
backbone structure of the protein, sorted in order of decreasing
probability.

The symbolic representation (in terms of the regions α, ε, $\alpha*$, $\varepsilon*$)
of the conformation of a protein is converted to a dihedral-angle
representation by randomly generating values of ϕ and ψ in each of the

assigned regions from appropriate probability distributions. A bivari-
ate (2D) Gaussian distribution parameterized on values of (ϕ,ψ) from
the known X-ray structures is used, together with standard techniques
for generating random numbers from Gaussian distributions. Several
such random structures are generated for each backbone prediction, and
the energy of each of them is minimized. The lowest-energy structure
is taken to represent the backbone prediction. The aforementioned
probabilities serve to reduce, to a manageable size, the set of conformations whose energies have to be minimized. This procedure has been
tested on the 36-residue avian pancreatic polypeptide, and the lowest-
energy structure [12] is compared to the X-ray structure [42,43] in
Fig. 1.

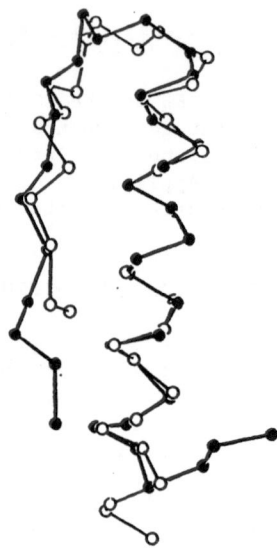

Figure 1. Superposition of computed structure of avian pancreatic
polypeptide (open circles) [12] on the X-ray structure (solid circles)
[42,43]. Only the C$^\alpha$ atoms are shown.

DIFFUSION EQUATION METHOD

Another method to surmount the multiple-minima problem is based on
the possibility of deforming the complex hypersurface in successive
stages so that higher-energy minima disappear, and only a descendant of
the global minimum remains. A reversal of the deformation procedure
then recovers the global minimum of the original potential function
[13].

This method is illustrated in Figure 2 by a simple one-dimensional
function with two minima. The original function, f(x), can be
deformed, in the first iteration, to f$^{[1]}$ (x) by adding its second
derivative, f''(x), which is zero at the inflection points, viz.

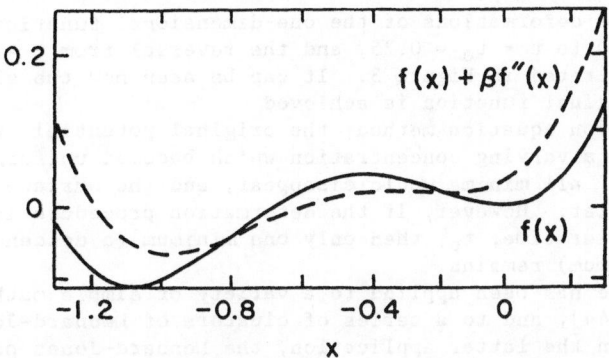

Figure 2. Original double-minimum potential energy curve f(x) (solid line) transformed, according to eq. 2, into a curve with only a single minimum (dashed line). The values of the transformed function at the inflection points do not change. The particular function used in this Figure is $f(x) = x^4 + ax^3 + bx^2$, with $\underline{a}$ and $\underline{b}$ equal to 2 and 0.9, respectively. When $\beta = 0.02$, one obtains $f(x) + \beta f''(x)$, which exhibits only one minimum [13].

$$f^{[1]}(x) = f(x) + \beta f''(x) = \left(1 + \beta \frac{d^2}{dx^2}\right) f(x) \qquad (2)$$

where β is a small positive constant. Repeated applications of this procedure lead to the following result in the Nth iteration:

$$f^{[N]}(x) = \left(1 + \frac{t}{N} \frac{d^2}{dx^2}\right)^N f(x) \qquad (3)$$

where t/N has been written for β, with the parameter t being positive. Destabilization of the surface is most effective when $N \to \infty$. Taking this limit, we may write

$$F(x,t) = \lim_{N \to \infty} \left(1 + \frac{t}{N} \frac{d^2}{dx^2}\right)^N f(x) = \exp\left(t \frac{d^2}{dx^2}\right) f(x) \qquad (4)$$

It can be shown that, equivalently, F(x,t) is a solution of the diffusion equation

$$\frac{\partial^2 F}{\partial x^2} = \frac{\partial f}{\partial t} \qquad (5)$$

where the parameter t takes on the meaning of "time", with the initial condition being $F(x,0) = f(x)$.

In higher dimensions, d^2/dx^2 is replaced by the Laplacian, $\Delta = \sum_{i=1}^{m} \partial^2/\partial x_i^2$, so that the diffusion equation becomes

$$\Delta F = \frac{\partial F}{\partial t} \qquad (6)$$

The successive deformations of the one-dimensional function of Figure 2 from t = 0 to t = t_o = 0.25, and the reversal from t = 0.25 to t = 0, is illustrated in Figure 3. It can be seen how the global minimum of the original function is achieved.

In the diffusion equation method, the original potential surface is the analogue of a varying concentration which becomes uniform as t → ∞. Thus, as t → ∞, all minima would disappear, and the surface would become uniformly flat. However, if the deformation procedure is stopped at an earlier time, t_o, then only one minimum (a descendant of the global minimum) remains.

This procedure has been applied to a variety of simple mathematical functions [13,44], and to a series of clusters of Lennard-Jones particles [14]. In the latter application, the Lennard-Jones potential function was expressed as a sum of Gaussians, thereby leading to an analytical solution of equation 6 in terms of the Fourier-Poisson integral. Calculations were carried out for various cluster sizes n = 5,6,7,...55. For n = 55, there are ~10^{45} local minima, the global minimum being the MacKay icosahedron [45]. This global minimum was found by the diffusion equation method [14] in ~400 seconds on an IBM

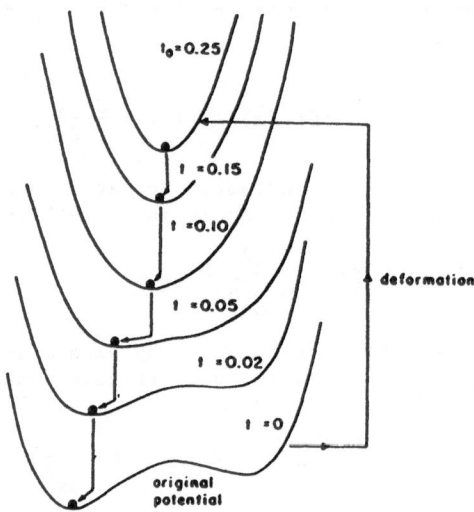

Figure 3. Illustration of the deformation of the original potential f(x) (the same as in Fig. 2), and of the reversing procedure. The deformation at t_o = 0.25 leads to the curve with the unique minimum that is achievable from any point of the space by a simple minimization. Then, the reversing procedure (shown by the arrows directed downward) is applied by considering a sequence of the deformed curves at t = 0.15, 0.10, 0.05, 0.02, and finally 0, where the original function is recovered. Each step of the procedure is followed by a minimization symbolized in the Figure by a ball moving downhill from the minimum position of the upper curve and always reaching the position of the minimum in the lower curve. In the final step, the global minimum is found [13].

3090 supercomputer. For a 55-particle cluster, there are $3n-6 = 159$
degrees of freedom, which is the number of independently variable
dihedral angles (internal coordinates) in a 26-residue polypeptide.
While the potential functions for treating polypeptides are more
complex than a Lennard-Jones potential, it is nevertheless hoped that
the diffusion equation method can be applied efficiently to large
polypeptides.

At present, this method is being used to treat the ECEPP/2
algorithm [33,34]; the Lennard-Jones 6-12 and hydrogen-bonding 10-12
components of ECEPP/2 are expressed as a sum of Gaussians, the
coulombic term is taken directly as $1/r$, and the intrinsic torsional
cosine-type terms are expressed as Tshebyshev polynomials in the
distance r between the two end atoms in the group of four that defines
the dihedral angle, for evaluation of the Fourier-Poisson integral
[15]. Initial applications are being made to terminally-blocked
alanine and to the pentapeptide Met-enkephalin [15].

CONCLUSION

It appears that progress is being made to surmount the multiple-
minima problem. The various procedures developed work efficiently for
polypeptides containing of the order of 5-20 residues. It is hoped
that the methods described here and elsewhere can provide accurate
structures, in a reasonable amount of computing time, for proteins containing of the order of 100 amino acid residues.

ACKNOWLEDGMENT

This work was supported by research grants from the National
Institute of General Medical Sciences, of the National Institutes of
Health (GM-14312), and from the National Science Foundation
(DMB84-01811).

REFERENCES

1. Anfinsen, C. B., Haber, E., Sela, M. and White, F. H., Jr. (1961)
    'The kinetics of formation of native ribonuclease during
    oxidation of the reduced polypeptide chain', Proc. Natl. Acad.
    Sci., U.S.A. 47, 1309-1314.
2. Scheraga, H. A. (1991) 'Experimental and theoretical aspects of
    protein conformation', in Theoretical Biochemistry and
    Molecular Biophysics, Vol. 2: Proteins, Ed. D. L. Beveridge
    and R. Lavery, Adenine Press, Guilderland, N.Y., p. 231-237.
3. Vásquez, M. and Scheraga, H. A. (1988) 'Calculation of protein
    conformation by the build-up procedure. Application to bovine
    pancreatic trypsin inhibitor using limited simulated nuclear
    magnetic resonance data', J. Biomolecular Structure & Dynamics
    5, 705-755.
4. Vásquez, M. and Scheraga, H. A. (1988) 'Variable-target-function
    and build-up procedures for the calculation of protein confor-
    mation. Application to bovine pancreatic trypsin inhibitor
    using limited simulated nuclear magnetic resonance data', J.
    Biomolecular Structure & Dynamics 5, 757-784.

5. Simon, I., Glasser, L. and Scheraga, H. A. (1991) 'Calculation of protein conformation as an assembly of stable overlapping segments: Application of bovine pancreatic trypsin inhibitor', Proc. Natl. Acad. Sci., U.S.A. 88, 3661-3665.

6. Li, Z. and Scheraga, H. A. (1987) 'Monte Carlo-minimization approach to the multiple-minima problem in protein folding', Proc. Natl. Acad. Sci., U.S.A. 84, 6611-6615.

7. Li, Z. and Scheraga, H. A. (1988) 'Structure and free energy of complex thermodynamic systems', J. Molec. Str. (Theochem). 179, 333-352.

8. Purisima, E. O. and Scheraga, H. A. (1986) 'An approach to the multiple-minima problem by relaxing dimensionality', Proc. Natl. Acad. Sci., U.S.A. 83, 2782-2786.

9. Purisima, E. O. and Scheraga, H. A. (1987) 'An approach to the multiple-minima problem in protein folding by relaxing dimensionality. Tests on enkephalin', J. Mol. Biol. 196, 697-709.

10. Lambert, M. H. and Scheraga, H. A. (1989) 'Pattern recognition in the prediction of protein structure. I. Tripeptide conformational probabilities calculated from the amino acid sequence', J. Comput. Chem. 10, 770-797.

11. Lambert, M. H. and Scheraga, H. A. (1989) 'Pattern recognition in the prediction of protein structure. II. Chain conformation from a probability-directed search procedure', J. Comput. Chem. 10, 798-816.

12. Lambert, M. H. and Scheraga, H. A. (1989) 'Pattern recognition in the prediction of protein structure. III. An importance-sampling minimization procedure', J. Comput. Chem. 10, 817-831.

13. Piela, L., Kostrowicki, J. and Scheraga, H. A. (1989) 'The multiple-minima problem in the conformational analysis of mole-cules. Deformation of the potential energy hypersurface by the diffusion equation method', J. Phys. Chem. 93, 3339-3346.

14. Kostrowicki, J., Piela, L., Cherayil, B. J. and Scheraga, H. A. (1991) 'Performance of the diffusion equation method in searches for optimum structures of clusters of Lennard-Jones atoms', J. Phys. Chem. 95, 4113-4119.

15. Kostrowicki, J. and Scheraga, H. A. (1991) work in progress.

16. Vásquez, M. and Scheraga, H. A. (1985) 'Use of buildup and energy-minimization procedures to compute low-energy structures of the backbone of enkephalin', Biopolymers 24, 1437-1447.

17. Glasser, L. and Scheraga, H. A. (1988) 'Calculations on crystal packing of a flexible molecule, Leu-enkephalin', J. Mol. Biol. 199, 513-524.

18. Dygert, M., Go, N. and Scheraga, H. A. (1975) 'Use of a symmetry condition to compute the conformation of gramicidin S', Macromolecules 8, 750-761.

19. Miller, M. H. and Scheraga, H. A. (1976) 'Calculation of the structures of collagen models. Role of interchain interactions in determining the triple-helical coiled-coil conformation. I. Poly(glycyl-prolyl-prolyl)', J. Polymer Sci.: Polymer Symposia, No 54, p. 171-200.

20. Gō, N. and Scheraga, H. A. (1973) 'Ring closure in chain molecules with C_n, I or S_{2n} symmetry', Macromolecules 6, 273-281.

21. Mirau, P. A. and Bovey, F. A. (1990) '2D and 3D NMR studies of polypeptide structure and function', Abstracts 199th April Amer. Chem. Soc. Meeting, Boston, POLY 58.

22. Okuyama, K., Tanaka, N., Ashida, T. and Kakudo, M. (1976) 'Structure analysis of a collagen model polypeptide, (Pro-Pro-Gly)$_{10}$', Bull. Chem. Soc. Japan 49, 1805-1810.

23. Gibson, K. D., Chin, S., Pincus, M. R., Clementi, E. and Scheraga, H. A. (1986) 'Parallelism in conformational energy calculations on proteins', in "Lecture Notes in Chemistry," Vol. 44, "Supercomputer Simulations in Chemistry," ed. M. Dupuis, Springer-Verlag, Berlin, 1986, p. 198-213.

24. Simon, I. (1985) 'Investigation of protein refolding: A special feature of native structure responsible for refolding ability', J. Theor. Biol. 113, 703-710.

25. Vonderviszt, F., Matrai, G. and Simon, I. (1986) 'Characteristic sequential residue environment of amino acids in proteins', Int. J. Peptide Protein Res. 27, 483-492.

26. Vonderviszt, F. and Simon, I. (1986) 'A possible way for prediction of domain boundaries in globular proteins from amino acid sequence', Biochem. Biophys. Res. Commun. 139, 11-17.

27. Cserzo, M. and Simon, I. (1989) 'Regularities in the primary structure of proteins', Int. J. Peptide Protein Res. 34, 184-195.

28. Tudos, E., Cserzo, M. and Simon, I. (1990) 'Predicting isomorphic residue replacements for protein design', Int. J. Peptide Protein Res. 36, 236-239.

29. Zimmerman, S. S., Pottle, M. S., Némethy, G. and Scheraga, H. A. (1977) 'Conformational analysis of the twenty naturally occurring amino acid residues using ECEPP', Macromolecules 10, 1-9.

30. Burgess, A. W. and Scheraga, H. A. (1975) 'Assessment of some problems associated with prediction of the three-dimensional structure of a protein from its amino-acid sequence', Proc. Natl. Acad. Sci., U.S. 72, 1221-1225.

31. Metropolis, N., Rosenbluth, A. W., Rosenbluth, M. N., Teller, A. H. and Teller, E. (1953) 'Equation of state calculations by fast computing machines', J. Chem. Phys. 21, 1087-1092.

32. Momany, F. A., McGuire, R. F., Burgess, A. W. and Scheraga, H. A. (1975) 'Energy parameters in polypeptides. VII. Geometric parameters, partial atomic charges, nonbonded interactions, hydrogen bond interactions, and intrinsic torsional potentials for the naturally occurring amino acids', J. Phys. Chem. 79, 2361-2381.

33. Némethy, G., Pottle, M. S. and Scheraga, H. A. (1983) 'Energy parameters in polypeptides. 9. Updating of geometrical parameters, nonbonded interactions, and hydrogen bond interactions for the naturally occurring amino acids', J. Phys. Chem. 87, 1883-1887.

12

34. Sippl, M. J., Némethy, G. and Scheraga, H. A. (1984) 'Intermolecular potentials from crystal data. 6. Determination of empirical potentials for O-H•••O=C hydrogen bonds from packing configurations', J. Phys. Chem. 88, 6231-6233.
35. Nayeem, A., Vila, J. and Scheraga, H. A. (1991) 'A comparative study of the simulated-annealing and Monte Carlo-with-minimization approaches to the minimum-energy structures of polypeptides: [Met]-Enkephalin', J. Comput. Chem. 12, 594-605.
36. Kirkpatrick, S., Gelatt, C. D., Jr. and Vecchi, M. P. (1983) 'Optimization by simulated annealing', Science 220, 671-680.
37. Vanderbilt, D. and Louie, S. G. (1984) 'A Monte Carlo simulated annealing approach to optimization over continuous variables', J. Comput. Phys. 56, 259-271.
38. Crippen, G. M. (1982) 'Conformational analysis by energy embedding', J. Comput. Chem. 3, 471-476.
39. Crippen, G. M. (1984) 'Conformational Analysis by scaled energy embedding', J. Comput. Chem. 5, 548-554.
40. Blumenthal, L. M. (1970) "Theory and Applications of Distance Geometry", Chelsea, New York, 97-99.
41. Kidera, A., Konishi, Y., Oka, M., Ooi, T. and Scheraga, H. A. (1985) 'Statistical analysis of the physical properties of the 20 naturally occurring amino acids', J. Protein Chem. 4, 23-55.
42. Blundell, T. L., Pitts, J. E., Tickle, I. J., Wood, S. P. and Wu, C. W. (1981) 'X-ray analysis (1.4-Å resolution) of avian pancreatic polypeptide: Small globular protein hormone', Proc. Natl. Acad. Sci., U.S.A. 78, 4175-4179.
43. Glover, I., Haneef, I., Pitts, J., Wood, S., Moss, D., Tickle, I. and Blundell, T. (1983) 'Conformational flexibility in a small globular hormone: X-ray analysis of avian pancreatic polypeptide at 0.98-Å resolution', Biopolymers 22, 293-304.
44. Kostrowicki, J. and Piela, L. (1991) 'Diffusion equation method of global minimization: Performance for standard test functions', J. Optimization Theory and Applications 69, 269-284.
45. Mackay, A. L. (1962) 'A dense non-crystallographic packing of equal spheres', Acta Cryst. 15, 916-918.

DISCUSSION

Kollman: Why do you divide the peptide map into 16 regions rather then fewer?

Scheraga: We used 16 conformational codes [29] to provide a reasonably fine division of the ϕ, ψ space of a single residue. If we were to use fewer regions, we would not have as precise a description of the computed conformation of each residue.

Kollman: Does the time-reversing procedure always end up in the global minimum?

Scheraga: There is no guarantee that the global minimum will always be reached. However, if, for example, we replace the potential function V by $[-\exp(-V/kT)]$ in the Fourier-Poisson integral, it becomes more likely that we will reach the global minimum. The lower that T is, the more likely it is that we will indeed reach the global minimum. From a practical point of view, one should explore the results with various small values of T. See reference 14 for details.

Kollman: Why don't you use a softer repulsive potential since you use only torsional angles as your independent degrees of freedom -- and your ϕ, ψ ECEPP energy maps have many X-ray-allowed conformations that are rather high in energy (> 8 kcal/mole)?

Scheraga: As we showed recently, in the Journal of Biomolecular Structure & Dynamics, 7, 319-419, 421-453 (1989); 8, 1109-1111 (1991), our ϕ, ψ ECEPP energy maps agree in large measure with the values of ϕ and ψ observed in highly-refined X-ray structures.

Weinstein: If you already know the global minimum, and it was reached before with another search procedure but with the same potential function, then why would you look for better functions if the diffusion equation procedure does not converge? Could it be that the global minimum simply "disappeared" during the transformation? Does this actually happen?

Scheraga: As far as your first question is concerned, we are continually trying to develop more and more efficient methods to locate the global minimum. The diffusion equation method is one of these new procedures. It is therefore natural to test it out first on a known system, i.e. one for which we already know the global minimum, before applying it to an unknown system.

As far as your question about the global minimum "disappearing" during the transformation is concerned, the reply that I gave to Kollman's question about always ending up in the global minimum also pertains to your question.

Finally, it does sometimes happen that the global minimum disappears. However, our new method of replacing V by [-exp(-V/kT)], referred to in my reply to Kollman, avoids this problem.

Warshel: What guarantees that the lowest minimum in the modified surface will be transformed to the lowest minimum in the real surface?

Scheraga: There is no guarantee. My reply to Kollman's question about always ending up in the global minimum also pertains to your question.

Mezey: Would you find any advantage in scaling the parameter β (the coefficient of the second-derivative term) in the adjusted potential function by making it dependent on some energy difference, for example, by the actual energy calculated at the given point minus the lowest energy which has been found so far in the procedure?

Scheraga: We actually do not add the second derivative to the original function. Equations (2)-(4), with β taken independent of the function f(x), were used simply to learn that the energy surface should be smoothed by solving the diffusion equation.

Perez: How does your smoothing procedure compare with simulated annealing? Moskowitz's procedure using simulated annealing seems to be very efficient in terms of computer time.

Scheraga: We have compared our MCM procedure with simulated annealing in runs on Met-enkephalin [35]. While simulated annealing is faster than MCM, we found that simulated annealing is (a) not reproducible from run to run, and (b) does not reach the global minimum. Consequently, we have not bothered to compare simulated annealing to the diffusion equation method.

While Moskowitz and others reported success with simulated annealing using various annealing strategies, but with a rather simplified force field compared to ECEPP, we have found that, in order to succeed, these annealing strategies seem to require foreknowledge of either the topology of the conformational space or of the region where the global minimum is located. They do not work reliably when the initial point of the Monte Carlo trajectory is chosen randomly, whereas MCM does [35].

Wilkie: When smoothing the potential curve using
$$f^{[1]}(x) = f(x) + \beta f''(x)$$
the rate at which minima fill up will depend on the curvature f''(x). Thus, which minimum is the sole remaining one will be determined by the value of the ratio
$$\frac{\text{well depth}}{\alpha\, f''(x)}$$
Do you know the value of α, since it may be critical to the

success of the diffusion equation method?

Scheraga: The answer to this question is similar to that which I gave in reply to Mezey's question. An estimate of the time to start the reversing procedure depends on the actual form of the potential, and it sometimes can be difficult to make such an estimate. However, one can examine the deformed surface at various times to determine the time at which only one minimum remains.

Pons: Could you comment on the use of increased dimensionality as a way for searching the conformational space, and how does it compare with the other methods that you described?

Scheraga: In our development of the procedure for relaxation of dimensionality from a high-dimensional space to a three-dimensional space, we have applied it so far only to single residues [8] and to Met-enkephalin [9]. We don't yet have enough experience with it to be able to make a detailed comparison with the other methods.

Löwdin: Could you comment about the temperature dependences of your protein structures?

Scheraga: Our empirical potential functions are parameterized on experimental data at essentially room temperature rather than at absolute zero. Therefore, peptide and protein structures, calculated with these potentials, are effectively room-temperature structures, but could also be used, for example, to explore conformational changes (denaturation) that occur as the temperature is raised because the potentials would not be expected to vary much with temperature.

COMPUTER MODELING OF CONSTRAINED PEPTIDE SYSTEMS

JAMES R. BLINN
KUO-CHEN CHOU
W. JEFFREY HOWE
GERALD M. MAGGIORA*
BORYEU MAO
JOSEPH B. MOON
Upjohn Laboratories
301 Henrietta Street
Kalamazoo, Michigan 49001
USA

ABSTRACT. Three approaches are described for dealing with conformational problems in constrained peptide systems. The first approach addresses the problem of *de novo* design of peptides where the structure of the binding or active site of the target protein is known. The method is based on a procedure, called GROW, which links together amino acid fragments from a library of amino acid conformations such that an "optimum fit" and solvation energy for the ligand are obtained. The second approach addresses the problem of modeling the surface loops of proteins. It is based upon the novel use of simulated annealing within the framework of a "random-tweak" type procedure [Shenkin *et al.*, *Proteins*, 1989]. The method has been tested on 6- and 8-residue loops, and yields results that agree well with experiment at relatively little computational cost. The third approach addresses both cyclic peptides and surface-loop modeling using a mass-weighted molecular dynamics (MWMD) procedure. It is shown that MWMD is clearly superior to normal-mass MD sampling at equivalent temperatures, and is numerically stable. Moreover, the sampling of individual torsional degrees of freedom is also seen to be essentially complete for the case studied here, *viz.* for pressinoic acid, a cyclic hexapeptide derived from vasopressin.

1. Introduction

The structural analysis of peptide systems, due to their great conformational flexibility, represents a significant challenge to experimentalists and theorists; difficulties persist even for constrained systems. The present work describes our efforts to address conformational questions in such constrained systems and covers peptides bound within the binding sites of proteins, peptides forming the surface loops of proteins, and cyclic peptides – the latter two cases being conceptually similar.

Although in all three cases the conformational space is significantly reduced from that of the corresponding unconstrained peptides, conformation analysis of such systems remains a difficult problem to treat theoretically, especially by systematic-search procedures. This is due to the combinatoric nature of conformational searching in systems with many degrees of freedom. Moreover, determination of the lowest-energy conformer is frustrated by the ever present multiple-minimum problem [1]. To overcome these difficulties, two approaches based upon Monte Carlo (MC) and molecular dynamics (MD) sampling procedures, respectively, have generally been

J. Bertrán (ed.), Molecular Aspects of Biotechnology: Computational Models and Theories, 17–38.
© 1992 *Kluwer Academic Publishers.*

employed [2,3]. MC-based procedures have the advantage that the choice of sample points is not determined by the forces acting on the particles, as is the case in MD-based sampling. Thus, passage through energy barriers does not present any serious difficulties. This can lead to a more efficient and thorough conformational-space sampling for longer-chain peptides, where intersection of the chain with itself can impede movement into certain "allowed" regions of conformational space. MD-based procedures, on the other hand, handle the coupled motions of particles quite naturally, a feature which is particularly useful when dealing with constrained systems. There exist, however, methods for dealing with coupled motions within the framework of MC-based sampling (Cf. [4-7]).

In the present work, it will be seen that the novel use of Monte-Carlo-based simulated annealing (MCSA) [8,9] and mass-weighted molecular dynamics (MWMD) procedures [10] can provide significant computational advantages over systematic searching for investigating the conformational properties of constrained peptide systems. Section 2 describes a novel method for "generating" peptides which fit within the binding sites of proteins of known structure. The initial formulation of the method, called GROW, described here is based upon the linking together of appropriate conformations of specific amino acid "fragments" obtained from a "conformational library" of structures through a systematic search procedure. A newer, computationally faster implementation of the method, based upon an MCSA procedure which samples the discrete conformational space contained within the conformational library, is also described. Section 3 describes an MCSA procedure procedure for modeling protein surface loops. The method addresses the coordinate-dependency problem, which plagues many MC-based procedures, in a novel and computationally feasible way that is reminiscent of the "random tweak" procedure developed by Shenkin *et al.* [11]. Section 4 describes a MWMD approach to conformational searching which also provides a computationally efficient means for sampling the conformational space of both surface loops and cyclic peptides. A clear advantage of the MWMD approach shown here over other MD-based sampling methods, particularly those employing elevated temperatures to enhance the sampling of regions separated by energy barriers, is the greater numerical stability afforded by the MWMD procedure without consequent loss in the overall extent of conformational-space sampling.

2. De Novo Design of Peptide Ligands (*Moon & Howe*)

With the advent of modern molecular biology, the possibility of expressing and characterizing the properties of a vast array of proteins with myriad functions now exists. In addition to their inherent biological interest, many of these proteins are attractive targets for the design of specific molecules which could inhibit or otherwise influence their behavior. Computational chemistry and molecular graphics methods have already been quite useful in the design of novel agents, but the vast array of possibilities can be bewildering, even to the most experienced of researchers.

Recently, a method for carrying out the *de novo* design of peptide ligands [12] has been developed. Basically, the method, called GROW, constructs (i.e., "grows") peptides one residue at a time by searching a library of residue templates for those templates which make the best overall energy contribution to the system. Initially, a systematic search procedure was used to generate a search tree from an initial "seed" structure, usually a peptide fragment. The tree is "pruned" at each branch in order to overcome the combinatoric explosion which otherwise results from such systematic-search procedures. The general features of GROW are illustrated in Figure 1.

SETUP:

(a) Interactive modeling: select site atoms
(b) Select seed position
(c) Specify control parameters

Template Library

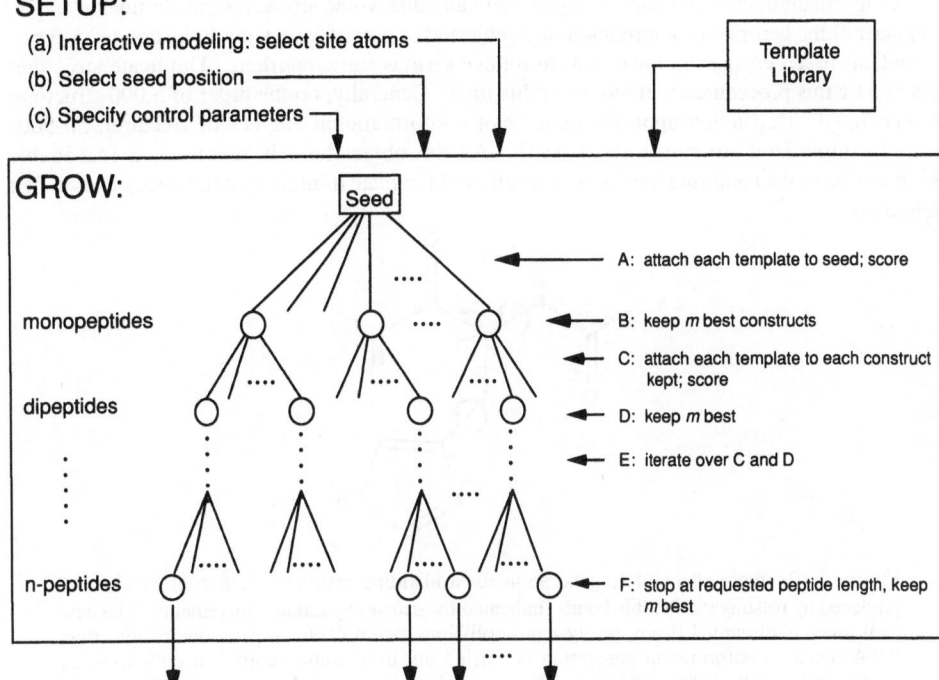

GROW:

Seed

monopeptides

A: attach each template to seed; score

B: keep *m* best constructs

C: attach each template to each construct kept; score

dipeptides

D: keep *m* best

E: iterate over C and D

n-peptides

F: stop at requested peptide length, keep *m* best

EVALUATE: Interactive modeling, binding energy estimation

Figure 1: Schematic of tree-based procedure for *de novo* peptide growth. After the user specifies a seed location, the growth begins. A: each template in the library is superimposed on the seed and scored; B: the highest scoring "m" constructs (specified by the user) are retained for the next level; C: each template in the library is attached to each of the constructs retained at the first level, and scored; D,E: this is repeated until F, the user-specified peptide length is achieved. These are also scored and the "m" best are retained for the final evaluation step, using more detailed methods.

GROW can operate basically in two growth modes: (1) *unrestricted*, where the "best" amino acid fragment is chosen at each step in the process, and (2) *restricted*, where the type of fragment substituted at each is pre-determined. Hybrid GROW runs which possess both restricted and unrestricted elements can also be carried out in a similar fashion.

2.1 TEMPLATE LIBRARY GENERATION

Generation of a suitable template library is crucial to the development of an effective and efficient GROW procedure. Each template is made up of an appropriate sidechain attached to an α-carbon flanked on either side by amide groups as illustrated in Figure 2 for the sidechain of a histidine residue. A set of conformations are generated by randomly perturbing the backbone φ-ψ and sidechain χ-angles as shown in the figure.

Conformations corresponding to highly sterically disfavored structures are eliminated and the remainder of the generated conformations are subjected to *partial* energy minimization (~15 cycles of gradient-based energy minimization) to relieve serious steric conflicts. Duplicate structures obtained by this procedure are removed at this time. Generally, on the order of 5,000 structures are generated. Depending upon the number of conformational degrees of freedom, 50-5000 partially minimized structures result [12]. As the object here is to obtain a reasonable discretization of the conformational space, location of template minima is unnecessary and, in fact, undesirable.

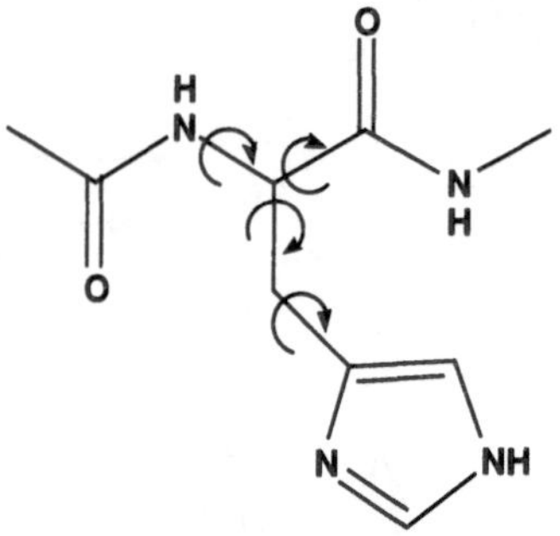

Figure 2. Example of template generation for a histidine residue. Each conformation is produced by rotating the flexible bonds (indicated by arrows) by random increments. The new conformer is discarded if any heavy atom collisions (atom center separations of less than 2.0 Å) exist. Conformation generation is carried out until some number of collision-free conformers, usually 5,000, are found. These are subjected to partial energy minimization and duplicate elimination, then stored in the template library.

2.2 SCORING PROCEDURE

Another critical element of GROW is the scoring procedure, which is based on the following equation

$$\text{Score} = -[V_{\text{ligand-recep}} - V_{\text{ligand}} - V_{\text{recep}}]$$

and includes an estimate of the solvent contribution to the ligand-receptor interaction. The general form of the potential-energy function used is similar to that found in AMBER [13] as implemented in MacroModel [14]. A notable exception is the Lennard-Jones terms which are "softened" to compensate for the discrete nature of the conformational space spanned by the set of conformations in the template set. Solvation energies are calculated based on the Nemethy-Scheraga hydration shell procedure [15]. Due to its computational speed, the surface area determination algorithm developed by Still and co-workers [16] is used. $V_{\text{ligand-recep}}$ contains intermolecular non-bonded energy contributions plus the solvation energy of the bound peptide-receptor complex. V_{ligand} takes account of the intramolecular bonded and non-bonded interactions and solvation energy of the unbound peptide. Intra-residue interaction energies, calculated during the template generation process, are simply retrieved from the template library rather than recalculated. V_{recep} is the solvation energy of the unbound receptor site. It should be emphasized that the above equation is used only for scoring the generated peptides; more detailed methods, such as free energy simulations [17-19], must be used if more accurate binding energies are desired.

2.3 VALIDATION OF GROW

A number of tests have been used to assess the reliability of the GROW procedure, and they are described in detail in a recent publication [12]. Here attention will be focused on a single example which, nonetheless, clearly illustrates the ability of the GROW procedure to accurately reproduce the binding site geometry of a ligand, namely the Upjohn inhibitor U-70531E [20], bound to the aspartyl protease rhizopuspepsin [21]. Figure 3 illustrates several steps of a restricted GROW run beginning with a seed obtained from the x-ray structure of the ligand-protein complex [21].

An analogous study of the HIV-1 protease inhibitor, MVT101, developed by Miller *et al.* [22] also yielded similar results [12]. In addition to the *de novo* generation of inhibitors, GROW has also generated peptide substrates for aspartyl proteases [12].

2.4 NEW DEVELOPMENTS

Since its initial development GROW has continued to grow! A number of significant developments have taken place which have improved the speed, flexibility, and robustness of the GROW procedure; the details will be presented in a future publication [23]. In the present work only a very brief outline of two important improvements will be described. The first involves replacement of the very computationally demanding systematic-search procedure used in early GROW studies with a "*discretized*" MCSA procedure for sampling the conformational space defined by the template library. This differs from usual MCSA conformational-space sampling procedures in that random sampling of a continuous dihedral-angle space is replaced by random sampling of the considerably more limited conformational space contained within the template library. As is the case in most MC methods applied to conformational problems, a form of the Metropolis algorithm [24] is used as a basis for accepting or rejecting structures generated by the MC sampling process. The temperature range considered is typically from 1000°K to 100°K, and is decremented by $T_{new} = 0.9 \cdot T_{old}$ for every 5,000 structures sampled. This approach has been shown to successfully reproduce the results obtained earlier by the systematic search procedure [12], with a 10- to 20-fold improvement in computational speed.

The second major improvement in GROW addresses the important question of how to locate the "seed". Seed location, not surprisingly, has a significant bearing both on what peptides will result from an unrestricted GROW run, and on the binding geometries of peptides obtained from either unrestricted or restricted growth. The sensitivity of the procedure to the initial seed placement arises from the constraint that the generated peptide must contain the user-placed seed fragment. Earlier studies with GROW generally located seeds based on crystallographic structures of bound ligands, which obviously cannot be used when only unliganded protein structures are available. Even in cases where crystallographic data for bound ligands does exist, its use introduces a bias into GROW runs. The solution to this sensitivity problem was an extension of the MCSA procedure in which the perturbations to a peptide can move it from the initial seed position.

When a peptide is perturbed by the replacement of one of its residues with a conformation from the template library, one end of the peptide must move to accommodate the newly introduced backbone ϕ and ψ angles. In the first implementation of the MCSA procedure, the end to be moved was chosen so that the seed position would remain anchored in place. The method was

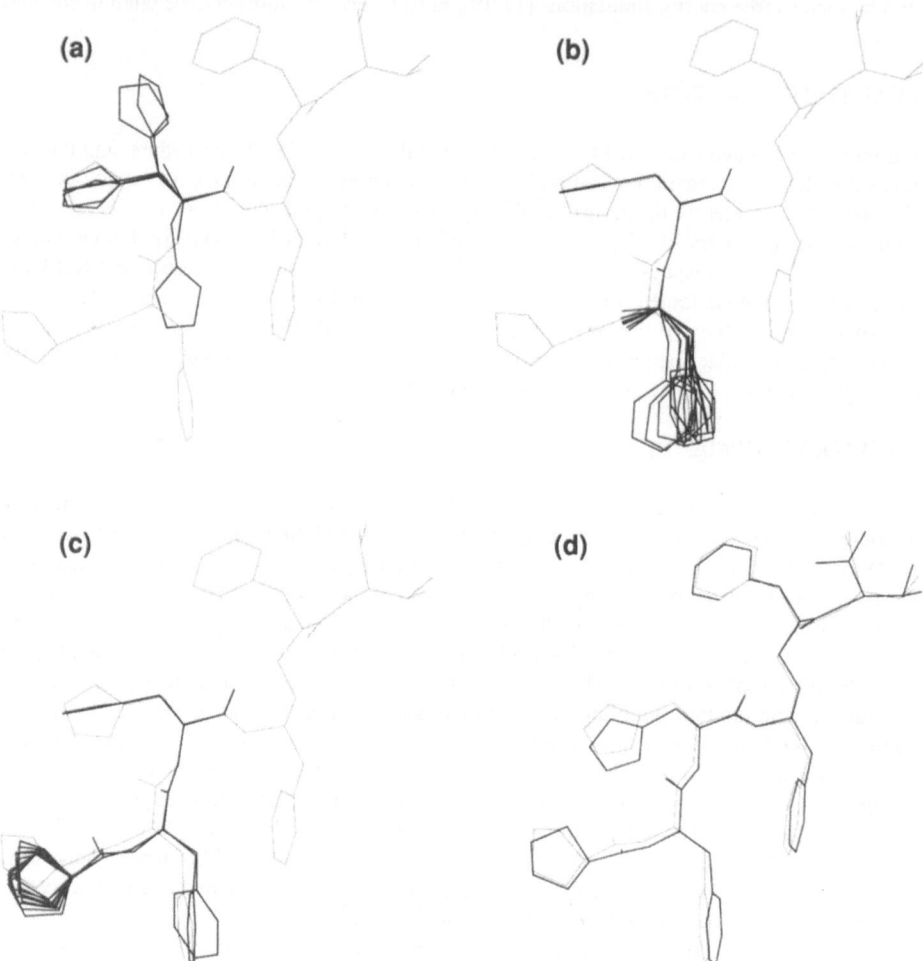

Figure 3. Steps in a restricted GROW run, in which the program was used to determine a binding geometry of the rhizopuspepsin inhibitor, U-70,531E. The crystallographic structure of the bound ligand (shaded) is shown to allow visual comparison with the conformations selected by GROW. However, the program used no information about this structure, except for its sequence and the location of one amide bond (seed position). (a) The 10 highest scoring histidine templates after the first stage of growth. Three regions were found which could accommodate the sidechain. (b) The ten highest scoring Phe–His dipeptide conformations found. Only one histidine conformation survived the second stage. (c) The ten best Pro–Phe–His tripeptide conformations after the third stage. (d) Overlay of the experimentally determined ligand structure (shaded) with the GROW's highest scoring final conformation (black) after both structures were minimized in the enzyme active site. The two structures differ by 0.6Å rms.

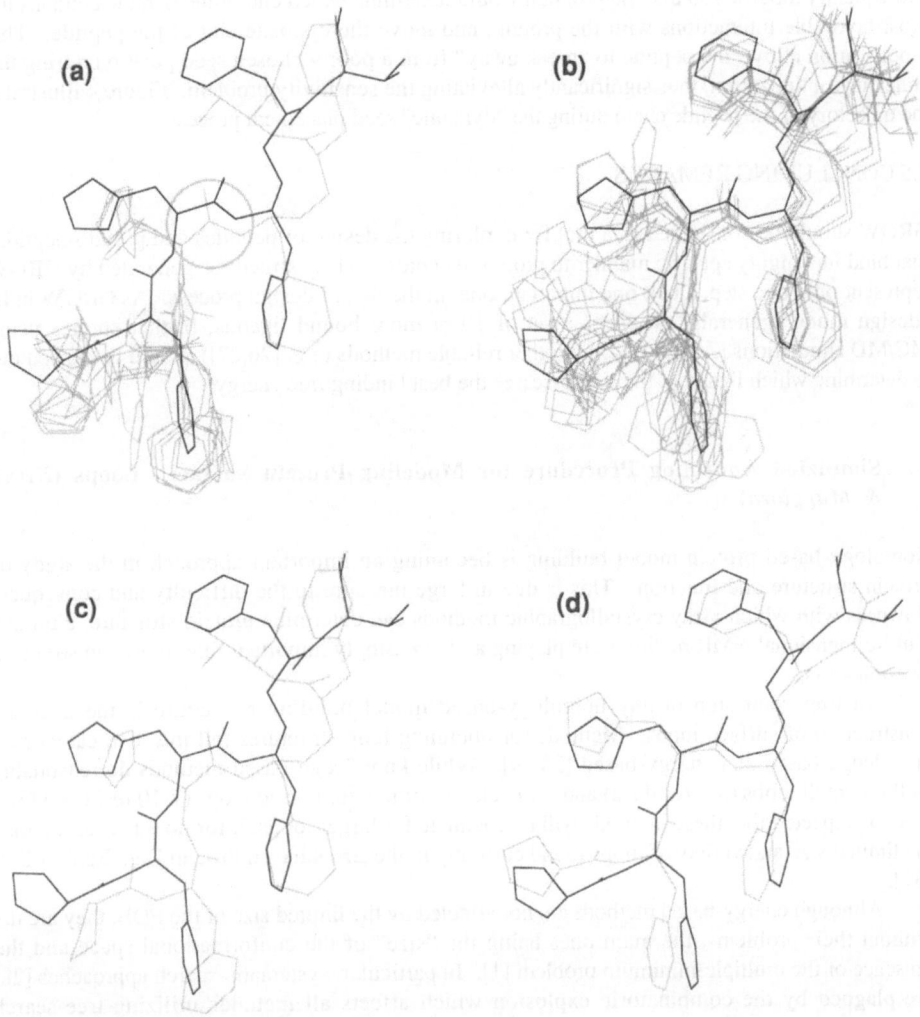

Figure 4 : Comparison of two MCSA implementations in GROW. The program was used to construct a binding geometry for U-70,531E to rhizopuspepsin using simulated annealing, starting from a poor seed position. For this example, the seed position was translated and rotated away from the x-ray structure (shown in black) of the bound ligand. (a) 20 ligand conformations during an annealing run in which the seed position (circled) is fixed in place. This seed position allows sampling of only one conformation of the c-terminal end of the peptide. (b) 20 ligand conformations generated from the same initial seed placement, during an annealing run in which the ligand is allowed to break away from the initial seed. Much broader sampling is evident. (c) The highest scoring structure form the annealing run shown in (a). The structure differs from the x-ray structure by 3.4Å RMS. (d) Highest scoring structure from the annealing run in (b). The deviation from the crystal structure is 1.4Å RMS before minimization, 0.8Å after minimization.

subsequently modified so that the program would determine which end of the peptide contains the most favorable interactions with the protein, and move the opposite end of the peptide. This modification allows the peptide to "break away" from a poorly chosen seed position during the annealing optimization, thus significantly alleviating the sensitivity problem. Figure 4 illustrates the trajectory of the peptide chain during the "dynamic" seed placement process.

2.5 CONCLUDING REMARKS

GROW should be considered as a tool for exploring the design of peptides and pseudo-peptides that bind in a highly specific manner to protein receptors. Thus, structures generated by GROW represent only one step, albeit one important one, in the ligand-design process. As GROW in its "design mode" generally produces a set of 10 or more bound ligands, further studies using MC/MD simulations [2,3,17-19,25] or other reliable methods (*e.g.* [26,27]) should be carried out to determine which ligand of the set possesses the best binding free energy.

3. Simulated Annealing Procedure for Modeling Protein Surface Loops (*Blinn & Maggiora*)

Homology-based protein model building is becoming an important approach in the study of protein structure and function. This is due in large measure to the difficulty and consequent slowness with which x-ray crystallographic methods can determine protein structure, although multi-dimensional NMR methods are playing an increasingly important role in protein structure determination.

An important step in any homology-based model building procedure is the accurate construction of surface loops. Methods for obtaining loop structures fall into two categories, knowledge-based and energy-based [28,29]. While knowledge-based methods do reasonably well for small loops (5-6 residues) and some classes of medium-sized loops (7-10 residues) [30], it is not expected that these methods will be adequate for larger loops, if for no other reason than the limited size and variety of loops found currently in the Brookhaven Protein Data Bank (PDB) [31].

Although energy-based methods are not affected by the limited size of the PDB, they are not without their problems, the main ones being the "size" of the conformational space and the presence of the multiple-minimum problem [1]. In particular, systematic-search approaches [28] are plagued by the combinatoric explosion which affects all methods utilizing tree-search procedures, even when effective pruning of the tree is employed. To overcome the difficulties of these procedures, Shenkin *et al.* [11] developed the "random tweak" procedure which produces sets of reasonable loop conformations that adequately "cover" the conformational space. The work reported here is related to this approach, the significant difference being that the current method utilizes an MCSA procedure [8,9] to overcome the multiple-minimum problem. As will be seen, loop structures determined in this way can closely approximate crystallographically determined ones for short and medium sized loops, but with relatively little computational effort.

The protein surface-loop modeling algorithm utilized is the following: (1) break the bond at one end, say the N-terminus, of the loop, (2) randomly perturb ϕ-ψ angles within the range of a given stepsize, (3) "reseal" the loop by varying the ϕ-ψ angles using a linear Lagrange multiplier procedure [11], a procedure which generally takes for 10-20 cycles, (4) evaluate the energy of the newly generated loop, (5) determine the acceptance or rejection of the loop using a Metropolis procedure [24], if the new conformer is rejected the parent conformation is used to generate a new randomly-perturbed conformation; if no acceptances are obtained after ~250 "cycles", the stepsize is reduced by 50%, and (6) after 10-20 acceptances at a given temperature (*e.g.* 2000°K) the stepsize is adjusted, and the system is "cooled" by a standard exponential cooling regimen, *e.g.* $T_{new} = 0.9 \cdot T_{old}$, until it reaches 100°K.

The potential-energy function employed in the present work is related to that found in AMBER [14] as implemented in MacroModel [15]; a dielectric constant given by $\varepsilon = r$ was used to partially correct for microenvironmental effects. As is the case with essentially all energy-based surface-loop modeling procedures, sidechains are truncated at the α-carbon so that all residues except glycine and proline are treated as alanine. Once the "primitive" loop has been fully optimized sidechains would be added; and numerous methods exist for carrying this out (see [28,29] for further details). Sidechains are not, however, considered here. The following two examples illustrate the exceptional performance of the current MCSA procedure on two loops, one of six and one of eight residues, taken from the immunoglobulin variable domain of 1IG2 [32]. Figure 5a shows five randomly-generated starting conformations for a 6-residue surface loop: the crystal structure is denoted by the dark line. Figure 5b portrays the loops after minimization by the MCSA procedure described here. As is clear from the figure, three of the five loops converged quite closely to the crystal structure, with an RMS difference of less than 0.5Å.

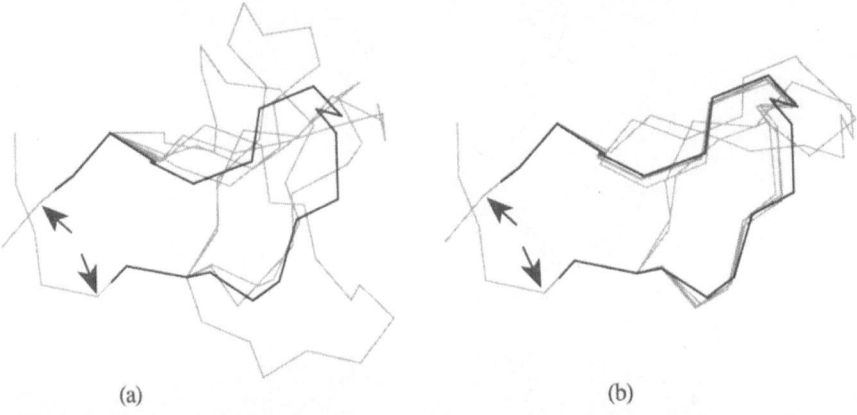

(a)   (b)

Figure 5. Application of the MCSA random-tweak procedure to a 6-residue surface loop from the immunoglobulin 1IG2 [32]. The two arrows indicate the points at which the surface is connected to the remainder of the protein. (a) The light lines depict five randomly-perturbed starting structures and the dark line the crystal structure. (b) The light lines depict the optimized structures obtained from the MCSA procedure and the dark line the crystal structure. Three of the starting structures converge to within 0.5Å of the crystal structure.

A more challenging case is that of the 8-residue surface loop illustrated in Figure 6a. Again the crystal structure is denoted by the dark line. The starting structure, indicated by the light gray line, was obtained by the random tweak procedure and is 2.6Å RMS from the crystal structure. The structure indicated by the thick, dark-gray line is the result of the MCSA procedure, and lies within 0.5Å RMS of the crystal structure.

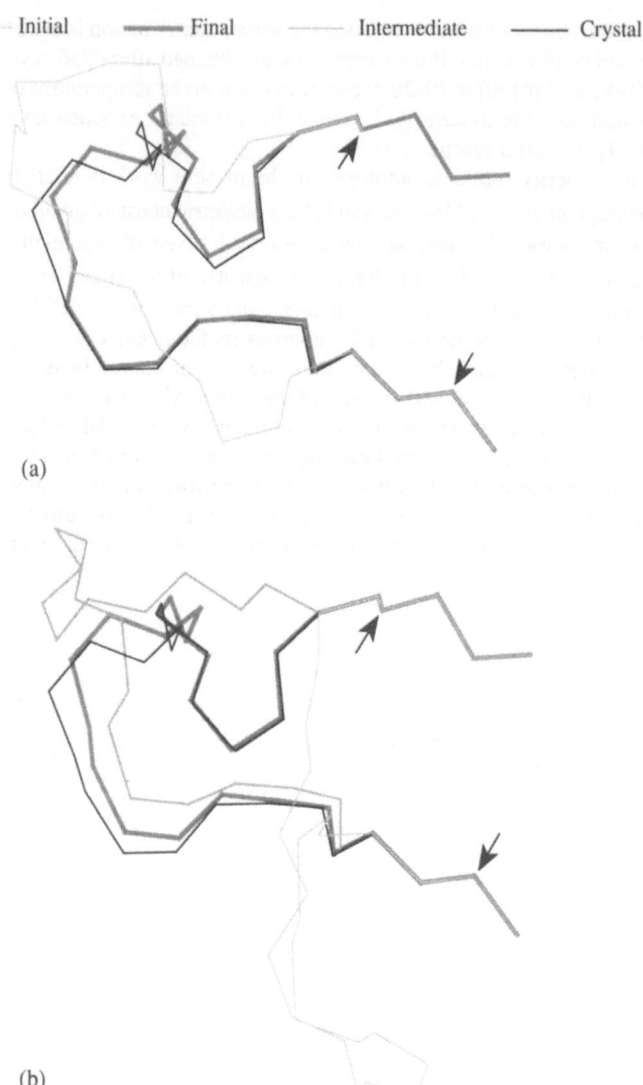

Figure 6. Application of the MCSA random-tweak procedure to an 8-residue surface loop from the immunoglobulin 1IG2 [32]. The two arrows indicate the points at which the surface loop is connected to the remainder of the protein. The legend at the top of the figure denotes the relationship of line quality to the type of structure, *viz.* 'Initial', 'Final', 'Intermediate', and 'Crystal'. (a) The light line denotes a randomly-generated surface loop 2.6Å RMS from the crystal structure, the darker line is the optimized structure obtained from the MCSA procedure, and the dark line the crystal structure. The latter two structures differ by 0.5Å RMS. (b) The light line denotes a randomly-generated surface loop 8.3Å RMS from the crystal structure given by the dark line. The two other lines denoted as 'Intermediate' and 'Final' represent, respectively, the structure obtained after the first pass of the MCSA procedure and that obtained after a second pass starting at the 'Intermediate' structure.

In order to examine the robustness of the procedure, a considerably more perturbed initial structure, denoted by the light gray line in Figure 6b, was generated with an RMS difference of 8.3Å from crystal structure denoted by the dark line. The 'Intermediate' structure (see legend in figure), which differs from the crystal structure by 2.6Å RMS, is the result of the first MCSA minimization. Applying the MCSA procedure to this structure yields a structure, denoted by the wider, dark-gray line in Figure 6b, which again lies close to the crystal structure, with an RMS deviation of only 0.4Å. *In all of the cases none of the converged structures were lower in energy than the crystal structure.*

All of the calculations described here were carried out on a VAX 8800 in a minimal amount of time (generally less than 20 minutes for the 6-residue case and 90 minutes for the 8-residue case). Thus, the MCSA procedure appears to be scalable to larger surface loops, at least to 14-16 residue loops and perhaps to even larger ones. This is currently under investigation. In addition, the use of simpler potential-energy functions, such as the PROSA potentials developed by Wilson and Doniach [33], is also under investigation. Finally, preliminary studies have shown that the current MCSA procedure can also be applied to cyclic systems, and further studies along this line are also being considered.

4. Mass-weighted Molecular Dynamics Conformational-space Sampling (*Mao, Maggiora & Chou*)

Sampling the conformational space of cyclic peptides presents interesting challenges. Although the conformational constraints imposed by cyclization reduce the effective conformational space accessible to the system, thoroughly sampling even this reduced space is difficult to accomplish using typical MC or MD procedures. In MC-based applications to constrained systems, such as cyclic peptides, the interdependence of the dihedral angles can present significant problems (*vide supra*).

In contrast to MC approaches, standard MD methods employ Cartesian coordinates, but the integration timescale is such that normally only a small region of the conformational space is sampled. The local conformational space can be sampled more efficiently when the scalar mass is replaced by a tensor [34]; on the other hand, increasing atomic velocities (*i.e.* conventional high-temperature simulations) increases the conformational sampling over potential-energy barriers, although a practical limit to this approach is reached relatively quickly. The difficulty encountered in such cases is that the the numerical integration becomes unstable [35], necessitating a reduction in integration time step size, lengthening the computational time.

Recently, mass-weighted molecular dynamics (MWMD) was investigated by Mao [10,36] as a means for overcoming this difficulty and improving conformational-space sampling. In MWMD simulations the velocity distribution of the particles is the same as that in normal MD simulations, whereas all the particle masses are numerically scaled by a common factor; this increases the momentum, kinetic energy, and temperature of the system. And although the potential-energy function remains unchanged, the total energy of the system increases due its increased kinetic energy. Hence, some of the potential-energy barriers that lie above the constant energy hyper plane in normal-mass MD now lie below it, and the system can explore regions of conformational space that heretofore were forbidden energetically [35]. In addition, the greater inertia of the par ticles in the system apparently enhances the likelihood of torsional barrier crossings within rela tively short periods of simulation. The procedure was shown [36] to provide essentially complete sampling of the dihedral conformational space for linear tetrapeptides (*vide infra*).

28

The procedure has subsequently been applied to pressinoic acid, a cyclic hexapeptide derived from vasopressin, and to a "tethered", pseudo-cyclic version of the same hexapeptide [37]. In these simulations, the mass-weighting factor is generally set at 10.0 and the atomic velocity distribution corresponds to that at 600°K. Covalent interactions such as bond stretching, valence-angle bending, and improper torsional-angle motions are constrained by harmonic potentials [36,37]. All of these features are easily programmed within CHARMM [38]. The computational time required to treat these systems was modest. For example, 100 ps of simulation for the tethered, pseudo-cyclic hexapeptide (consisting of 66 atoms in CHARMM's extended-atom representation), required approximately 50 m of CPU time on an IBM 3090J.

The results on both constrained systems showed that the sampling of the conformational space of individual dihedral angles is dramatically improved over that obtained from conventional MD simulations. Figure 7 depicts the dihedral angles of pressinoic acid investigated in the current study, a total of 10 φ-ψ and 5 χ-angles, although all dihedral angles including those associated with sidechains were treated in the study. Figure 8 illustrates the sampling of three of these angles, the φ-ψ pair, 8 and 9, of the asparagine residue and the χ-angle, 14, of the disulfide bond. The darker lines correspond to MWMD and the lighter ones to normal-mass MD. From the figure it is clear that the extent of sampling is considerably better for MWMD than for normal-mass MD.

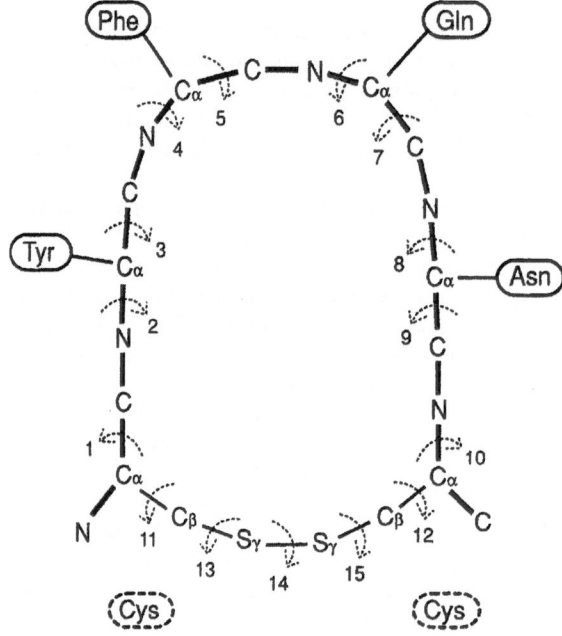

Figure 7. Schematic depiction of pressinoic acid. Each curve arrow designates one of the dihedral angles studied. Dihedral angles 1-10 represent backbone φ-ψ angles and 11-15 represent sidechain χ-angles of the disulfide linkage between the two terminal cysteine residues. (*Reprinted with permission from* [37])

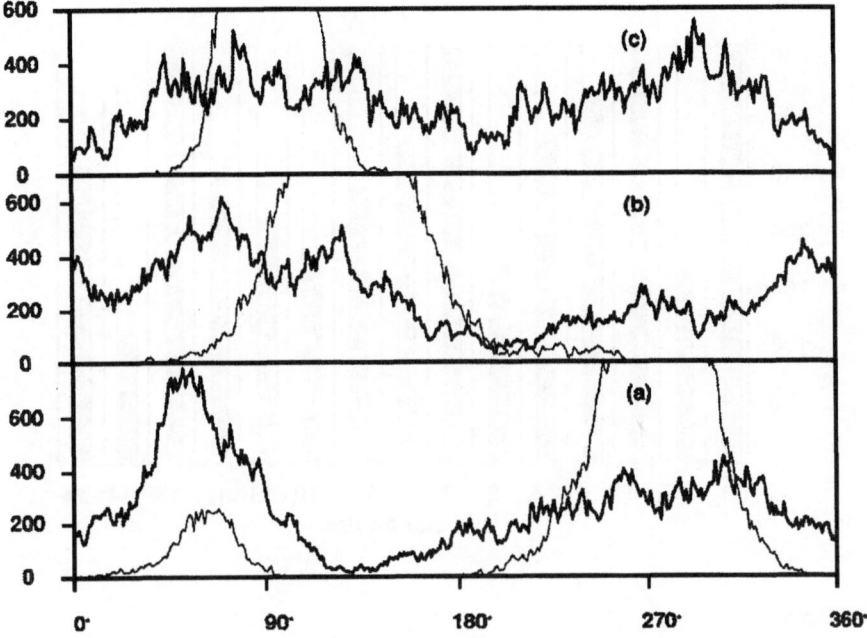

Figure 8. Distribution of dihedral-angle values sampled over the full 360°-range for angles (a) 8, (b) 9, and (c) 14. For each angle the 360°-range is divided into 1° bins and the number of times a given angle "visits" a particular bin during a simulation is plotted against the bin number. Curves shown as thin lines are obtained from normal-mass MD simulations while those shown as thick lines are obtained from MWMD simulations with a mass-weighting factor of 10. All simulations were carried out at a temperature of 600°K. (*Reprinted with permission from* [37])

Two parameters, the *range R* and *coverage C*, were defined to characterize the sampling quantitatively [37]. R is defined as the number of 1°-regions of the total 360° dihedral-angle space sampled during an MWMD simulation; thus, $1 \leq R \leq 360$. C is given by the ratio, expressed as percent, of the area under the dihedral angle sampling distribution curve to the area of the rectangle formed by the value of the dihedral angle which is sampled the maximum number of times and the full 360°-range; thus, $0.278\% \leq C \leq 100\%$. Figure 9 shows the R and C parameters computed from curves such as those in Figure 8 for all backbone dihedral angles in pressinoic acid. In terms of the R and C parameters, it is clear from the histograms in Figure 9 that MWMD is clearly superior to normal-mass MD as a means for sampling conformational space.

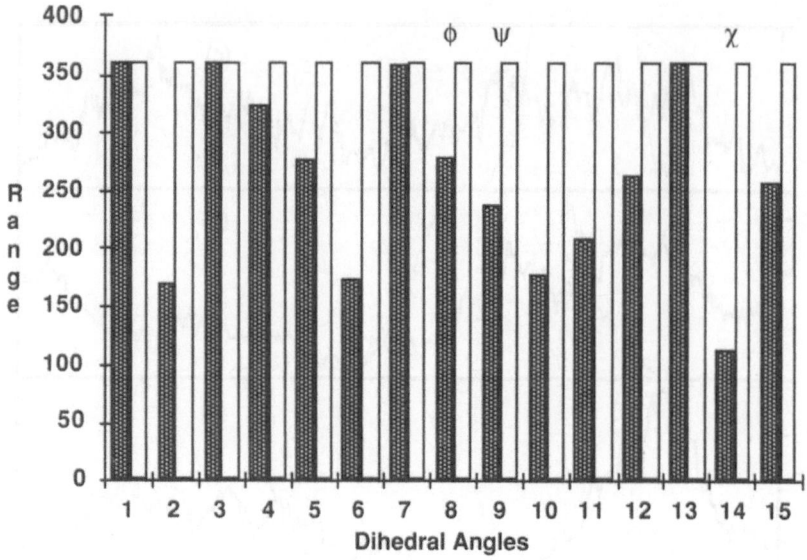

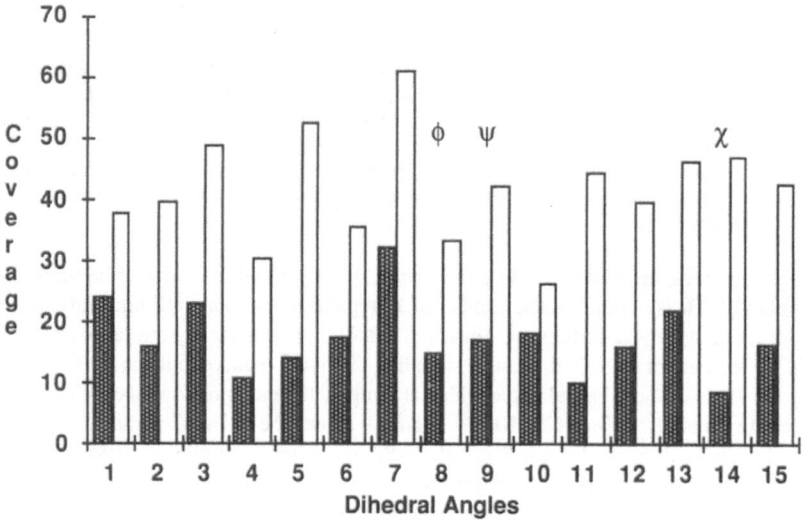

Figure 9. Histograms illustrating the range **R** and coverage **C** parameters for normal-mass and MWMD simulations of pressinoic acid. The dark bars correspond to normal-mass MD and the shaded bars to MWMD. The ϕ, ψ, and χ angles corresponding respectively to dihedrals 8, 9, and 14 (see also Figure 8) are indicated on the figure.

The results in Figure 10 for a tethered, pseudo-cyclic peptide show that the conformational space of individual ϕ-ψ angles are well sampled by the MWMD procedure. Moreover, Figure 11 shows the scope of the backbone motion of the tethered loop. As is clear from the figure, the physical 3-D space of the loop about the fixed points is also sampled.

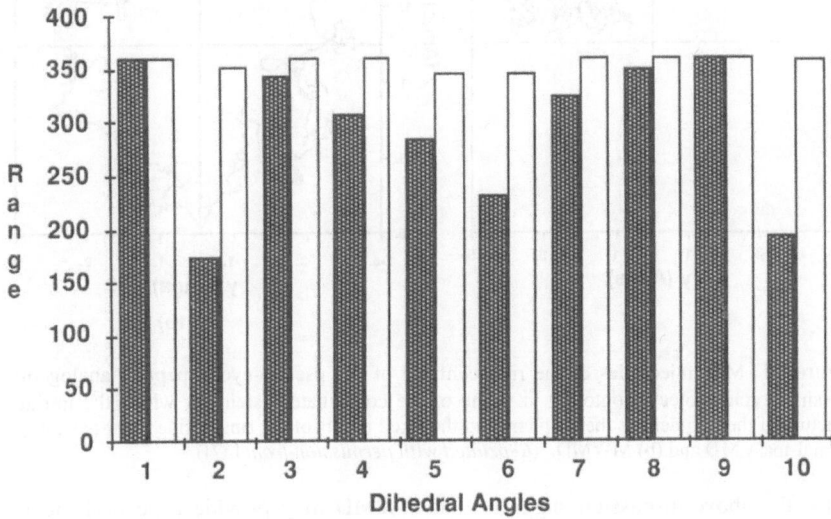

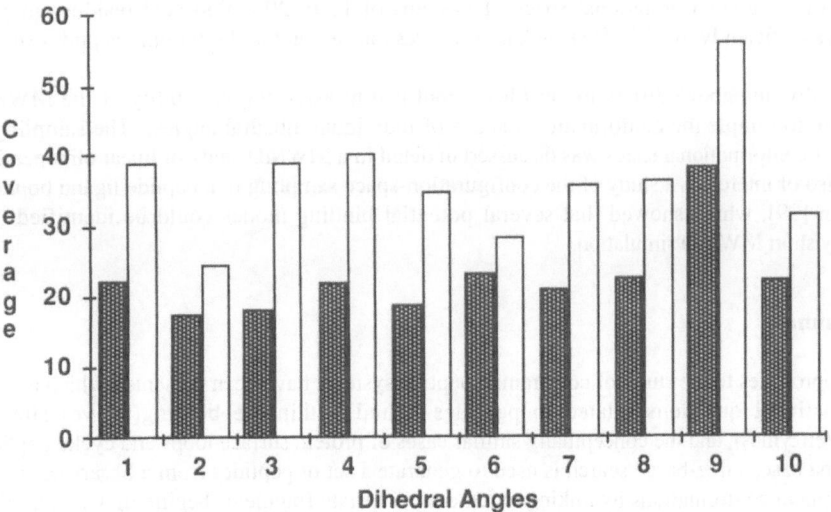

Figure 10. Histograms illustrating the range R and coverage C parameters for normal-mass and MWMD simulations of the pseudo-cyclic analog of pressinoic acid. The dark bars correspond to normal-mass MD and the shaded bars to MWMD.

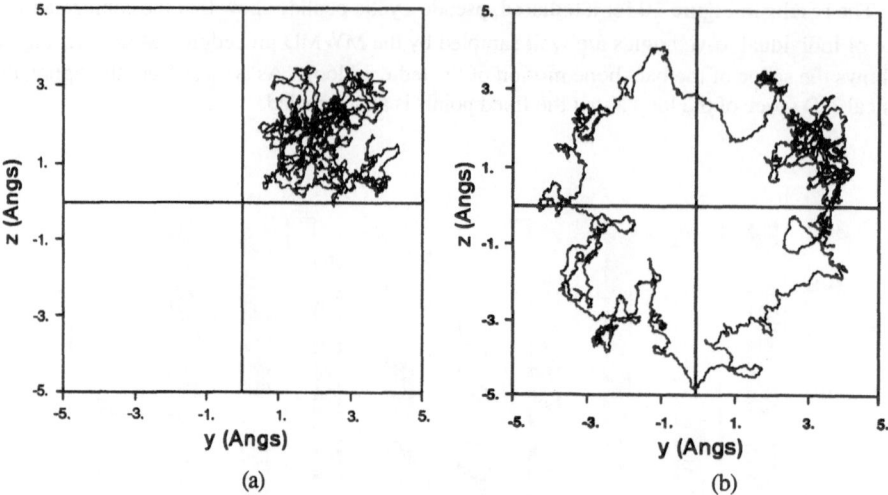

Figure 11. MD trajectories of the ring centroid of the pseudo-cyclic peptide analog of pressinoic acid projected onto the yz-plane of the coordinate system for which the initial structure of the ring defines the xy-plane and the fixed points of the ring define the x-axis: (a) normal-mass MD and (b) MWMD. (*Reprinted with permission from* [37])

Thus, the above discussion suggests that MWMD may provide a general means for conformational-space sampling in constrained peptide systems. Extension of the procedure to longer peptides is currently under investigation. Based on current findings, however, it is expected that the conformational space of systems of 15 to 20 amino acid residues may be sampled as efficiently by MWMD simulations as was the case in the shorter oligopeptides studied thus far.

Finally, the above discussion has been confined to assessing the ability of the MWMD procedure to sample the conformational space of individual dihedral angles. The sampling of molecular conformational states was discussed in detail in a MWMD study of linear oligopeptides [36]. Also of interest is a study of the configuration-space sampling of a peptide ligand bound to a protein [39], which showed that several potential binding modes could be identified in a relatively short MWMD simulation.

5. Summary

Three approaches to the study of constrained peptide systems have been presented which address conformational questions related to peptides bound within the binding(active) sites of proteins(enzymes), and the conceptually similar cases of protein surface loops and cyclic peptides. In the first case, a tree-based search is used to generate a set of peptides from a library of amino-acid fragment conformations by linking together appropriate fragments beginning with an initial "seed". Fragments are chosen both to optimize the fit of the peptide to the binding(active) site and to maximize an estimate of its binding free energy. GROW has already demonstrated the ability to generate biologically active peptide inhibitors and substrates, and to generate conformers that match corresponding crystal structures to within 1Å RMS. A recent modification, which replaces the tree search with an MCSA algorithm, significantly reduces computational times. And, at the option of the user, this method also permits the original seed location to move, resulting in a much broader exploration of the binding site.

In the second case, an MCSA procedure for modeling protein surface loops is described which extends, in a novel way, the random-tweak method of Shenkin et al. [11]. Preliminary tests of the procedure on 6- and 8-residue surface loops of 1IG2 [32] indicate that the method produces loops whose structure are in close agreement to the corresponding crystal structure. Moreover, the method is seen to be both robust and computationally efficient.

In the third case, an MWMD procedure is shown to provide a highly efficient means for sampling the conformational space of cyclic and constrained peptides, as illustrated by pressinoic acid and a "tethered" pseudo-cyclic peptide analog meant to model a protein surface loop. The results described here are also in accord with several other studies of small linear peptides [10, 35-37]. In all instances examined thus far, MWMD is seen to be superior to normal-mass, high-temperature MD simulations as a means of conformational-space sampling. Moreover, MWMD is not plagued by numerical instabilities which present significant problems in typical high-temperature MD sampling studies.

As shown by the present work, MC- and MD-based sampling procedures can deal with flexible molecular systems of modest dimension, such as the constrained 6- to 8-residue peptides considered here, in a computationally tractable manner. Extension of these approaches to larger systems is currently under investigation, and it is not unreasonable to assume that they should be scalable to systems of approximately twice the size of those examined to date. Further extension to even larger systems is, however, problematic and it is quite likely that significant practical computational limitations will be encountered. How this computational bottleneck can be dealt with is the subject for detailed future investigations, investigations that will begin to merge with those of the related, but still unsolved, protein-folding problem.

References

[1]   Gibson, K.D. and Scheraga, H.A. (1988) in M.H. Sarma and R.H. Sarma (eds.), *Structure and Expression*, Vol. 1, From Proteins to Ribosomes, Adenine Press, Guilderland, New York, pp. 67-94.

[2]   Kalos, M.H. and Whitlock, P.A. (1986) M*onte Carlo Methods*, Vol.I, John Wiley & Sons, New York.

[3]   DiNola, A., Berendsen, H.J.C., and Edholm, O. (1984) *Macromolecules* **7**, 2044-2050.

[4]   Noguti, T. and Go, N. (1985) *Biopolymers* **24**, 527-546.

[5]   Go, N. and Scheraga, H.A. (1970) *Macromolecules* **3**, 178-187.

[6]   Chang, G., Guida, W.C., and Still, W.C. (1989) *J. Amer. Chem. Soc.* **111**, 4379-4386.

[7]   Li, Z. and Scheraga, H.A. (1987) P*roc. Natl. Acad. Sci. USA* **84**, 6611-6615.

[8]   Kirkpatrick, S., Gellatt, C.D., and Vecchi, M.P. (1983) *Science* **220**, 671-680.

34

[9]   Wilson, S.R. and Cui, W. (1990) *Biopolymers* **29**, 225-235.

[10] Mao, B. and Friedman, A.R. (1990) *Biophys. J.* **58**, 803-805.

[11] Shenkin, P.S., Yarmush, D.L., Fine, R.M., Wang, H., and Levinthal, C. (1987) *Proteins* **1**, 2053-2085.

[12] Moon, J.B. and Howe, W.J. (1991) *Proteins*, in press.

[13] Weiner, S.J., Kollman, P.A., Case, D.A., Singh, U.C., Ghio, C., Alagona, G., Profeta, S., Jr., and Weiner, P. (1984) *J. Amer. Chem. Soc.* **106**, 765-784.

[14] Mohamadi, F., Richards, N.G.J., Guida, W.C., Liskamp, R., Lipton, M., Caufield, C., Chang, G., Hendrickson, T., and Still, W.C. (1990) *J. Comp. Chem.* **11**, 440-467.

[15] Ooi, T., Oobatake, M., Nemethy, G., and Scheraga, H.A. (1987) *Proc. Natl. Acad. Sci. USA* **84**, 3086-3090.

[16] Hasel, W., Hendrickson, T., and Still, W.C. (1988) *Tetrahedron Comput. Methodol.* **1**, 103-116.

[17] McCammon, J.A. and Harvey, S.C. (1987) *Dynamics of Proteins and Nucleic Acids*, Cambridge University Press, Cambridge.

[18] Brooks, C.L., III, Karplus, M., and Pettitt, B.M. (1988) *Proteins: A Theoretical Perspective of Dynamics, Structure, and Thermodynamics*, John Wiley & Sons, New York.

[19] Beveridge, D.L. and DiCapua, F.M. (1989) in W.F. van Gunsteren and P.K. Weiner (eds.), *Computer Simulation of Biomolecular Systems*, ESCOM, Leiden, pp. 1-26.

[20] Sawyer, T.K., Pals, D.T., Mao, B., Maggiora, L.L., Staples, D.J., deVaux, A.E., Schostarez, H.J., Kinner, J.H., and Smith, C.W. (1988) *Tetrahedron* **44**, 661-673.

[21] Suguna, K., Padlan, E.A., Smith, C.W., Carlson, W.D., and Davies, D.R. (1987) *Proc. Natl. Acad. Sci. USA* **84**, 7009-7013.

[22] Miller, M., Schneider, J., Sathyanarayana, B.K., Toth, M.V., Marshall, G.R., Clawson, L., Selk, L., Kent, S.B.H., and Wlodawer, A. (1989) *Science* **246**, 1149-1152.

[23] Moon, J.B. and Howe, W.J. (1992) Manuscript in preparation.

[24] Metropolis, N., Rosenbluth, A.R., Rosenbluth, M.N., Teller, A.H., and Teller, E. (1953) *J. Chem. Phys.* **21**, 1087-1092.

[25] van Gunsteren, W.F. (1989) in W.F. van Gunsteren and P.K. Weiner (eds.), *Computer Simulation of Biomolecular Systems*, ESCOM, Leiden, pp. 27-59.

[26] Gilson, M. and Honig, B. (1988) *Proteins* **4**, 7-18.

[27] Warshel, A. and Creighton, S. (1989) in W.F. van Gunsteren and P.K. Weiner (eds.), *Computer Simulation of Biomolecular Systems*, ESCOM, Leiden, pp. 121-138.

[28] Maggiora, G.M., Mao, B., Chou, K.C., and Narasimhan, S.L. (1990) *Methods of Biochemical Analysis* **35**, 1-86.

[29] Maggiora, G.M., Narasimhan, S.L., Granatir, C.A., Blinn, J.R., and Moon, J.B. (1991) in S.J. Formosinho, I.G. Csizmadia, and L.G. Arnaut (eds.), *Theoretical and Computational Models for Organic Chemistry*, Kluwer Academic Publishers, Dordrecht, The Netherlands, pp. 137-158.

[30] Tramontano, A., Chothia, C., and Lesk, A.M. (1989) *Proteins* **6**, 382-394.

[31] Bernstein, R., Koetzle, T.F., Williams, G.J.B., Meyer, E.F., Jr., Brice, M.D., Rodgers, J.R., Kennard, O., Shimanouchi, T., and Tasumi, M. (1977) *J. Mol. Biol.* **112**, 535-542.

[32] Marquart, M., Deisenhofer, J., and Huber, R. (1980) *J. Mol. Biol.* **141**, 369-391.

[33] Wilson, C. and Doniach, S. (1989) *Proteins* **6**, 193-209.

[34] Bennett, C.H. (1975) *J. Comp. Phys.* **19**, 267-279.

[35] Mao, B. (1991) Manuscript in preparation.

[36] Mao, B. (1991) *Biophys. J.* **60**, 611-622.

[37] Mao, B., Maggiora, G.M., and Chou, K.C. (1991) *Biopolymers* **31**, 1077-1086.

[38] Brooks, B.R., Bruccoleri, R.E., Olafson, B.D., States, D.J., Swaminathan, S., and Karplus, M. (1983) *J. Comp. Chem.* **4**, 187-217.

[39] Mao, B. (1991) *Biophys. J.*, in press.

DISCUSSION

Question (Professor P.A. Kollman):

"What is the consequence of mass-weighted molecular dynamics sampling with ε = 80? Will MWMD preferentially sample higher-energy regions of conformational space? Will it work with other ε?"

The simulations were carried out with ε = 80; the procedure also works with other ε values. In MWMD, high-energy regions of dihedral-angle conformational space are not preferentially sampled (see [36] in text).

"In the simulated-annealing-based loop modeling, when do sidechain contributions become important?"

While the examples provided here show that it is not necessary in all cases to include sidechains, there certainly are situations where sidechain interactions are significant, *e.g.*, when H-bonding between a sidechain group and the remainder of the protein is important (see *e.g.* [30] in text). We have carried out some very preliminary studies using the rather simple effective (PROSA) potentials developed by Wilson and Doniach (see [33] in text). The results were, unfortunately, rather inconclusive: although the location of sidechains was reasonable (*i.e.* hydrophobic sidechains tended to be buried and hydrophilic ones exposed) the corresponding backbone geometries obtained were not well accounted for.

Question (Professor P.G. Mezey):

"The mass-scaled molecular dynamics approach allows you to formally exaggerate momentum in order to lead the system out of local potential energy 'traps' on potential-energy surfaces. By keeping velocities, and hence directions, constant, the freedom in the theoretically possible motions is reduced, based on those motions accessible within the non-scaled molecular dynamics at lower energy."

The velocities of individual particles are not fixed. Rather, the same velocity distribution is maintained for comparison purposes; thus, the directions in which the particles move are not necessarily more restricted.

Question (Dr. A. Taymaz):

"First, in this computational modeling, what sort of computer programs have you used?"

The programs used in the GROW and simulated annealing procedures were developed locally; CHARMM was used for the MWMD studies (see [38] in text).

"Second, what is a typical CPU time for typical samples?"

As we have used a number of different machines for the calculations describe here, it is difficult to give a simple answer to your question - suffice it say that most calculations took less than an hour, and that is none of the calculations was computationally burdensome or intractable.

Question (Dr. M. Pons):

"Shouldn't we be concerned about using forcing potentials that were not present when the force field was developed, but keeping the rest of the parameters?"

Yes, but only to the extent that such potentials affect the sampling of particular degrees of freedom, which in our studies were the dihedral angles; the results from a study of linear peptides showed (see [36] in text) that this is not the case, at least for linear peptides.

Question (Dr. J.J. Perez):

"From your experience modeling protein loops, how important is the effect of the rest of the protein when you modeled it?"

In our studies the protein in the region of the loop is explicitly treated, although it is held constant during the simulated annealing. It is critical to take some account of the environment in which the loop resides, but how much of the protein needs to treated is not clear. Also, I should point out that we dealt with the simplest case, that is we treated the loops one at a time. In the more realistic case when all loops must be treated, interactions between near loops would present additional computational difficulties that we did not address.

"Are the results expected to be similar if you model a constrained peptide?"

By constrained I assume you mean a cyclic peptide. We have carried out some preliminary studies on cyclic peptides, which indicate that the simulated annealing approach described here will produce comparable results to those obtained for protein surface loops.

"What information can you expect from your loop modeling studies?"

I'm not sure what you mean by your question. What we hope to gain is a procedure for calculating surface loops that is both reliable and computationally tractable.

Question (Professor H.A. Weinstein):

"Just a short comment to emphasize the importance of the rest of the protein structure in determining the conformation of the loop. In exploring the role of the protein secondary and tertiary structure surrounding the calcium binding loop in Calbinden D_{9k}, we found that removal of the flanking helices or changes in their tertiary structures yielded loop

conformations that differed much from that known from the crystal structure. One of these 'new' minima does, in fact, correspond to a loop found in a different protein that diverged from the class of Ca^{2+}-binding proteins through mutations that changed the tertiary structure. That loop lost its ability to bind Ca^{2+}. Thus, the rest of the protein structure seemed to play a determinant role in maintaining the functional conformation of the loop, and should be included in some form in any exploration of possible structures."

I fully agree with you, but would add the additional point that it may not be necessary to include all the protein in preliminary efforts to obtain reasonable surface loops as additional, and more complete structural optimizations will be performed before the model-building process is completed. Nevertheless, we included a significant portion of the protein in our calculations —a sphere of ~10 Å radius which "surrounds" the loop being modeled.

PEPTIDE CONFORMATIONAL POTENTIAL ENERGY SURFACES AND THEIR RELEVANCE TO PROTEIN FOLDING*

Andras Perczel
Institute of Organic Chemistry
Eötvös University
Budapest
Hungary

Wladia Viviani
Laboratoire de Chimie Theorique
Universite de Nancy I
Vandoeuvre-les-Nancy
France

Imre G. Csizmadia
Department of Chemistry
University of Toronto
Toronto, Ontario
Canada

ABSTRACT

This paper reviews the background and outlines the utility of a 3D → 1D transformation of peptide conformation. Although this transformation leads to a linearized notation of protein secondary and tertiary structures that may be used for an objective description and classification of protein folding*, nevertheless, the method is intended to be descriptive and is not meant to be predictive.

---

*In the present paper "protein folding" is used exclusively to describe the static aspect of a primary structure of a protein, folded to some particular secondary and tertiary structure. In other words a particular "protein folding" is taken here to be the synonym of a particular "protein conformation".

J. Bertrán (ed.), Molecular Aspects of Biotechnology: Computational Models and Theories, 39–82.
© 1992 Kluwer Academic Publishers.

MULTIDIMENSIONAL CONFORMATIONAL ANALYSIS AND POTENTIAL ENERGY SURFACES

Multidimensional Conformational Analysis (MDCA), which is an intuitive conceptional tool of every organic chemist, allows one to predict from the topology of the component potential energy curves (PEC) the topology of the potential energy surface (PES) if the molecular system is ideal[1,2]. This is illustrated in Figure 1 when the PEC's have three fold periodicity. In this case the $3 \times 3 = 9$ minima are energetically degenerate.

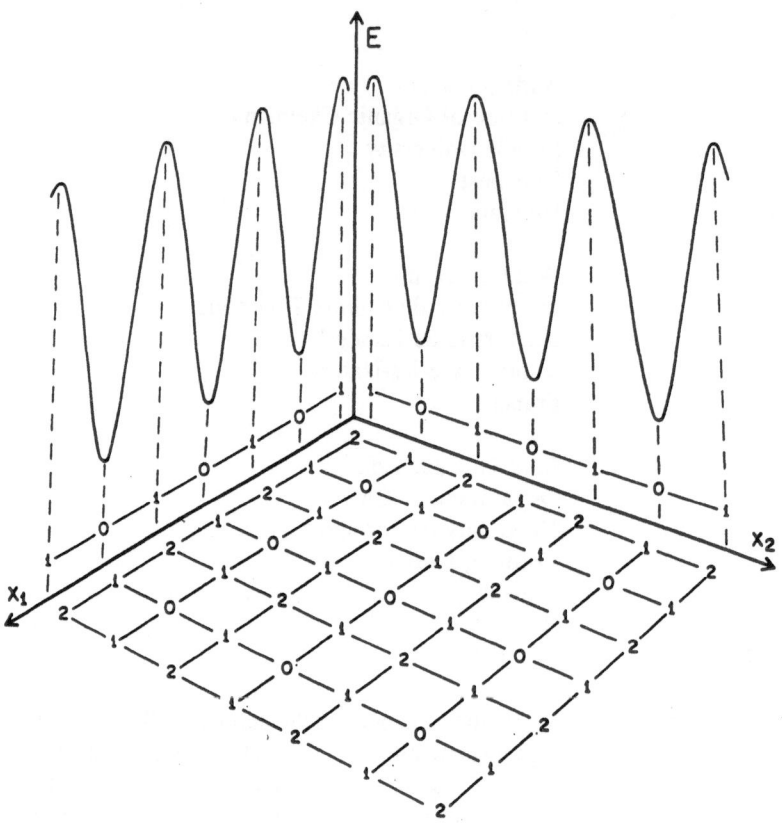

Figure 1: A schematic representation of how the ideal potential energy surface topology may be predicted from potential curves. The component potential energy curves show triple degeneracy.

This is the case which is operative for two CH_3 rotors as may be occurring in propane and in molecules with two equivalent CH_3 groups (e.g. CH_3OCH_3, CH_3-$(CH_2)_2$-CH_3 etc.) as illustrated by Figure 2.

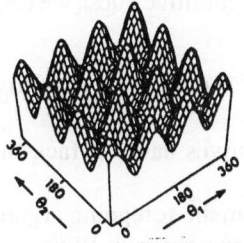

Figure 2: A pseudo three dimensional representation of the conformational potential energy surface: $E = E(\theta_1, \theta_2)$ of propane showing the full cycle of rotation ($360° \times 360°$).

If, on the other hand, the component PEC's continue to have three minima but these minima are energetically non-degenerate then the resultant PES will have 9 non equivalent minima. Such a hypothetical case is illustrated in Figure 3.

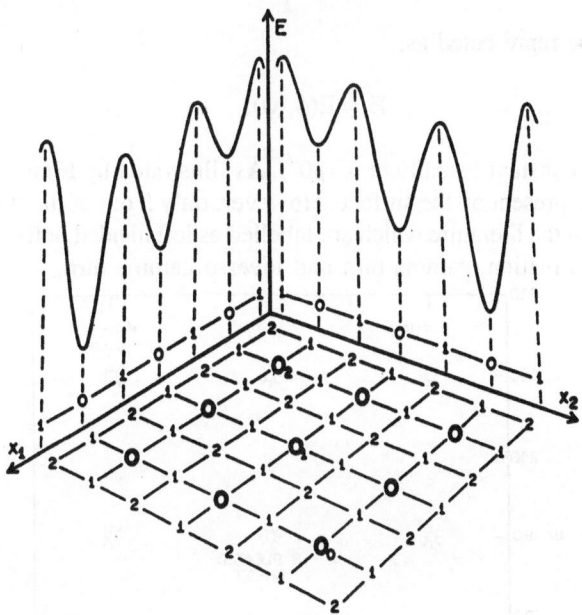

Figure 3: A schematic representation of how the ideal potential energy surface topology may be predicted from potential curves. The component potential energy curves do not show degeneracy.

In the case of Figure 1 it was possible to make a statement that all 9 minima have the same energy value; in the case of Figure 3 we can make an analogous statement that all 9 minima have different energy values. However, we are not in the position to predict what the energy spectrum of these nine minima might be and what the relative stability of these minima could be. Nevertheless, by making an intuitive guess we can suggest an order for the relative stabilities of the diagonal elements:

$$E(0_2) > E(0_1) > E(0_0) \quad [1]$$

if the component potential energy curves have, in fact, the shape as shown in Figure 3.

What is important to note from the foregoing arguments is that the PES for a single peptide unit (I) is analogous to the one given in Figure 3:

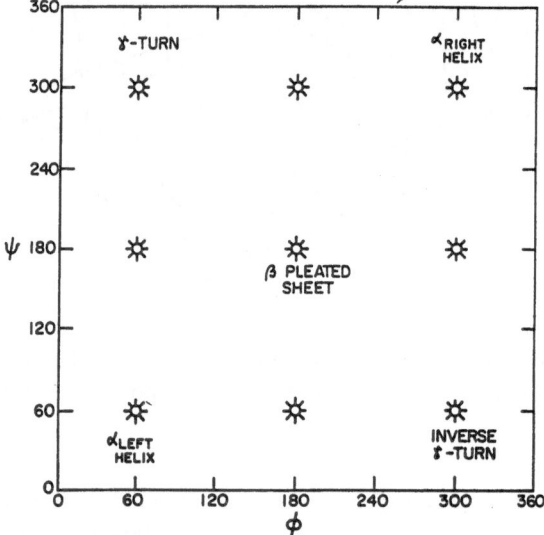

I

and its PES may be represented as,

$$E = E(\phi, \psi), \quad [2]$$

assuming that ω is constant (usually $\omega = 180^\circ$). As illustrated by Figure 4, 9 minima have been expected to be present on the surface. However, only 5 out of the 9 minima have been recognized earlier in the literature which are labelled as left handed helix right handed helix, extended like conformation, gamma turn and inverse gamma turn.

Figure 4: Idealized PES topology for a single amino acid residue indicating the five minima that already have been identified in the protein literature. (The idealized location of the minima are specified by open stars.)

In Figure 4, shown above, both ϕ and ψ vary between $0°$ and $360°$. However, protein chemists have adopted a range for both ϕ and ψ that runs between $-180°$ and $+180°$ covering both clockwise and counter-clockwise rotations, which may be labelled as standard (STD):

$$-180° \leq \phi_{STD} \leq +180°$$

$$-180° \leq \psi_{STD} \leq +180°$$

[3]

We feel that our representation is more useful as topological (TOP) relationships can be recognized with a greater ease:

$$0° \leq \phi_{TOP} \leq 360°$$

$$0 \leq \psi_{TOP} \leq 360°$$

[4]

More important is the fact that apart from the central minimum (the conformation associated with β-pleated sheet) the minima occur in pairs. Thus the remaining unassigned 4 minima (c.f. Figure 4) could be regarded as 2 pairs of minima. Preliminary study indicated that, apart from the β conformation, the most important, that is the energetically most favoured, conformations for the L-enantiomer are at the extreme right and lower right and for the D-enantiomer the most favoured conformations are at the upper left and extreme left. The topological relationship of these two families are illustrated in the following SCHEME

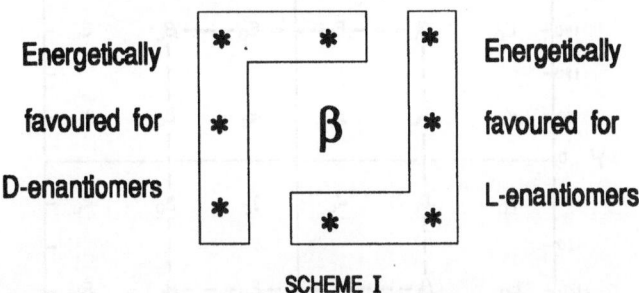

Energetically favoured for D-enantiomers   β   Energetically favoured for L-enantiomers

SCHEME I

Consequently, the greek symbols associated with the extreme right are subscripted by L and the greek symbols associated with the extreme left are subscripted by D. Thus instead of saying that the conformation that is capable of producing a right handed helix (α_{RIGHT}; sometime as α_R denoted) we say it is the α_L conformation. Similarly, instead of the conformation that is capable of generating a left handed helix (α_{LEFT}; sometime referred to as α_L) we might say it is the α_D conformation.

In order to refer to the as of yet unassigned conformations, the mid-point at the top is labelled as δ_D and the mid-point at the bottom is labelled as δ_L. The mid-point at the left is labelled as ε_D and the mid-point at the right is labelled as ε_L.

Utilizing the labels used previously[2], to denote the location of the minima, we obtain the following arrangement:

$$\gamma_D \qquad\qquad \delta_D \qquad\qquad \alpha_L$$

$$\varepsilon_D \qquad\qquad \beta_{DL} \qquad\qquad \varepsilon_L \qquad\qquad \text{SCHEME II}$$

$$\alpha_D \qquad\qquad \delta_L \qquad\qquad \gamma_L$$

For glycine where no chiral centre exists, the β_{DL} conformation is to be located at the geometrical centre. For L-amino acids β_{DL} becomes β_L and its position is shifted towards the lower right hand corner. For D-amino acids β_{DL} becomes β_D and its position is shifted towards the upper left hand corner of the idealized topological scheme (c.f. SCHEME II) which represents only a different cut of the PES as illustrated by the broken lines in Figure 5.

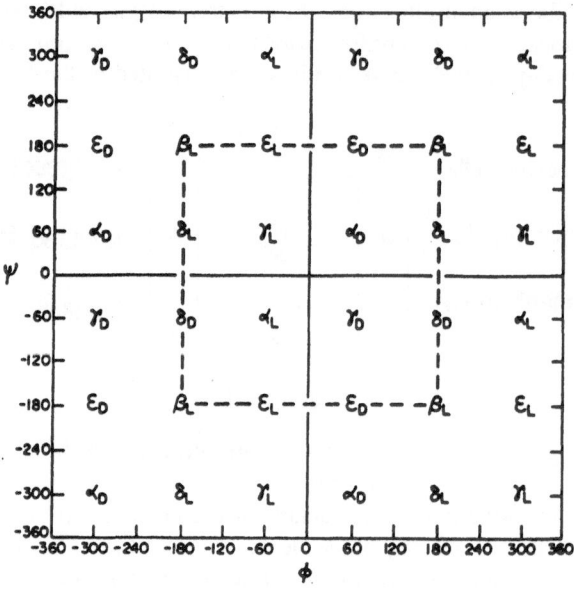

Figure 5: Idealized PES topology for a single amino acid residue involving two complete cycles of rotation in both ϕ and ψ (the location of the minima are specified by their names in terms of subscripted greek letters).

In a PES, associated with an ideal molecular system, minima, saddle points and maxima occur in a predictable regular pattern. It is customary to denote these critical points with the number of negative eigen-values of the Hessian matrix, with elements:

$$H_{ij} = \frac{\partial^2 E}{\partial x_i \partial x_j}$$ [5]

where $\{x_i, x_j\}$ are any pair of the total of n variables including $\{\phi, \psi\}$.

The number of negative eigen-values of the Hessian is usually referred to by the index (λ) of the critical point. For ordinary surfaces n varies between 0 and 2 ($0 \leq \lambda \leq 2$):

$$\lambda = 0 \quad \text{for minima}$$
$$\lambda = 1 \quad \text{for saddle points} \quad [6]$$
$$\lambda = 2 \quad \text{for maxima}$$

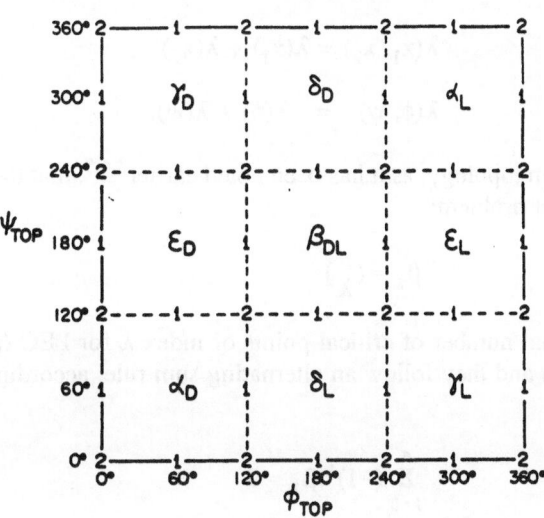

Fig. 6: The topology of an idealized two-dimensional (2D) Ramachandran map containing the **a-priori** predicted nine minima for a single amino-acid residue (...-CONH-CHR-CONH-...). The horizontal and vertical dashed lines represent low lying mountain ridges that separate the nine distinctly different catchment regions[3] labelled by the greek letters. (Note that the topologically (TOP) useful regions of ϕ and ψ are given in a $0°$ to $360°$ range.) Numerals indicate the expected location of transition states (λ=1) and maxima (λ=2)

For potential energy hypersurfaces (PEHS):

$$0 \leq \lambda \leq n \qquad [7]$$

for minima $\lambda = 0$, for maxima $\lambda = n$ and in between are located the saddle points with a variety of indices ranging from 1 to (n-1).

Figures 1 and 3 are ideal surfaces for hypothetical systems where the critical points were labelled with their indices: 0, 1 and 2. Figure 6 again shows an ideal surface as applied to a single peptide residue.

In this figure the minima are not labelled by 0 but by the greek letters introduced earlier α_L, β_L γ_L etc., but critical points of higher indices are denoted by their λ values: 1 and 2. There are two points to note about this figure. First of all, the minima are separated from each other by mountain ridges containing maxima and saddle points. Each valley, contains a single minimum and these valleys are normally referred to, after Mezey[3], as catchment regions. Secondly, in Figure 6, as in Figures 1 and 3, the indices of the PES may be calculated from the indices of the appropriate PEC if Mezey's criteria are fulfilled[3]:

$$\lambda(x_1, x_2) = \lambda(x_1) + \lambda(x_2) \qquad [8]$$

or

$$\lambda(\phi, \psi) = \lambda(\phi) + \lambda(\psi). \qquad [9]$$

It is well known from topology, as it has been noted earlier[3,4,5], that the Betti numbers (β_λ) for a conformational problem:

$$\beta_\lambda = \binom{n}{\lambda} \qquad [10]$$

represent the minimal number of critical points of index λ for PEC (n=1), for PES (n=2) and for PEHS (n$\geq$3) and they follow an alternating sum rule, according to Morse theory[6]:

$$\sum_{\lambda=0}^{n} (-1)^\lambda \beta_\lambda = 0 \qquad [11]$$

just like the actual number of critical points (N_λ):

$$\sum_{\lambda=0}^{n} (-1)^\lambda N_\lambda = 0 \qquad [12]$$

where

$$N_\lambda \geq \beta_\lambda \qquad [13]$$

On the basis of the foregoing for an ideal surface, one might expect that the Betti numbers, β_λ, (that originate from mathematics) and the actual number of critical points N_λ (that incorporates chemistry) may be functionally related to each other.

$$N_\lambda = f(\beta_\lambda) \tag{14}$$

The full functional relationship introduced by Mezey[7,8] may be written as

$$N_\lambda = (\prod_{i=1}^{n} m_i)\, \beta_\lambda \tag{15}$$

Each m_i may be associated with the number of bonds eclipsing during a full cycle (i.e. from $0°$ to $360°$) of internal rotation (torsion). For a couple of CH_3 - rotations or for that matter, rotation about a chiral centre (C_α) of a peptide residue (i.e. about ϕ and ψ), $m = m_1 = m_2 = 3$:

$$N_\lambda = (\prod_{i=1}^{2} m_i)\, \beta_\lambda = m^2 \beta_\lambda = 3^2 \beta_\lambda = 9\beta_\lambda \tag{16}$$

THE PROBLEM OF PROTEIN FOLDING

A molecular system, of course, is not always ideal. Sometimes N_λ cannot be calculated as simply as equation [16] implies. In Mezey's papers[7,8]

> "precise sufficient conditions are given for the validity of these
> predictions in terms of curvatures, that are generalized force
> constants of interactions (eqs. 6a and 6b), and also in terms
> of first derivatives describing the approximate alignment of
> mountain ridges and valley floors with the coordinate directions
> (eq. 31), that are the constraint on the extent of interactions
> between motions along different internal coordinates."

In general, we might say that whenever excessive attractive or excessive repulsion interactions occur we may anticipate deviation from ideal behaviour. It is generally observed that excessive attractive interactions and excessive repulsive interactions may create or annihilate critical points. This principle is illustrated schematically by Figure 7.

Our accumulated experience indicates that hydrogen bonding do not qualify as excessive attractive interactions therefore in peptide conformations new minima are never created (a hydrogen bond may only stabilize an otherwise legitimate minimum). However, repulsive interactions in peptides may be excessive enough to annihilate certain minima. For the annihilation of critical points the selection rules for annihilation are those as published earlier[9] and illustrated schematically in Figure 8. The upshot of all of this is that in the case of peptides and proteins $N_0 = 9$ which appears to represent an upper bound for the number of possible minima for a single amino acid residue.

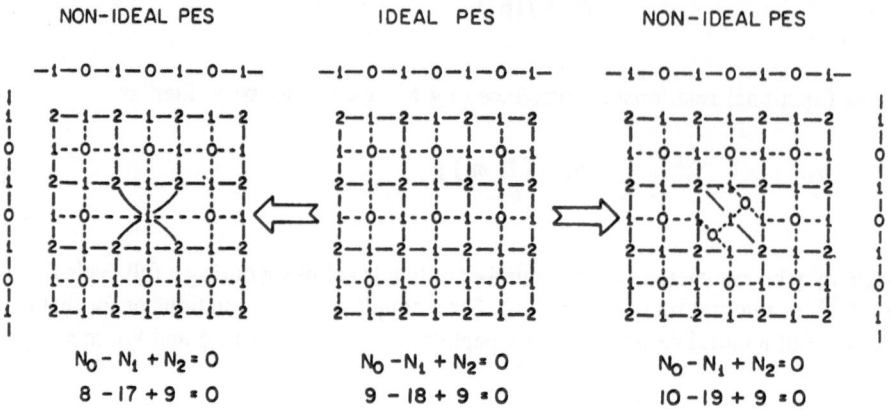

Figure 7: A schematic illustration for the creation (from centre to right) and annihilation (from centre to left) of critical points with respect to the ideal PES.

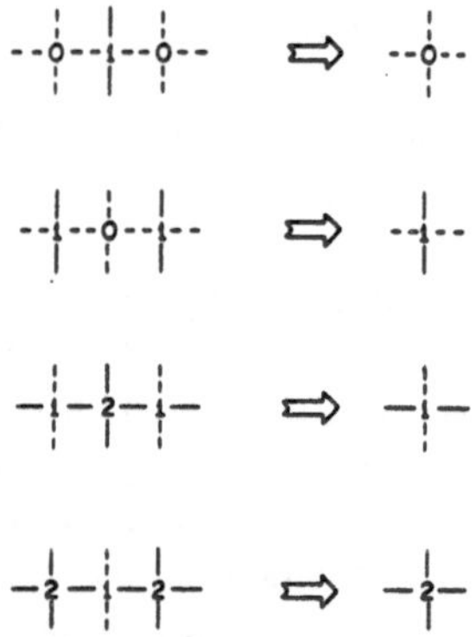

Figure 8: Selection rules for the collapse of three critical points to one.

It has been established from first principles that the idealized 2D-Ramachandran PES (c.f. Figure 6) has to have nine minima. It is therefore an obvious question to ask whether all these nine conformations are actually occurring in proteins. We therefore carried out an analysis of 258 proteins[29] with known x-ray structure[30]. These proteins contained 56,495 amino acid residues with well defined ϕ and ψ angles. The minima were identified with the aid of the nine minima[2] of Ac-L-Ala-NHMe determined by ECEPP/2 method allowing a $\pm 40°$ tolerance[29] in the ϕ and ψ values. The results are summarized in Table I. A number of conclusions may be drawn from this table.

1)   The "not-assigned" (N/A) conformations are quite large indicating that Ac-L-Ala-NHMe may not be as good a model to mimic a single amino acid residue in a protein than hitherto might have been believed.

2)   Glycine has the greatest number of N/A cases implying that the alanine derivative cases that has a side chain may be a much better model to all amino acid residues with side chains than to glycine residues that have no side chain (perhaps glycine should be modelled with glycine).

3)   Since glycine is achiral instead of 9 only 5 unique conformations occur. This means to say that the α_L conformation must be the same as α_D and similarly might be expected for the other three pairs: $\gamma_L = \gamma_D$ $\delta_L = \delta_D$ and $\varepsilon_L = \varepsilon_D$. The actual finding is not all that far from expectation

$$\alpha_L = 850 \qquad 631 = \alpha_D$$
$$\gamma_L = 79 \qquad 160 = \gamma_D$$
$$\delta_L = 62 \qquad 45 = \delta_D$$
$$\varepsilon_L = 388 \qquad 324 = \varepsilon_D$$

Undoubtedly, the actual degeneracy is lost in the 1799 NA conformations.

4)   Phenylalanine (Phe) has no γ_D conformation and Proline has no ε_D and γ_D conformation. All other amino acid residues do occur in all the possible 9 conformations.

It is important to emphasize, nevertheless, that all 9 conformations do occur in proteins as demonstrated by Table I.

This is not a new discovery; in fact it has been noticed by Professor Scheraga and his co-workers in 1977. This discovery led Professor Scheraga[10] to divide to total conformational domain into 12 regions. With the aid of the new, topologically more meaningful cut, it is possible to redraw Scheraga's diagram from the (ϕ_{STD}, ψ_{STD}) to the (ϕ_{TOP}, ψ_{TOP}) representation. Such a redrawn picture is shown in Figure 9. One can readily see that some of the 12 minima Scheraga[10] identified (A, C, D, E, F, G, A*, C*, D*, E*, F*, G*) are in fact identical conformations (cf E=E*, D=G*, G=D*). SCHEME III shows the equivalence between the two sets of notation.

Some might be convinced, that Scheraga intended nothing more than to create merely a practical subdivision of the (ϕ, ψ) map into 12 regions. The present authors, however tend to believe that, in fact Scheraga's work has shown an incredible insight into the problem, since after taking the appropriate $(\phi_{STD}, \psi_{STD}) \rightarrow (\phi_{TOP}, \psi_{TOP})$ transformation one obtains a one-to-one correspondence between the 9 catchment regions of the ideal conformational potential energy surface and those 12 regions of Scheraga (c.f. SCHEME III).

Table I. Frequency of occurrence of the nine backbone conformations of various amino acids in proteins.

Amino Acid	Backbone Conformation										
	γ_L	β_{DL}	α_L	δ_L	ε_L	δ_D	α_D	ε_D	γ_D	N/A*	Total
Gly	79	460	850	62	388	45	631	324	160	1799	4798
Ala	147	612	2593	78	793	43	54	12	36	526	4894
Val	368	918	1620	56	907	53	14	8	8	427	4379
Ile	222	570	976	35	549	15	14	3	4	260	2648
Leu	294	467	2159	98	953	37	51	14	11	432	4516
Pro	113	4	936	3	1170	1	1	0	0	242	2470
Phe	182	387	788	75	392	27	31	2	0	222	2106
Trp	40	160	392	27	158	12	3	1	10	90	893
Tyr	146	444	565	76	439	15	36	3	3	221	1948
Ser	177	868	1703	161	952	49	70	24	28	641	4673
Thr	175	706	1252	65	764	49	15	12	14	435	3487
Cys	76	362	483	20	323	25	19	5	3	176	1492
Met	58	180	365	13	138	7	13	2	1	75	852
Lys	191	438	1769	88	619	40	71	12	21	529	3778
Arg	109	292	723	57	303	10	56	3	5	222	1780
His	76	197	622	75	194	18	52	4	5	227	1470
Asp	301	223	1245	146	446	47	68	12	21	470	2979
Asn	275	277	866	145	375	30	286	9	20	484	2767
Gln	111	251	1392	40	401	27	26	12	14	315	2589
Gln	86	252	897	36	373	18	42	6	6	260	1976

* Not assigned conformations

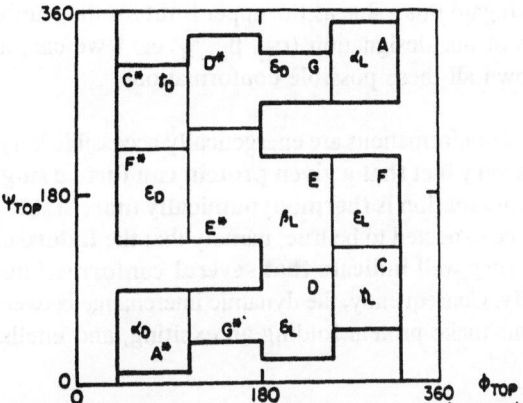

Figure 9: Correlation of Scheraga's twelve categories $(A,A^*, C,C^*, D,D^*, E,E^*, F,F^*$ and $G,G^*)$ with the currently recognized nine minima $(\alpha_L, \alpha_D, \beta_L, \gamma_L, \gamma_D, \epsilon_L, \epsilon_D)$ on the (ϕ_{TOP}, ψ_{TOP}) coordinate system. [Note that $D^* = \delta_D = G$, $E^* = \beta_L = E$ and $G^* = \delta_L = D$.]

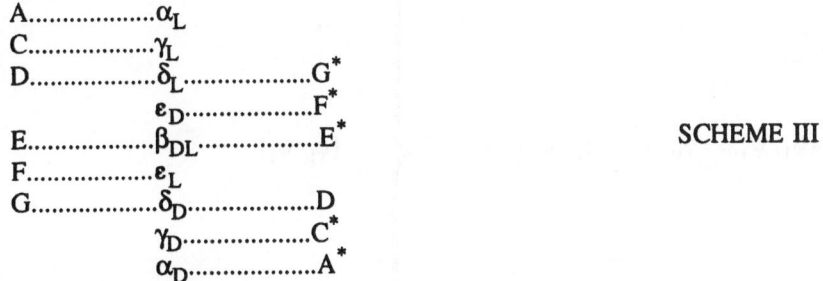

SCHEME III

It should now be emphasized that every amino acid may occur, at least in principle, in $3^2 = 9$ conformations and therefore the number of <u>a priori</u> stable conformations (N_0) for a protein containing **n** amino acids is:

$$N_0 = (3^2)^n = 3^{2n} = 9^n \qquad [17]$$

For a protein consisting of 100 amino acid residues n = 100 and therefore the value of N_0 is a very large number:

$$N_0 = 9^{100} = 2.65 \times 10^{95} \qquad [17a]$$

Although N_0 represents an astronomically large number of conformations nevertheless it should be noted that

(i)   it is finite
(ii)   we now know its magnitude

(iii) we may regard this value as the upper limit for the number of conformations

(iv) in terms of our designation (α_L, β_L, γ_L etc.) we can, at least in principle, write down all these possible conformations

Obviously, not all protein conformations are energetically accessible but many conformations might be. However, **the very fact that a given protein can form a single crystal indicates that one particular conformation is thermodynamically more stable than all the others.** The converse may also be expected to be true, namely that the **failure of a protein to form a single crystal could very well indicate that several conformations may have similar thermodynamic stability.** Consequently, the dynamic interchange between these energetically accessible conformations make protein folding an exciting, and intellectually challenging subject.

The mathematical implications of studying only backbone and backbone/side chain interactions is the partitioning of the 2n-dimensional space to **n** two-dimensional subspaces:

$$\{\phi_1, \psi_1,...,\phi_i, \psi_i,...,\phi_n, \psi_n\} \rightarrow \left\{ \begin{array}{c} \{\phi_1, \psi_1\} \\ \cdot \\ \cdot \\ \{\phi_i, \psi_i\} \\ \cdot \\ \cdot \\ \{\phi_n, \psi_n\} \end{array} \right. \qquad [18]$$

Consequently, one can replace the study of the overall protein PEHS: $E = E(\underline{x})$ by n regular and more manageable PES: $E = E(\phi_i, \psi_i)$, where $1 \leq i \leq n$, (i.e. number of amino acid residues in the peptide chain):

$$E(\phi_1, \psi_1,...,\phi_i,\psi_i,...,\phi_n,\psi_n) \rightarrow \{E(\phi_1, \psi_1),...,E(\phi_i,\psi_i),..E(\phi_n,\psi_n)\} \qquad [19]$$

These surfaces are expected to have a topology outlined in the previous section (cf Figure 5). Therefore, anything we may have learned from such 2D-Ramachandran maps can only be related to protein conformations only very carefully.

Turning back to Figure 10, in which it is indicated that dipeptides are the smallest units that might mimic nearest - neighbour interactions, this idea suggests a different partitioning of the vector space:

$$\underline{x} = (\phi_1, \psi_1, ... \phi_n, \psi_n) \qquad [20]$$

The great advantage in partitioning the energy of interactions into these three components is that we may be able to model these three types of interactions that occur in proteins by smaller peptides. The minimum sized peptides that are capable of modelling such interactions are specified in Figure 10.

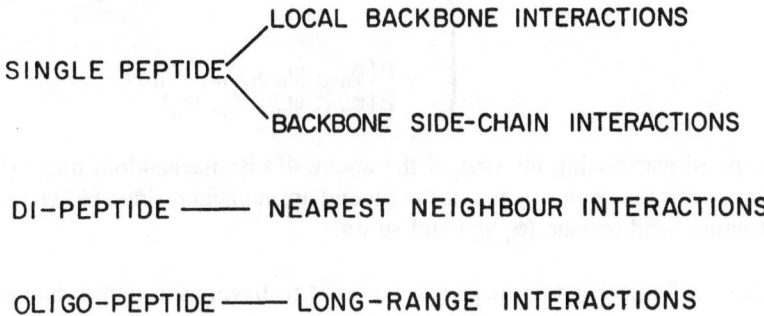

Figure 10: Minimum sized peptides and types of interactions that might be modelled.

Wittingly or unwittingly, during the past 40 years, ever since the time of Pauling and Corey[11,12], protein chemists have simplified their approach to the study of protein folding by separating, at least mentally, the three types of interactions outlined above. This philosophy implied the conviction that first we have to understand the problem of backbone conformation in the absence of nearest neighbour and long-range interactions (such as electrostatic attraction or repulsion, dipole-dipole stabilization or destabilization and H-bonding) before we can gain a full comprehension of the overall problem. Of course, even such a basic notion as the concept of the Ramachandran's map[13] has the above approximation at its very core when related throughout the years to protein secondary structure.

In view of the above, the backbone conformational problem of a protein can be viewed in terms of a hypothetical conformational potential energy hypersurface in which nearest neighbour and long-range interactions are eliminated. As a mathematical consequence, this traditional idea of conceptual partitioning implies that there must exist a conformational potential energy hypersurface (PEHS) of a protein $E=E(\underline{x})$ (where the variables are torsional angles, $\underline{x} = (\phi_1, \psi_1,..., \phi_n, \psi_n)$, may be defined according to the IUPAC-IUB convention) that be partitioned, mathematically, in different ways.

Thus, the partitioning of the PEHS of 2n independent variables to the PES of n-pairs of independent variables (ϕ_i, ψ_i where $1 \leq i \leq n$) may be superseded by the partitioning to (n-1) PEHS with four independent variables (e.g. $\phi_i, \psi_i, \phi_{i+1}, \psi_{i+1}$) that is to (n-1) 4D

Ramachandran maps:

$$E = E\{\phi_1, \psi_1, ..., \phi_i, \psi_i, ..., \phi_n, \psi_n\} \rightarrow \begin{cases} E\{\phi_1, \psi_1, \phi_2, \psi_2\} \\ E\{\phi_2, \psi_2, \phi_3, \psi_3\} \\ \cdot \\ \cdot \\ \cdot \\ E\{\phi_{i-1}, \psi_{i-1}, \phi_i, \psi_i\} \\ E\{\phi_i, \psi_i, \phi_{i+1}, \psi_{i+1}\} \\ \cdot \\ \cdot \\ \cdot \\ E\{\phi_{n-2}, \psi_{n-2}, \phi_{n-1}, \psi_{n-1}\} \\ E\{\phi_{n-1}, \psi_{n-1}, \phi_n, \psi_n\} \end{cases} \quad [21]$$

With this type of partitioning the first of the above 4D-Ramachandran maps (i.e.: $E\{\phi_1, \psi_1, \phi_2, \psi_2\}$) reveals the conformation of the second animo acid residue (ϕ_2, ψ_2) in the field of the first amino acid residue (ϕ_1, ψ_1) and so on.

These 4D-Ramachandran maps are expected to have, as an upper bound,

$$N_0 = (3^2)^2 = 9^2 = 81 \quad [22]$$

distinctly different minima. The topology of this 4D-Ramachandran maps with their 81 minima is depicted in a 2D projection in Figure 11.

$$
\begin{array}{ccc|ccc|ccc}
\gamma_D\gamma_D & \gamma_D\delta_D & \gamma_D\alpha_L & \delta_D\gamma_D & \delta_D\delta_D & \delta_D\alpha_L & \alpha_L\gamma_D & \alpha_L\delta_D & \alpha_L\alpha_L \\
\gamma_D\epsilon_D & \gamma_D\beta_L & \gamma_D\epsilon_L & \delta_D\epsilon_D & \delta_D\beta_L & \delta_D\epsilon_L & \alpha_L\epsilon_D & \alpha_L\beta_L & \alpha_L\epsilon_L \\
\gamma_D\alpha_D & \gamma_D\delta_L & \gamma_D\gamma_L & \delta_D\alpha_D & \delta_D\delta_L & \delta_D\gamma_L & \alpha_L\alpha_D & \alpha_L\delta_L & \alpha_L\gamma_L \\
\\
\epsilon_D\gamma_D & \epsilon_D\delta_D & \epsilon_D\alpha_L & \beta_L\gamma_D & \beta_L\delta_D & \beta_L\alpha_L & \epsilon_L\gamma_D & \epsilon_L\delta_D & \epsilon_L\alpha_L \\
\epsilon_D\epsilon_D & \epsilon_D\beta_L & \epsilon_D\epsilon_L & \beta_L\epsilon_D & \beta_L\beta_L & \beta_L\epsilon_L & \epsilon_L\epsilon_D & \epsilon_L\beta_L & \epsilon_L\epsilon_L \\
\epsilon_D\alpha_D & \epsilon_D\delta_L & \epsilon_D\gamma_L & \beta_L\alpha_D & \beta_L\delta_L & \beta_L\gamma_L & \epsilon_L\alpha_D & \epsilon_L\delta_L & \epsilon_L\gamma_L \\
\\
\alpha_D\gamma_D & \alpha_D\delta_D & \alpha_D\alpha_L & \delta_L\gamma_D & \delta_L\delta_D & \delta_L\alpha_L & \gamma_L\gamma_D & \gamma_L\delta_D & \gamma_L\alpha_L \\
\alpha_D\epsilon_D & \alpha_D\beta_L & \alpha_D\epsilon_L & \delta_L\epsilon_D & \delta_L\beta_L & \delta_L\epsilon_L & \gamma_L\epsilon_D & \gamma_L\beta_L & \gamma_L\epsilon_L \\
\alpha_D\alpha_D & \alpha_D\delta_L & \alpha_D\gamma_L & \delta_L\alpha_D & \delta_L\delta_L & \delta_L\gamma_L & \gamma_L\alpha_D & \gamma_L\delta_L & \gamma_L\gamma_L
\end{array}
$$

Figure 11: A 2D projection of an ideal 4D Ramachandran map. $E=E(\phi_1, \psi_1, \phi_2, \psi_2)$ with no coupling between the (ϕ_1, ψ_1) and (ϕ_2, ψ_2) subspaces. The symbols γ_L α_L, etc. designate the conformation of a dipeptide from left to right; $HN\text{-}CHR(\gamma_L)\text{-}CONH\text{-}CHR'(\alpha_L)CONH\cdots$. [Note that without coupling 9*9=81 minima may be expected as shown in this 2D projection of the 4D map. For non-ideal cases the 81 minima may be reduced.]

CONFORMATIONS OF SINGLE PEPTIDE UNITS

The diamides, derived from a single amino acid unit, such as N-formyl-alanine-amide (II) and N-acetyl-alanine-N'-methyl amide (III)

are the two simplest complete molecules that could mimic an amino acid residue (I) within a protein, which we may be referred to as a "single peptide unit". The molecular structure of such systems have been investigated both experimentally[14-20] and theoretically [20-23] and it is generally believed that the γ_L (or C_7^{equ}) conformation of all the cases studied is the global minimum on the PES. (Note that in the case of glycine, where R=H, by necessity $\gamma_L = \gamma_D$ or $C_7^{equ} = C_7^{ax}$ due to the loss of chirality).

In our own work, formyl glycine-amide[2], formyl-L-alanine-amide[2] and formyl-L-valine-amide[23] have been investigated in detail. (The formyl-D-alanine-amide was also studied[2] and it proved to have a PES that was enantiomeric to the PES of the L-alanine derivative) In addition to these three amino acid derivatives a partial study has also been conducted on formyl-L-serine amide.[24] The above four amino acids represent the four prototypes of single peptide units as far as the nature of the ligand attached to the α carbon is concerned. These four types, associated with the four side-chains, are summarized in SCHEME IV.

Family Type	Side chain	Example
1	-H	GLYCINE
2	-CH3	ALANINE
3	-CH2Q	SERINE
4	-CHQ'Q"	VALINE

Q = -OH
Q" = Q' = -CH3   SCHEME IV

The topology of the glycine and L-alanine derivatives are shown in SCHEME V:

$$\begin{array}{ccc} \gamma_D & \delta_D & \alpha_L \\ \cdot & \beta_L & \cdot \\ \alpha_D & \delta_L & \gamma_L \end{array} \qquad \begin{array}{ccc} \gamma_D & \delta_D & \cdot \\ \varepsilon_D & \beta_L & \cdot \\ \alpha_D & \delta_L & \gamma_L \end{array}$$

SCHEME V

N-formyl-glycine amide   N-formyl-L-alanine amide

As it can be seen from SCHEME V, in contrast to the general topology, given in SCHEME II, instead of 9 only 7 minima were found; 2 minima were annihilated. The selection rules[9], presented schematically in Figure 8, had to be applied several times[2] in order to explain the annihilation of 2 minima. With the annihilation of the 2 minima the topology predicted the creation of a new mountain ridge with a single maximum in the general area where the two minima α_L and ε_L (in the case of the L-alanine derivative) are expected to be located. The PES generated previously[22] for the L-alanine derivative can be replotted to encompass two full cycles of rotation along each of the two torsional angles:

$$-360° \leq \phi \leq +360°$$

$$-360° \leq \psi \leq +360°$$

[23]

This new plot, shown in Figure 12, which covers the same domain as Figure 5, clearly indicates the emergence of a new mountain ridge.

at the expected locations of the α_L and ε_L minima. This figure also indicates that inter-conversion of conformations in the topological units (c.f. equation [3]), that are the four quadrants of the PES, is far easier than within the cut, indicated by broken lines, at the centre corresponding to the IUPAP-IUB convention (c.f. equation [4]).

In contrast to alanine, where the $-CH_3$ side chain had only one conformation, in the Valine derivative (IV)

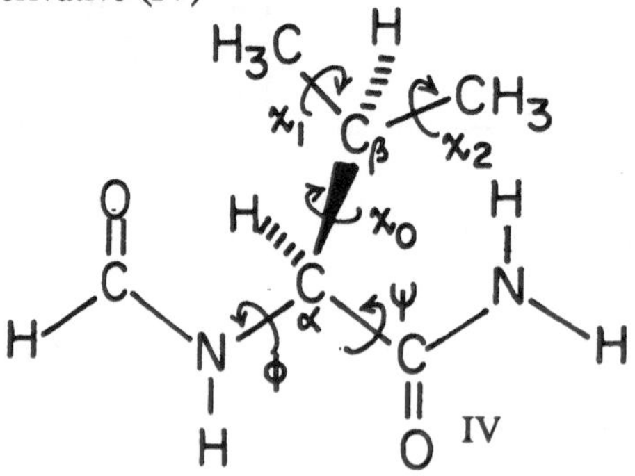

the iso-propyl side chain may have several conformations. The most prominent ones are depicted in SCHEME VI.

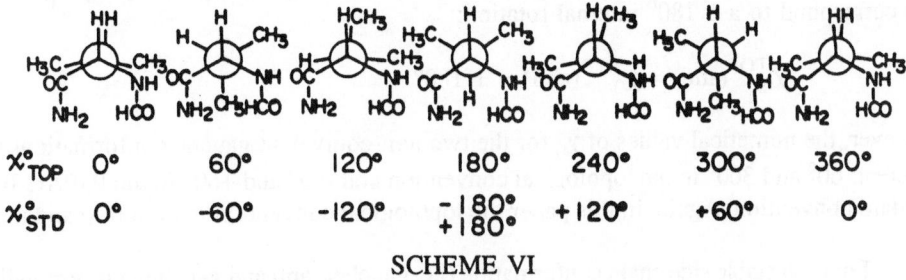

$$\chi^{\circ}_{TOP} \quad 0^{\circ} \qquad 60^{\circ} \qquad 120^{\circ} \qquad 180^{\circ} \qquad 240^{\circ} \qquad 300^{\circ} \qquad 360^{\circ}$$

$$\chi^{\circ}_{STD} \quad 0^{\circ} \qquad -60^{\circ} \qquad -120^{\circ} \qquad \begin{matrix}-180^{\circ}\\+180^{\circ}\end{matrix} \qquad +120^{\circ} \qquad +60^{\circ} \qquad 0^{\circ}$$

SCHEME VI

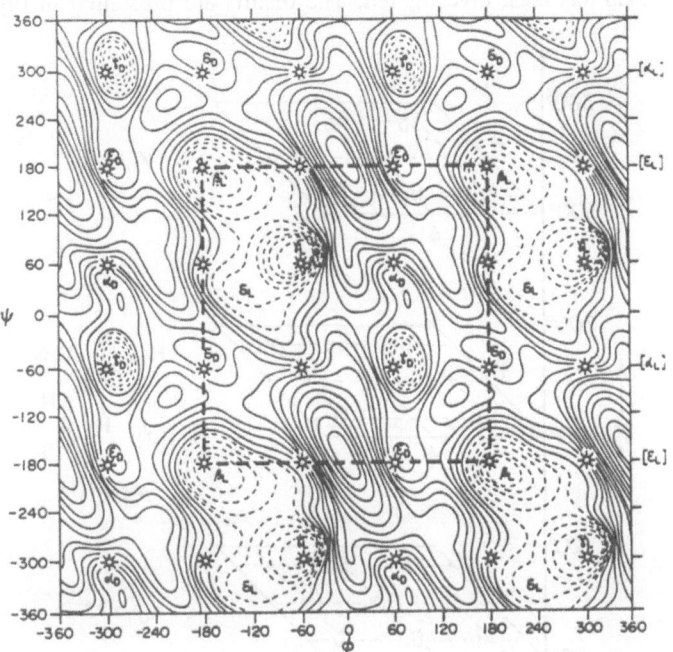

Figure 12: Contours of conformational potential energy surface, $E=E(\phi,\psi)$, for N-formyl-L-alanine-amide computed at the 3-21G-set level of SCF theory. The plot involves two complete cycles of rotation in both ϕ and ψ like in Figure 5. Locations of the actual minima are specified by their names in terms of subscripted greek letters. The anticipated location of minima, according to idealized topology, are specified by open stars. The central square (dotted lines) gives the conventional IUPAC-IUB cut while the four quadrants specify the topologically more useful cuts. This plot is based on the plot published by J.A. Pople et al[22]. The authors gratefully acknowledge Professor Pople's permission to use his results in the present context.

58

Here, again, for the torsional angle (χ_0) of the iso-Pr group, the topological convention (χ_0^{TOP}) is favoured over the IUPAC-IUB standard (χ_0^{STD}) convention. In both conventions the conformation in which the two hydrogens associated with C_α and C_β, i.e. $H-C_\alpha-C_\beta-H$, are anti correspond to a $\pm 180^\circ$ internal rotation:

$$\chi_0^{TOP}(\text{anti}) = \chi_0^{STD}(\text{anti}) = \pm 180^\circ \qquad [24]$$

However, the numerical values of χ_0 for the two non-equivalent gauche conformations are different: 60° and 300° in the topological convention and -60° and $+60^\circ$ in the IUPAC-IUB standard convention. Again, in this paper the topological convention is preferred and used.

For each stable side chain conformation (i.e. gauche$_1$, anti and gauche$_2$, corresponding roughly to 60°, 180° and 300° of χ_0 values respectively) the topology of the backbone conformational PES has been investigated. The results are presented in Figure 13.

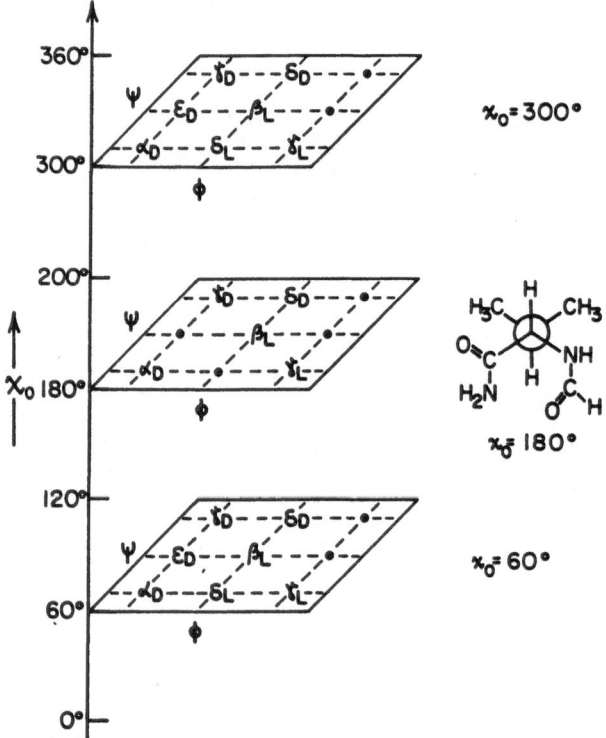

Figure 13: The topology of backbone conformational PES, $E=E(\phi, \psi)$ as the function of iso-propyl, $H-C^\alpha-C^\beta-H$, torsional angle (χ_0). At every entry the solid dot indicates the annihilation of a particular minimum predicted to have a predicted existence by multidimensional conformational analysis (MDCA).

Now we are in the position to compare the backbone conformations for the N-formyl glycine, L-alanine and L-valine amides (c.f. Figure 14) with the idealized conformations.

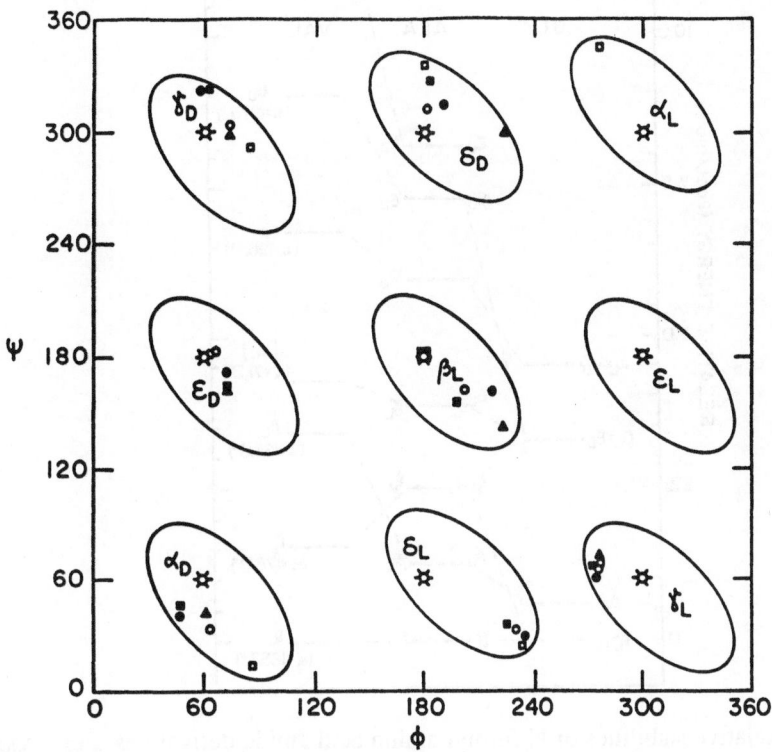

Figure 14: A scatter of ab initio SCF points on the 2D-Ramachandran map, $E=E(\phi,\psi)$, including conformations of N-formyl glycine amides (open squares) N-formyl-L-alanine amide (open circles) and N-formyl L-valine amide at $\chi_0 = 60^\circ$ (solid squares), at $\chi_0 = 180^\circ$ (solid triangles) and at $\chi_0 = 300^\circ$ (solid circles) in relation to the idealized conformations (open stars).

It is interesting to note that one may draw an ellipse of a relatively small area around the ideal conformations and these ellipses contain all computed conformations of a variety of amino acid derivatives. This observation indicates that although the computed conformations are different from the ideal, the differences, however, are not very large implying that the idealized amino acid residue topology may be applicable to real single peptide units with some predictable errors. The computed energy spectrum for these three amino acid derivatives are shown in Figure 15.

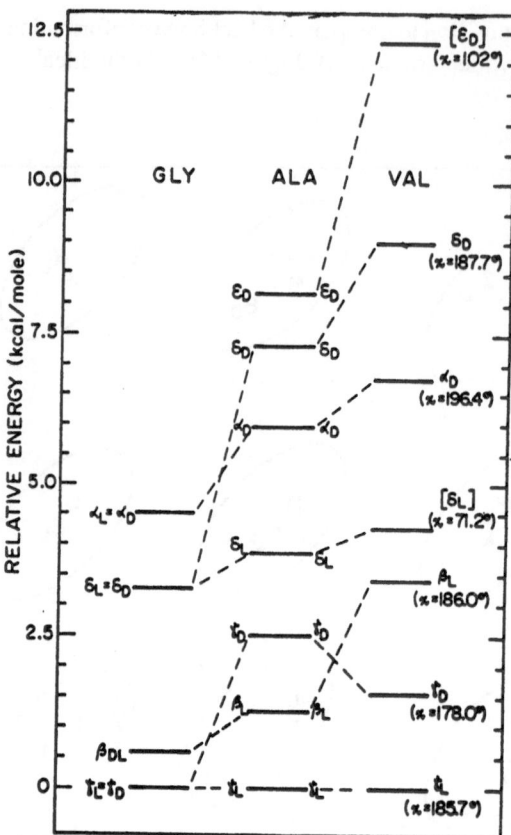

Figure 15: Relative stabilities of N-formyl amino acid amide derivatives: For - Xxx-NH$_2$ for Xxx = glycine (GLY), alanine (ALA) and valine (VAL).

Although it was not possible to always take the same side-chain conformation (χ_0) in the valine derivative as the ε_D and δ_L conformations do not exist in the valine derivative at χ_0=180°. Nevertheless the ever increasing spread of the energy spectrum indicates the increasing spatial requirement of the R group on going from R=H via R=CH$_3$ to R=CH(CH$_3$)$_2$.

Local geometry changes, while backbone conformations are being varied, is a topic that should not be neglected. Professor Lothar Schäfer and his group have made[20,21,25-27] some noticeable advances in this respect on a series of compounds of the type: Ac-amino acid-NHMe including glycine, L-alanine and L-serine. The results are summarized in the following table.

Table II: The variation of bond angle about the α carbon of a single peptide unit as the function of backbone conformation.

Amino acid	γ_L			β_L		
	ϕ	ψ	∢ N-C$^\alpha$-CO	ϕ	ψ	∢ N-C$^\alpha$-CO
Gly	-83	+71	111.9	±180	±180	107.9
Ala	-85	+73	109.5	-166	+167	106.4
Ser*	-84	+72	109.6	-168	-174	106.1
	-84	+63	111.3	-174	+170	105.6

* Two different side-chain conformations were studied.

CONFORMATIONS OF DI-PEPTIDE UNITS

Dipeptides have been studied less extensively than single peptide units. We ourselves made only an initial attempt to optimize at least one conformation of For-Gly-Gly-NH$_2$ (V).

V

We started with a geometry that was anticipated to be correct for a $\gamma_L\gamma_L$ conformation. Since the γ_L conformation is the most stable one for a number of single peptide units, we presumed the $\gamma_L\gamma_L$ conformation to be the most stable for a dipeptide. After some lengthy optimization the geometry has migrated to what might be labelled as the $\gamma_L\varepsilon_L$ conformation. However, this partially optimized structure has turned out to be an inflexion point at best and the final optimized structures in fact became a $\gamma_L\delta_L$ conformation. This ab initio "rollercoaster" on the 4D Ramachandran PEHS is illustrated by the next table.

Table III. Ab initio conformational "rollercoaster" on the 4D Ramachandran map (PEHS) associated with For-Gly-Gly-NH$_2$

Phases	Conf.	ϕ_1	ψ_1	ϕ_2	ψ_2
Initial guess	$\gamma_L\gamma_L$	-80°	$+70^\circ$	-80°	$+70^\circ$
Partially optimized geometry	$\gamma_L\varepsilon_L$	-83.9°	$+70.2$	-90.8°	$+173.5$
Final Optimized geometry	$\gamma_L\delta_L$	-84.6°	$+71.1^\circ$	-128.3°	$+23.3^\circ$
Ab initio single peptide units[*]	$(\gamma_L)(\delta_L)$	-83.9°	$+67.8^\circ$	-126.0°	$+25.5^\circ$
Idealized position	$(\gamma_L)(\delta_L)$	-60°	$+60^\circ$	-180°	$+60^\circ$

* Taken from Table 3 of Reference 2.

It should perhaps be mentioned that the first phase of the optimization (i.e. $\gamma_L\gamma_L \rightarrow \gamma_L\varepsilon_L$) was carried out in Toronto (Canada) using the MONSTERGAUSS Program System. The second phase of the optimization (i.e. $\gamma_L\varepsilon_L \rightarrow \gamma_L\delta_L$) was carried out by Professor Lothar Schäfer's group in Fayetteville (USA) using the TEXAS Program System.

Finally, we must recognize that of dipeptide conformations have not been investigated very widely by ab initio chemists in the past, so we are eagerly exploring this territory jointly with Professor Lothar Schäfer and his group. At this time we can offer only an empirical "database" that we have obtained for Ac-L-Ala-L-Ala-NHMe using Scheraga's ECEPP/2 method. From the 4D-Ramachandran map (c.f. Figures 11 and 12) it is clear that we expect, under ideal conditions, the existence of $9^2 = 81$ minima. However, the "database" given in Table IV recognizes only 75 distinctly different minima (6 minima were annihilated). Nevertheless, this is still the most extensive set of minima since instead of the $9^2 = 81$ legitimate minima (idealized case) one may anticipate only $7^2 = 49$ minima (ab initio case) and certainly the 75 minima located on the empirical surface is relatively close to the idealized case.

Furthermore, one should say that dipeptide derivatives (such as V) may be extended to oligopeptide derivatives

$$\text{For-(Amino Acid)}_n\text{-NH}_2$$
or $\qquad$ $\text{Ac -(Amino Acid)}_n\text{-NHMe}$

One such attempt has been made by Schäfer and coworkers[28] on the derivative of pentaglycine: For-(gly)$_5$-CONH$_2$. They studied the poly β_{DL} (β pleated sheet) and poly γ_L conformations:

HCO-HN-Gly-Gly- Gly-Gly- Gly-CONH$_2$

β_{DL}-β_{DL}-β_{DL}-β_{DL}-β_{DL}

γ_L - γ_L - γ_L - γ_L - γ_L

SCHEME VII

Interestingly enough, as shown in SCHEME VII, the N-C$^\alpha$-CO bond angles are different for the two conformations (c.f. Structure VI):

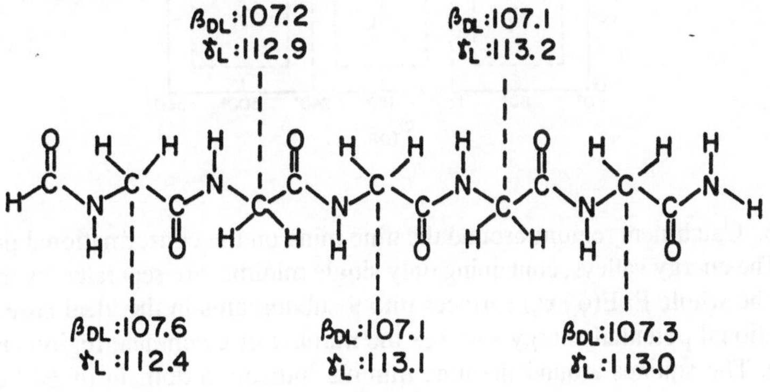

β_{DL}:107.2 β_{α}:107.1
δ_L:112.9 γ_L:113.2

β_{DL}:107.6 β_{DL}:107.1 β_{DL}:107.3
γ_L:112.4 γ_L:113.1 γ_L:113.0

VI

THE UTILITY OF THE 4D-RAMACHANDRAN MAP

The 2D-Ramachandran map (c.f. Figures 4, 5 and 6) under ideal conditions has 9 minima and the 4D-Ramachandran map (Figures 11) similarly under ideal conditions is expected to have 9^2=81 minima. The idealized pattern of the 2D-map (Figures 5 and 6) and 4D map (Figures 11) has all these minima in ± 120° intervals. However, the non-idealized but computed pattern of an ideal molecular system doesn't show such regularity. In the case of the 2D system [2] the ECEPP/2 molecular mechanics method has indeed shown the existence of the 9 minima even if some of these minima were energetically very shallow. However, in the 4D system 6 minima have been annihilated as 12 minima collapsed pairwise to 6 minima. The following SCHEME shows the range of ϕ and ψ values obtained for the γ_L conformation. The first three rows are for Ac-L-Ala-NHMe and the last two rows show the ranges obtained for Ac-L-Ala-L-Ala-NHMe where at least one amino acid residue had a γ_L conformation while the other amino acid was allowed to assume variable (X) conformations.

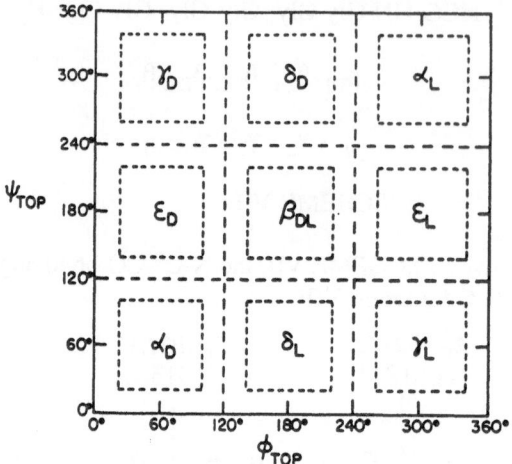

Figure 16: Catchment regions around the nine mina on the conformational potential energy surface. The energy valleys, containing only single minima, are separated by mountain-ridges dividing the whole $E=E(\phi_i, \psi_i)$ surfaces into 9 subdomains in the ideal case. For non-ideal conformational potential energy surfaces the number of catchment regions might be fewer than nine. The squares around the nine minima indicate a domain of ±40 degrees.

SCHEME VIII

IDEALIZED	$\phi=-60.0^\circ$	$\psi=+60.0^\circ$
Ab Initio SCF	$\phi=-84.4^\circ$	$\psi=67.7^\circ$

ECEP/2

γ_L	$\phi = -80.4^\circ$	$\psi=+75.8^\circ$
$X\gamma_L$	$-81.6^\circ < \phi < -79.3^\circ$	$72.7^\circ < \psi < 76.4^\circ$
$\gamma_L X$	$-80.3^\circ < \phi < -76.2^\circ$	$73.8^\circ < \psi < 81.2^\circ$

In order to create a system of classification for denoting various conformations of the amino acid residues in proteins one must choose a primary standard for conformations. The IDEALIZED angles might be easy to use but they are not necessarily close enough to the actual conformation. Figure 17 clearly indicates that the actual location of δ_L is quite far from the idealized location of δ_L (denoted by an open star); the deviation in the case of δ_L is much larger than that of γ_L (c.f. SCHEME VIII).

Clearly, the most appropriate choice, at this time, is the family of dipeptide (V) conformations for which the "database" (in the case of R=CH$_3$) is given in Table IV. These data are ECEPP/2 results obtained on Ac-L-Ala-L-Ala-NH-Me, but in the absence of a better "database" we shall use these conformations (TABLE IV) as primary standard for all amino acid residues, irrespectfully of the nature of the R-group.

TABLE IV

Molecular mechanics (ECEPP/2) results[#] for the 81 conformations of Ac-L-Ala-L-Ala-NHCH

Type			Torsional angle (deg)				E (kcal mol⁻¹)	SDB	LDB
			$\phi(1)$	$\psi(1)$	$\phi(2)$	$\psi(2)$			
γ_L	γ_L		−79.82	75.16	−79.91	75.73	−3.109	8	41
α_L	γ_L		−68.30	−39.21	−81.62	72.67	−2.941	14	62
α_L	ϵ_L		−68.28	−39.18	−81.62	72.67	−2.941	14	62
α_L	α_L		−70.55	−29.25	−70.01	−32.55	−2.915	1113	8103
ϵ_L	α_L		−80.24	73.86	−74.81	−35.74	−2.728	40	149
γ_L	α_L		−80.28	73.83	−74.78	−35.72	−2.728	40	149
α_L	δ_L		−63.28	−33.69	−113.95	41.30	−2.701	129	435
γ_L	δ_D		−78.12	75.49	−159.45	−56.04	−2.389	2	16
ϵ_L	δ_D		−78.11	75.45	−159.52	−56.02	−2.389	2	16
γ_L	δ_L		−80.27	73.79	−151.51	42.00	−2.381	4	24
ϵ_L	δ_L		−80.24	73.83	−151.51	41.98	−2.381	4	24
β_L	γ_L		−154.78	157.46	−80.07	76.01	−2.272	20	83
γ_L	ϵ_L		−79.31	76.50	−74.14	145.98	−2.136	50	219
ϵ_L	α_D		−63.85	109.75	54.48	40.64	−2.129	29	90
γ_L	β_L		−79.52	76.91	−153.47	157.92	−2.125	38	113
α_L	β_L		−72.71	−33.24	−153.94	156.52	−2.041	31	55
ϵ_L	γ_L		−74.69	141.04	−80.09	76.19	−1.951	65	209
δ_L	γ_L		−150.56	43.68	−80.68	75.55	−1.885	7	13
β_L	α_L		−155.00	157.76	−73.15	−35.60	−1.654	81	232
γ_L	γ_D		−76.18	81.22	55.59	44.49	−1.633	9	19
γ_L	α_D		−76.22	81.20	55.59	44.48	−1.633	9	19
β_L	β_L		−154.88	158.41	−154.50	158.16	−1.622	246	735
δ_D	γ_L		−157.88	−57.86	−79.63	76.36	−1.459	2	5
β_L	ϵ_L		−154.57	158.19	−74.88	145.62	−1.396	285	665
δ_L	β_L		−150.58	42.83	−155.59	156.72	−1.337	14	37
ϵ_L	β_L		−73.34	140.99	−154.84	159.06	−1.335	230	493
δ_L	α_L		−149.01	43.63	−73.83	−35.17	−1.270	32	184
β_L	δ_L		−155.05	158.32	−150.16	44.52	−1.190	10	25
ϵ_L	ϵ_L		−75.13	142.13	−74.91	145.22	−1.077	534	1339
δ_L	δ_L		−148.97	43.43	−151.42	44.97	−0.998	1	12
δ_D	β_L		−157.99	−57.00	−153.74	158.63	−0.932	3	11
α_D	γ_L		54.46	44.94	−80.38	76.29	−0.875	12	36
δ_L	ϵ_L		−151.10	45.11	−75.56	143.05	−0.817	75	266
δ_D	ϵ_L		−158.69	−57.29	−76.40	145.34	−0.639	9	12
β_L	δ_D		−154.37	157.17	−158.17	−57.51	−0.624	1	17
δ_D	α_L		−157.50	−57.54	−72.90	−37.76	−0.616	10	34
α_D	β_L		54.36	44.51	−155.77	158.57	−0.581	26	72
α_L	δ_D		−73.65	−32.03	−157.37	−57.05	−0.575	10	66
δ_L	δ_D		−149.54	44.51	−158.45	−57.15	−0.317	4	9
δ_D	δ_L		−157.97	−56.73	−149.37	46.55	−0.316	3	6
α_D	α_D		56.39	37.04	54.24	40.30	−0.175	28	77

66

TABLE (continued)

Type		Torsional angle (deg)				E	SDB	LDB
						(kcal mol^{-1})		
		$(\phi(1))$	$\psi(1)$	$\phi(2)$	$\psi(2)$			
α_L	α_D	-71.76	-34.90	54.85	45.11	-0.139	39	138
β_L	α_D	-153.95	156.41	54.39	46.31	-0.118	12	35
α_D	ϵ_L	54.57	44.63	-76.37	142.62	0.095	76	217
α_D	δ_L	54.35	44.66	-151.87	46.23	0.164	5	18
α_D	α_L	54.70	44.02	-74.82	-34.70	0.199	22	59
δ_L	α_D	-150.55	48.30	54.46	46.55	0.321	4	23
δ_D	δ_D	-158.87	-57.43	-157.84	-57.57	0.327	29	4
α_D	δ_D	54.51	41.91	-160.87	-57.16	0.369	1	4
δ_D	α_D	-158.67	-57.24	55.05	45.27	0.869	0	4
ϵ_D	γ_L	62.54	-172.69	-79.26	73.33	1.325	1	8
ϵ_D	ϵ_L	62.53	-172.68	-79.26	73.35	1.325	1	8
γ_L	ϵ_D	-80.08	74.92	63.73	-175.87	1.501	1	5
ϵ_L	ϵ_D	-80.08	74.92	63.72	-175.85	1.501	1	5
α_L	ϵ_D	-71.53	-35.03	63.57	-174.48	2.101	24	42
ϵ_D	α_L	62.72	-170.96	-68.50	-35.66	2.184	15	43
ϵ_D	β_L	63.42	-174.07	-153.75	156.59	2.546	9	16
δ_L	ϵ_D	-150.81	47.31	63.28	-174.80	2.563	1	3
β_L	ϵ_D	-155.32	156.27	63.51	-175.45	2.570	9	31
ϵ_D	δ_L	63.10	-172.17	-147.16	45.48	2.801	1	1
δ_D	ϵ_D	-158.16	-57.19	63.88	-174.92	3.365	0	1
γ_D	α_L	76.47	-64.70	-77.37	-36.13	3.624	5	22
δ_D	γ_L	77.26	-64.68	-80.31	75.33	3.853	0	5
γ_D	δ_L	76.76	-63.97	-149.93	39.32	3.939	1	6
ϵ_D	δ_D	63.64	-174.49	-157.39	-57.45	3.962	0	1
ϵ_L	γ_D	-78.90	77.60	78.25	-63.63	4.023	3	8
ϵ_D	α_D	63.22	-173.44	56.03	45.08	4.351	1	2
α_L	γ_D	-71.22	-36.07	77.84	-64.28	4.626	1	16
γ_D	ϵ_L	77.19	-63.48	-74.47	147.56	4.718	2	13
γ_D	δ_D	77.60	-63.57	-155.85	-56.37	4.724	1	3
α_D	ϵ_D	55.92	51.62	64.81	-173.96	4.963	0	4
β_L	γ_D	-154.52	157.92	77.75	-64.10	4.988	1	3
γ_D	β_L	77.72	-63.55	-153.20	156.48	5.003	1	5
δ_L	γ_D	-150.82	45.74	77.14	-64.36	5.263	0	2
α_D	γ_D	53.26	50.67	76.17	-63.35	5.314	3	10
δ_D	γ_D	-157.85	-56.13	78.49	-64.02	5.795	1	4
γ_D	α_D	77.65	-63.82	54.94	46.01	5.941	0	7
ϵ_D	ϵ_D	63.17	-173.23	64.71	-174.81	6.575	2	2
γ_D	ϵ_D	77.43	-64.25	63.53	-175.54	8.745	0	1
ϵ_D	γ_D	62.51	-172.06	80.12	-63.73	8.806	0	1
γ_D	γ_D	77.64	-62.60	78.40	-63.86	11.320	0	4
							3517	15108

\# Torsion angles ($\phi(1)$, $\phi(1)$, $\phi(2)$, $\psi(2)$, in degrees and energy (E) in kcal/mol units
* Degenerate minima

The question of accuracy in assigning a given experimental conformation to one of the nine minima has been a problem. It turned out that a ± 30° or a ± 40° tolerance is φ and ψ from the theoretically pinpointed minima leads to a rather high (nearly 80% for ± 30° and over 90% for ± 40°) in the assignments. Figure 21 shows squares around the IDEALIZED minima with sides of 2 * 40° = 80° on the 2D-Ramadiandran map

Similarly, one may construct 4D-hypercubes with sides of 80° about the theoretical minima pinpointed by the ECEPP/2 method (c.f. Table IV). The volume of such a hypercube is given in the following equations:

$$\text{Volume} = (\phi_i - \phi_1)(\psi_i - \psi_1)(\phi_{i+1} - \phi_2)(\psi_{i+1} - \psi_2) \qquad [25]$$

in which ϕ_i and ψ_i are the torsional angles of the i-th and ϕ_{i+1} and ψ_{i+1} are the torsional angles of the (i+1)-st amino acid residues in a protein under investigation. Similarly, ϕ_1, in which ϕ_i and ψ_i are the torsional angles of the i-th and ϕ_{i+1} and ψ_{i+1} are the torsional angles of the (i+1)-st amino acid residues in a protein under investigation. Similarly, ϕ_1, ψ_1, ϕ_2, ψ_2 are the corresponding torsional angles listed in Table IV. If the angle differences are all 80°, as discussed above, then the overall volume allocated for the 81 minima is simply

$$V = 81*(80°)^4 \qquad [26]$$

which represents only 18% of the total volume of $(360°)^4$. Hence, the allowed geometrical variations have been narrowed down considerably in the configuration space.

If the ϕ_i ψ_i pair falls within such a 40°-hypercube then the assignment is confirmed. However, a numerical value, may be associated with the match, which is calculated in the form of a well defined deviation (dev):

$$\text{dev} = \tfrac{1}{4}[(\phi_i - \phi_1)^2 + (\psi_i - \psi_1)^2 + (\phi_{i+1} - \phi_2)^2 + (\psi_{i+1} - \psi_2)^2] \qquad [27]$$

Table IV clearly indicates the practicality of the assignment. Since a dipeptide maybe fitted two ways to a given amino acid unit within a protein chain as shown below:

protein   * - * - (i-1) - (i) - (i+1) - * - *

First fit of dipeptide   1 - 2   [28]

Second fit of dipeptide   1 - 2

therefore each fit will lead to a particular assignment. In the example given in SCHEME IX the conformational assignment (using dev < 30°) was uncertain for the 37-th amino acid (γ_L or ε_L) and the 62-nd amino acid (ε_L or β_L); all the other conformations were doubly reconfirmed by both the first and second fitting of the dipeptide.

The following SCHEME shows the explicit form of this linearized notation of backbone conformation for Cytochrome b_5 (2B5C).

10

Ala-Val-Lys-Tyr-Tyr-Thr-Leu-Glu-Gln-Ile-Glu-Lys-His-Asn-Asn-Ser-Lys-

δ_L- ε_L- ε_L- ε_L- β_L- ε_L- α_L- α_L- α_L- α_L- α_L- α_L- ε_L- δ_L- β_L- α_L- α_L-

20   30

Ser-Thr-Trp-Leu-Ile-Leu-Hys-Tyr-Lys-Val-Tyr-Asp-Leu-Thr-Lys-Phe-Leu-

β_L $-\varepsilon_L$ $-\beta_L$ $-\beta_L$ $-\varepsilon_L$ $-\beta_L$ $-\alpha_D$ $-\alpha_D$ $-\varepsilon_L$ $-\varepsilon_L$ $-\beta_L$ $-\gamma_L$ $-\gamma_L$ $-\alpha_L$ $-\alpha_L$ $-\alpha_L$ $-\alpha_L$ -

40   50

Glu-Glu-Hys-Pro-Gly-Gly-Glu-Glu-Val-Leu-Arg-Glu-Gln-Ala-Gly-Gly-Asp-

α_L $-\alpha_L$ $-\varepsilon_L$ $-\alpha_L$ $-\alpha_L$ $-\varepsilon_D$ $-\alpha_L$ $-\alpha_L$ $-\alpha_L$ $-\alpha_L$ $-\alpha_L$ $-\alpha_L$ $-\delta_L$ $-\varepsilon_L$ $-\alpha_D$ $-\beta_L$ $-\gamma_L$ -

γ_L ε_L

60

Ala-Thr-Glu-Asp-Phe-Glu-Asp-Val-Gly-Hys-Ser-Thr-Asp-Ala-Arg-Glu-Leu-

δ_D $-\alpha_L$ $-\alpha_L$ $-\alpha_L$ $-\alpha_L$ $-\alpha_L$ $-\alpha_L$ $-\delta_L$ $-\alpha_D$ $-\varepsilon_L$ $-\varepsilon_L$ $-\alpha_L$ $-\alpha_L$ $-\alpha_L$ $-\alpha_L$ $-\alpha_L$ $-\alpha_L$

β_L

70   80

Ser-Lys-Thr-Phe-Ile-Ile-Gly-Glu-Leu-Hys-Pro-Asp-Asp-Arg-Ser-Lys-Ile

α_L $-\alpha_L$ $-\alpha_L$ $-\delta_L$ $-\varepsilon_L$ $-\alpha_L$ $-\varepsilon_D$ $-\beta_L$ $-\varepsilon_L$ $-\varepsilon_L$ $-\alpha_L$ $-\alpha_L$ $-\alpha_L$ $-\alpha_L$ $-\alpha_L$ $-\alpha_L$ $-\alpha_L$

SCHEME IX

The utility of the linearized notation of secondary and tertiary structure can be demonstrated by comparing two proteins. As an example, the linearized notation for two proteins labelled as 2C2C and 3C3C in the Brookhaven Protein Data Bank[30] is given in Figure 22. In this figure the oxidized (2C2C) and reduced (3C2C) forms of the Rodospirillum Cytochrome C2 are compared. Although, one might have anticipated only small quantitative conformational changes, as the result of the $Fe^{3+} \rightarrow Fe^{2+}$ reduction, a qualitative change did occur at the third amino acid unit where an asparagine residue switched from a γ_L to an ε_L conformation (c.f. Figure 22). This switch between γ_L and ε_L conformations represent nearest neighbour transformation in accordance with SCHEME II.

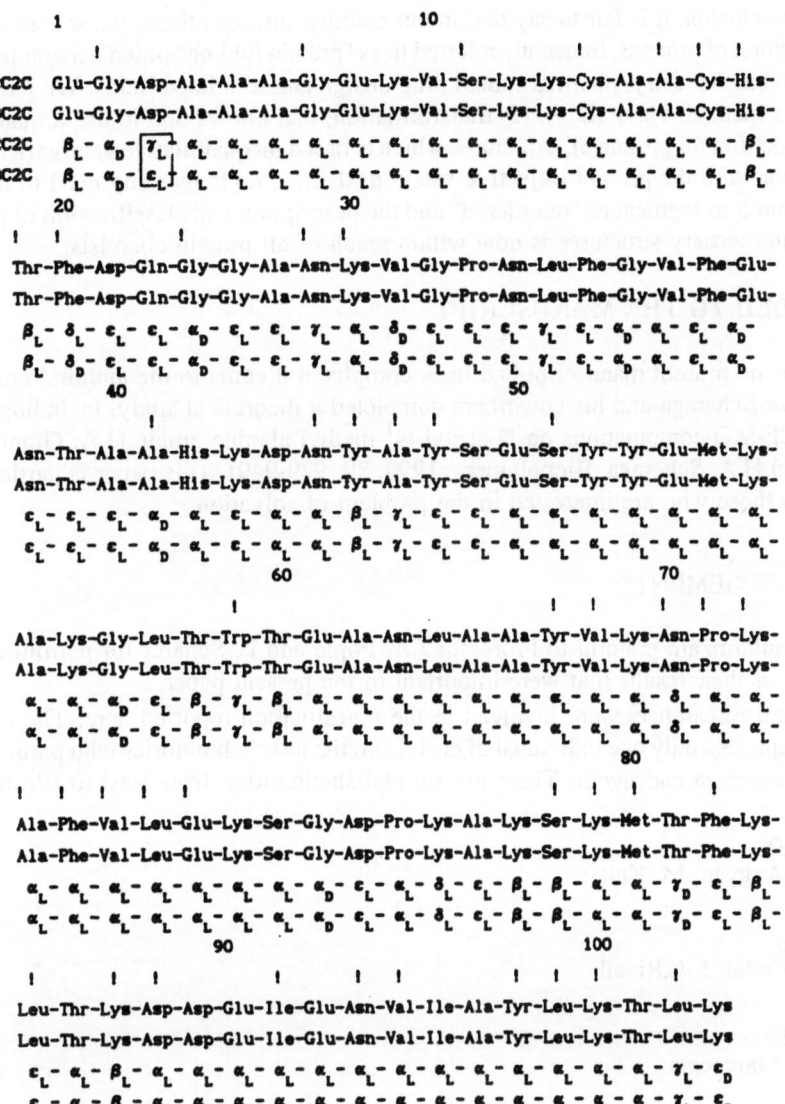

Figure 17: The linearized notation of the primary and secondary structures of 2C2C (Rodospirillum C2 oxidized form), 3C2C (Rodospirillum C2 reduced form). For numbering of amino acids see Louie et al.[31]

In conclusion, it is fair to say that in our century, among others, the secondary and tertiary structural of proteins, frequently referred to as "protein folding" posed a major problem to medicine, biochemistry, pharmaceutical drug design and to related fields. We have now formulated a method, via a 3D → 1D transformation, that allows one to use a qualitative linearized notation for protein 3D structures which is based on quantitative geometrical data. It appears that with the present **objective method**, there is no longer any need to refer to a particular protein segment as "unordered" and the description and classification of protein secondary and tertiary structures is now within reach of all protein chemists.

NOTE ADDED TO THE MANUSCRIPT

After the present manuscript had been completed it came to the authors' attention that Professor Scheraga and his coworkers completed a theoretical study, including large basis set SCF-MO computations on N-acetyl-N^1-methyl alanine amide [J.A. Grant, R.L. Williams and H.A. Scheraga, Biopolymers, **1990**, 30, 929-949]. This paper is particularly important to those who are interested in the problem of solvation.

ACKNOWLEDGEMENTS

The authors are grateful to Professor J.A. Pople and L. Schäfer for permitting the use of some of their results that were important to the present paper.

Several researchers were involved in the investigation reported here. The present co-authors represent only one individual of each from the three laboratories who participated in this joint research endeavour. These are (in alphabetic order, from East to West):

from Hungary:
    J.G. Ángyán, M. Kajtár

from France:
    R. Daudel, J.-L.Rivail

from Canada:
    J.-F. Marcoccia

This research was supported in part by a grant from the Institute of Science Management and Informatics, Hungary.

The authors thank IBM-France and the French CNRS for funding as well as for the generous allocation of computer time within the framework of the GS (Groupment Scientifique) "Modelisation Moleculaire". Many of the ab initio computations were carried out at CIRCE (Orsey, France).

The continued financial support of the NSERC of Canada as well as the Canadian NRC are gratefully acknowledged.

REFERENCES

1.a   I.G. Csizmadia, "General and Theoretical aspects of the thiol group" in "The chemistry of the thiol group" a member of the series "The chemistry of functional groups" edited by Saul Patai, John Wiley and Sons **1974**, pp 1-109 (c.f. particularly pages 36-41 including Figures 23 and 24 as well as Table 20).

1.b   I.G. Csizmadia, Multidimensional Theoretical Stereochemistry and Conformational Potential Energy Surface Topology in "New Theoretical Concept for Understanding Organic Reactions" **1989** by J. Bertrán (ed.) D. Reidel Publishing Co. pp.1-31.

2.   A. Perczel, J.G. Angyan, M. Kajtar, W. Viviani, J.-L. Rivail, J.-F. Marcoccia and I.G. Csizmadia, J. Am. Chem. Soc., **1991** 113, 6256

3.   P.G. Mezey Potential Energy Hypersurfaces Elsevier Science Publishers **1987** p. 227

4.   M.R. Peterson "Determination of Critical Point Geometries of Conformational Energy Hypersurfaces" University of Toronto, Ph.D. Thesis **1980**.

5.   M.R. Peterson, I.G. Csizmadia and R.W. Sharpe, J. Mol. Struct., THEOCHEM **1983**, 94, 363.

6.   M. Greenberg, "Lectures on Algebraic Topology", W.A. Benjamin New York, **1967**, pp. 99-103.

7.   P.G. Mezey, Chem. Phys. Letters **1981**, 82, 100

8.   P.G. Mezey, Chem. Phys. Letters **1982**, 86, 562.

9.   J.G. Angyan, R. Daudel, A. Kucsman and I.G. Csizmadia, Chem. Phys. Letters, **1987**, 136, 1

10.   S.S. Zimmerman, M.S. Pottle, G. Nemethy and H.A. Scheraga, Macromolecules, **1977** 10, 1.

11.   L. Pauling and R. Corey, Proc. Nat. Acad. Sci. U.S.A., **1951**, 37, 729.

12.   L. Pauling, R. Corey and H. Branson, Proc. Nat. Acad. Sci. U.S.A., **1951**, 37, 205.

13.   G.N. Ramachandran, C. Ramakrishnan and Sasisekharan, J. Mol. Biol., **1963**, 7, 95.

14.   Y. Koyama, T. Shimanouchi, Biopolymers **1971**, 10 1059.

15.   M. Avignon, P. V. Huong, J. Lascombe M. Marraud and J. Néel, Biopolymers, **1969**, 8 69.

16. M. Avignon and P.V. Houng, Biopolymers, **1970**, 9, 427.

17. M. Avignon, C. Garrigou-Lagrange and P. Bothorel, Biopolymers, **1973**, 12, 1651.

18. Y. Grenie, M. Avignon, C. Garigou-Lagrange, J. Mol. Struct. **1975**, 24, 293.

19. A. Balázs, J. Phys. Chem. **1990**, 94, 2754.

20. L. Schäfer, C. Van Alsenoy and J.N. Scarsdale, J. Chem. Phys. **1982**, 76 1439.

21. J.N. Scarsdale, C. Van Alsenoy, V.J. Klimkowski, L. Schäfer and F.A. Momany, J. Am. Chem. Soc. **1983**, 105, 3438.

22. T. Head Gordon, M. Head Gordon, M.J. Frish, C. Brooks III, and J.A. Pople, Int. J. Quantum Chem. Quantum Biology Symposium, **1989**, 16, 311.

23. W. Viviani, J.-L. Rivail, A. Perczel and I.G. Csizmadia, to be published.

24. A. Perczel, R. Daudel, J.G. Ángyán and I.G. Csizmadia, Can. J. Chem. **1989**, 68 1182.

25. a) V.J. Klimkowski, L. Schäfer, F.A. Momany and C. Van Alsenoy, J. Mol. Struct. **1985**, 124 143. b) L. Schäfer, V.J. Klimkowski, F.A. Momany, H. Chuman and C. Van Alsenoy, Biopolymers, **1984**, 23 2335.

26. K. Siam, V.J. Klimkowski, C. Van Alsenoy, J.D. Ewbank and L. Schäfer, J. Mol. Struct., **1987**, 152 261.

27. K. Siam, S.Q. Kulp, J.D. Ewbank, L. Schäfer and C. Van Alsenoy, J. Mol. Struct. **1989**, 189 143.

28. L. Schäfer, Chem. Design Autom. News, July **1990**, 5, 10.

29. A. Perczel, M. Kajtàr, J.F. Marcoccia and I.G. Csizmadia, J. Mol. Struct. (THEOCHEM) **1991** 232, 291-319.

30.a F.C. Bernstein, T.F. Koetzle, G.J.B. Williams, E.F. Mayer, Jr. M.D. Brice, J.R. Rodgers, O. Kennard, T. Shimanouchi and M. Tasumi, J. Mol. Biol. **1977**, 112, 535-542.

30.b E.E. Abola, F.G. Bernstein, S.H. Bryant, T.F. Koetzle and J. Weng in Crystallographic Database- Information Content, Software System, Scientific Applications, eds. F.H. Allen, G. Bergerhoff and R. Sievers, Data Commission of the International Union of Crystallography, Bonn/Cambridge/Chester, **1987**, 107-132

31. G.V. Louie, W.L.B. Hutcheon and G.D. Brayer, J. Mol. Biol., **1988** 199, 295-314.

DISCUSSION

COMMENTS OF PROF. KOLLMAN

How many local minima? Seven? What are their relative energy in JACS paper? Basis set and correlation dependence of higher energy minima? I am concerned on the accuracy of your survey of protein structures; you should redo with only the highest resolution structure to see if all the minima are found with these.

LECTURER: I.G. Csizmadia

The following table, taken from our J.Am.Chem. Soc. paper (Ref. 2), shows the relative energy of the 7 minima we have found by gradient optimizing all internal coordinates (internal forces were less than $5*10^{-4}$ mdyn/Å).

Ab initio SCF (3-21G) results[a] for
For-L-Ala-NH$_2$ conformations

BB[b]	ϕ_{STAND}	ψ_{STAND}	χ_1	ϕ_{TOP}	ψ_{TOP}	E	ΔE_{rel}
α_D	63.8	32.7	60.6	63.8	32.7	-412.465 293	5.93
α_L		not found					
β_L	-168.4	170.9	60.0	191.6	170.9	-412.472 746	1.25
γ_D	73.9	-56.7	60.5	73.9	303.3	-412.470 720	2.52
γ_L	-84.4	67.7	64.7	275.6	67.7	-412.474 738	0.00
δ_D	-178.6	-44.0	58.4	181.4	316.0	-412.463 100	7.30
δ_L	-127.8	30.0	58.7	232.2	30.0	-412.468 678	3.80
ϵ_D	67.6	-178.1	64.9	67.6	181.9	-412.461 724	8.16
ϵ_L		not found					

[a] Torsion angles (ϕ, ψ, χ_1) in degrees, energy (E) in hartrees, and energy difference (ΔE_{rel}) in kilocalories per mole. [b] BB = backbone conformation.

These minima may or may not correspond to the results of a particular MM. For example some MM method might give you 5 minima (α_L, β_L, γ_L, α_D and γ_D) and ignores some of the higher energy minima (δ_D, γ_D, ϵ_L and ϵ_D), yet we have only two minima, α_L and ϵ_L, annihilated at the ab initio level. Higher level of ab initio theory may alter the situation but it does appear to us that the nine minima we are talking about:

$$\gamma_D \quad \delta_D \quad \alpha_L$$
$$\epsilon_D \quad \beta_L \quad \epsilon_L$$
$$\alpha_D \quad \delta_L \quad \gamma_L$$

are legitimate minima and some of them may appear on a PES or some of them may disappear from the PES depending on the situation.

For the a-posteriori control the selection of the proteins were made as suggested by McGregor et al (J.Mol.Biol. 1987, 198, 295-310). Consequently two databases were created, the first set contained 78 protein sequences with entities being determined to a resolution of 2.A or better. This is labelled as the small data base of proteins (SDB-PROTEINS). In order to find recognizable conformational examples in proteins for the poorly populated dipeptide conformations, which were predicted by the theory as higher energy minima, a large data base of proteins (LDB-PROTEINS) with some 250 proteins was also introduced. All protein structures were taken from the Brookhaven Protein Data Bank[30]. The members of the two databases are given below, maintaining the notation of the Brookhaven Protein Data Bank.

SDB-PROTEINS

1BP2, 1CC5, 1CCR, 1CPV, 1CRND, 1CTF, 1ECA, 1FB41, 1FBJ1, 1FC2c, 1FDX, 1GCR, 1HIP, 1HMQ, 1INSa, 1LH1, 1LZ1, 1LZT, 1MBD, 1NXB, 1PCY, 1PP2, 1PPD, 1PPT, 1SBT, 1SN3, 1TGSz, 2ABX, 2ACT, 2ALP, 2APP, 2AZA, 2CAB, 2CCY, 2CDV, 2CTS, 2CYP, 2EST, 2GN5, 2INSa, 2LHB, 2LZM, 2OVO, 2PAD, 2PKAa, 2RHE, 2SGA, 2SNS, 2SOD, 351C, 3C2C, 3DFR, 3ICB, 3RP2, 3RXN, 3SGBe, 3TLN, AADH, 4APE, 4ATCa, 4CYT, 4DFR, 4FXN, 4HHBa, 5CPA, 5LDH, 5RXN, 3PGM, 1FB4h, 1FBJh, 1FC2d, 1INSb, 1TGSi, 2INSb, 2PKAb, 3SGBi, 4ATCb, 4HHBb.

LDB-PROTEINS

4APE, 2APP, 1APRA, 2ACT, 1ACX, 2ADKD, 1AGA, 2ALP, 3WGA, 4ADHB, 5ADH, 6ADH, 6ADH, 7ADH, 2TAA, 5API, 6API, 6API$_L$, 1ABP, 1AAT, 2ATC, 5ATC, 5ATC$_a$, 2AZA, 1AZU, 2BCL, 2ABX, 4ATC, 1CPVB, 2CPVB, 3CPVB, 3ICB, 1CAP, 2CAB, 2CAB, 1CACC, 3CPA, 4CPA, 4CPA, 5CPA, 1CPB, 1CAR, 1CAR, 1PTE, 1C4S, 2CHA, 7CAT, 8CAT, 4CAT, 2C4S, 4CAT, 2C4S, 5CHA, 2GCH, 1CTXA, 1CHG, 1CTS, 4CTS, 2CTS, 3CTS, 2CNA, 3CNA, 1CRN, 1GCR, 1CN1, 2B5C, 2B5CA, 156BA, 3CYT, 4CYT, 1CYCJ, 1CCR, 2CCY, 2CYP, 2C2C, 3C2C, 1CY3, 2CDV, 1CC5, 155CD, 351C, 451C, 1CPPA, 3DFR, 4DFR, 4DFR, 1ANA, 2ANA, 1BNAD, 2BNA, 2BNA, 3BNA, 3BNA, 4BNA, 5BNA, 5BNA, 6BNA, 7BNA, 7BNA, 8BNA, 8BNA, 1ZNA, 2GN5, 1ESTC, 2EST, 2EST, 2EBX, 1ECD, 1ECO, 1ECA, 1ECN, 2FD1, 1FDXF, 3FXC, 3FXNA, 4FXNA, 1FX1, 1GBP, 1GAP, 2GAP, 1GCN, 1PGI, 1GP1, 2GRS, 1HCO, 1GPDA, 2GPD, 3GPD, 1HMG, 1HRB, 1HRB, 1HMQ, 1HMZ, 1HR3, 1HDS, 2MHB, 2DHB, 2HHB, 3HHB, 4HHB, 2HCO, 1HHO, 1FDH, 2YHX, 1HKG, 1HBS, 2LHB, 1HIP, 1HYAA, 1HYA, 2HYA, 3HYA, 3HYA, 4HYA, 4HYA, 1FBJ, 1MCP, 2MCP, 1FB4, 3FAB, 1MCG, 1REIC, 2RHE, 1FC1, 1FC2, 1PFC, 1IG2, 1INS, 2INS, 2PKA, 2KAI, 2KAI, 1KGA, 1KES, 4LDH, 3LDH, 3LDHA, 5LDH, 1LDX, 1LH1, 2LH1, 1LH2, 2LH2, 1LH3, 2LH3, 1LH4, 2LH4, 1LH5, 2LH5, 1LH6, 2LH6, 1LH7, 2LH7, 2LZM, 1LYZB, 2LYZB, 3LYZB, 5LYZB, 6LYZB, 7LYZC, 1LZT, 8LYZ, 9LYZ, 1LZHB, 2LZHD, 1LYM, 1LZ1, 1LZ2, 1CTF, 2MDH, 1MLT, 2MT2, 1MBSE, 1MBN, 2MBN, 3MEN, 1MBD, 1MBO, 1MB5, 1MHR, 1NXB, 1NXB, 1SN3, 1OVO, 2OVO, 1PPT, 1PADB, 2PAD, 2PADB, 9PAP, 4PADC, 5PADB, 5PADB, 6PADE, 1PPD, 1PEP, 3PGK, 2PGK, 2PGK, 3PGM, 1BP2, 2BP2, 3BP2, 1P2P, 1P2P, 1PP2, 1PCY, 2PCY.

In the SDB we also found, as reported in our J.Mol.Struct. paper (Ref. 29), that all nine minima $(\alpha_L\ \beta_L\ \gamma_L\ \delta_L\ \varepsilon_L\ \alpha_D\ \gamma_D\ \delta_D\ \varepsilon_D)$ were present in the various amino acid residues of those proteins that had resolution of 2.0Å or better. If you demand, let's say an order of magnitude better resolution (i.e. less than 0.2 Å); we probably have to wait until protein structures will be determined in a statistically significant number by X-ray crystallographers to such an accuracy.

COMMENTS OF PROF. ADNAN TAYMAZ

Modifying potential energy curves (PEC) and the potential energy surfaces (PES), you are just taking angular variables such as ψ and ϕ. However, the potential energy between nuclei are also space dependent such as the radial position.

LECTURER: I.G. Csizmadia

Yes, it is true that we focus our attention on the Energy as a function of these torsional angles of ϕ and ψ. however, it is understood that <u>all internal variables</u> (bond length, bond angles) are optimized to be at their minima.

COMMENTS OF PROF. P.G. MEZEY (Comment No.1)

In many of the examples you have shown, nine minima predicted by the "backbone only" ideal model yet in your ab initio calculation the minimum in the ε region disappeared. This is not unexpected, since in this high energy region the catchment region of a minimum is expected to be shallow. Interestingly, the minimum in the α region also disappeared. I suspect this is due to the chirality differences of various centres.

Would you say that side chain interactions are responsible not only for the disappearance but also for the appearance of some minima?

LECTURER: I.G. CSIZMADIA

The disappearance of certain minima is due to the shape of a major mountain in the PES which in turn is controlled by the substitution at the chiral centre. The following sketch for this mountain is taken from the ab initio Ramachandran map (Figure 12) of this paper:

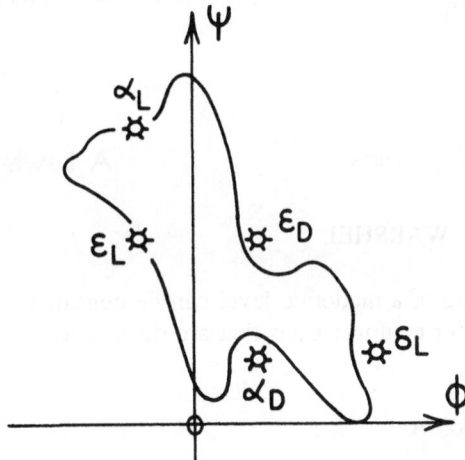

It is clear that the substituent (and its conformation) is in fact influencing the shape of this mountain and some minima will disappear as a result of such surface modification.

COMMENTS OF PROF. P.G. MEZEY (Comment No.2)

I suspect that for many of the features [the presence or the lack of catchment regions of minima within the ε region] and also in the other regions, the dominant effects causing these features might be related the side chain/backbone interactions and not necessarily the conformational preferences within a pure, "side chain - less" backbone model.

LECTURER: I.G. CSIZMADIA

As it is indicated in the answer to the previous question, due to the shape of the mountain, for a given side chain attached to the chiral centre, some minima will have intrinsic instability. [For example for For-L-Ala-NH$_2$ the α_L and ε_L minima were annihilated.] However, certain side-chain/backbone interaction may restabilize these intrinsically unstable minima. We have found that in For-Ser-NH$_2$ an intramolecular Bronsted complex (H-bonded complex) will stabilize the α_L conformation (see our Can.J.Chem. paper Ref. 24) and we might expect an intramolecular Lewis Complex (charge transfer complex) to stabilize either the α_L or ε_L conformation

A Bronsted Complex   A Lewis Complex

COMMENTS OF PROF. A. WARSHEL

It is hard to see how ab initio of a moderate level can be considered as better predictions than force field approaches for conformations that are determined by steric repulsions

LECTURER: I.G. CSIZMADIA

It would be unscientific to claim that the results obtained by the use of a 3-21 G basis set represent a primary standard. However, it is equally unscientific to pretend that the existing force field methods give us the absolute truth. Scherage et al [Journal of Biomolecular Structure and Dynamics Vol. 7 Issue No. 3 (1989) 421-453] compared CHARMM, AMBER and ECEPP and although they agreed in certain gross features they were sufficiently different in details so it is hard to make a choice. At this stage of history, the ab initio and force field methods should be presented side by side until a reliable and therefore universally acceptable method will emerge.

COMMENTS OF PROF. MAGGIORA (Comment No.1)

Why not check out the existence and depth of minima of several amino acids by higher order methods?

LECTURER: I.G. CSIZMADIA

That will be the first thing we shall do as soon as we will get some computer time.

COMMENTS OF PROF. RIVAIL

1) It is obvious that ab initio SCF results do not include dispersion energy which is an important contribution to intermolecular interactions. Nevertheless intramolecular interactions are pretty well predicted by ab initio SCF results as shown by Lothar Schäfer's work predicting gas phase geometries confirmed by microwave spectroscopy. The question is, what can be considered as intramolecular or intermolecular in a protein. I would propose the conjecture that a peptidic unit can be treated as a molecule (like in I.G. Csizmadia's approach) and that interactions between peptidic units can be represented as intermolecular.

2) Another important contribution comes from induction forces which, of course, are well represented by ab initio SCF computations. I wonder whether they are represented to sufficient accuracy in most of present days force fields.

COMMENTS OF PROF. H. WEINSTEIN

I must point out that the idea that the conformation surrounding a single peptide bond determines protein folding, is still an article of faith. Consequently, comparing crystals structures with the results of ab initio calculations of such a minimal unit may be misleading in that it ignores interactions among main chain and side chain elements along the sequence - the most obvious examples being (i, i+4) hydrogen bonds in helices, and hydrogen bonds across β-strands to from β-sheets. The local confirmation at a certain residue will be dependent on such interactions, and these may be more important than the "self energy' in determining such local conformations.

I would hope that an evaluation of this problem through theoretical criteria, not numerical correlations, will be attempted by theoretical chemists to establish the feasibility and possible nature of the "smallest unit" determining the secondary and/or tertiary structure.

LECTURER: I.G. CSIZMADIA

Considering your "article of faith" remark, surely nobody is suggesting that studying a diamide derived from a single amino residue one can predict the secondary and tertiary structure of proteins. What is suggested here, however is that the conformational PES of such diamides are not entirely irrelevant to protein 3D structures (see also the answer to Prof. Löwdin's question). The main point of this paper was that various MM methods predict only 5 conformations for a single amino acid diamides, yet there are 9 legitimate conformations (some of which might be annihilated). The comparison to X-ray data was only used to illustrate that indeed all 9 conformations do occur in proteins. As far as your quest for smallest useful unit is concerned the paper does present a classification which is shown below:

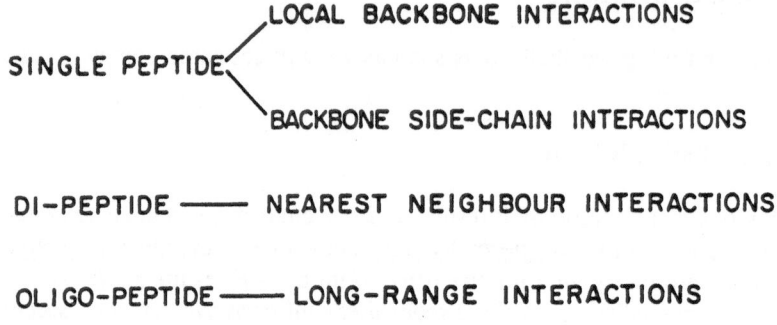

COMMENTS OF PROF. MAGGIORA (Comment No.2)

What is the "message" of β-turns?

LECTURER: I.G. CSIZMADIA

Every enzyme carries a "message" in its 3D structure. The message is really a command that may be regarded as a sentence. For example:

> "Take a given reactant (substrate) and convert it to a predetermined product utilizing a prespecified reaction mechanism."

The enzyme really and truly must have such a 3D structure as it has in order to obey this command. The particular 3D structure only partially determined by the repetitive units such as $(\alpha_L)_n$ the right handed α helix or $(\beta_L)_n$ the β pleated sheet and the actual detailed conformation must come from the "in between" or "unordered" structural elements such as the various turns including β turns. To use a poetic expression if the 3D structure of a protein is the "message" then the various turns (including the β turns) are the vocabulary to describe such a message. (About 32% of the residues are involved in recognized β turns. The actual percentage however might be higher.)

In our J.Mol.Struct. paper (Ref. 29) we have speculated that the classical Type I β-turn actually may be associated with three different conformations:

$$\alpha_L \alpha_L \quad \alpha_L \gamma_L \quad \alpha_L \delta_L$$

Yet there might be as many as 6 distinctly different β-turns with $\alpha_L X$ conformations:

$$\begin{array}{cc} \alpha_L \delta_D & \alpha_L \alpha_L \\ \alpha_L \beta_L & \alpha_L \varepsilon_L \\ \alpha_L \delta_L & \alpha_L \gamma_L \end{array}$$

In our most recent, as yet unpublished, paper we have already demonstrated that the β-turn of $\alpha_L \delta_L$ conformation has intrinsic stability so its existence is not due to external hydrogen bonding, involving solvent or interaction with far away peptide bonds in the protein chain.

COMMENTS OF PROF. TAIMAZ

In your computational calculation of potential energy between molecular fragments what sort of mathematical function are you using in the Hamiltonian?

LECTURER: I.G. CSIZMADIA

Our Hamiltonian operator is the usual Hamiltonian in which the potential energy operator includes all attractive and repulsive electrostatic interactions.

COMMENTS OF PROF. P.O. LÖWDIN

I believe that your "ab initio" approach refers to a restricted Hartree-Fock calculation with fixed nuclei in the Coulombic Hamiltonian and that, in that approach, the energy surface of the dipeptides may be of great importance. We know, however, that the biological function of e.g. an enzyme is strongly related to its tertiary structure. Could you comment on how you would explain the importance of the di-peptides for understanding the tertiary structures?

LECTURER: I.G. CSIZMADIA

We are keenly aware of the moderate basis set-RHF limitation. However, this limitation is not by personal choice but it is imposed upon us by the scarcity of computer time.

As far as the tertiary structure of a protein is concerned one must do calculations on olygopeptides, at least, to study the problem. However, when a protein is folded into a particular structure a number of the ϕ, ψ angles of dipeptides will be altered. So, although you can never predict the tertiary structures of proteins by studying dipeptides, nevertheless if you do have a particular tertiary structure you can analyze them in terms of dipeptide conformations. Thus our method is not predictive but descriptive. It is our hope, however, that this objective descriptive method will allow chemists to classify 3D-protein structures.

COMMENTS OF PROF. MAGGIORA (Comment No.3)

In the presentations of your "linear" conformational code it may be useful to include the designations of secondary-structure such as given by Kabsch & Sander's algorithm.

LECTURER: I.G. CSIZMADIA

The definition of the primary structure of proteins is an objective description which is based on the factual information of the sequence of amino acids that make up the protein molecule. By contrast, as of today, the definition of the secondary structures of proteins is a subjective description as it requires an extensive pattern recognition from the researcher. Since a unique pattern recognition system is required to formulate an objective description, the classification of protein secondary structures has turned out to be a rather elusive goal for quite some time.

During the folding process many proteins share highly similar structures; the so called common secondary structural elements. Usually the remaining part of the backbone is commonly treated as "unordered" (or coil) conformations. In order to give a precise description of the 3D structure of proteins, the "unordered" conformations or the so-called coil groups (a general class containing non typical secondary structural elements) have become an endless target of scientific investigation. Using different types of algorithms the deconvolution of that general class of conformations into subgroups has been a promising approach. Kabsch et al.[a] succeeded in creating seven groups of major conformational units by fixing the backbone conformations and relying on the H-bond types. Based on the C^α distances Levitt et al.[b] have created several models. Also on the basis of inter chain C^α distances Richardes et al.[c] recommended similar models. More recently Rooman et al.[d] created a procedure for the automatic definition of recurrent local structures in proteins. Using general equivalencies Sali et al[e] reported a method that is useful in such classifications.

---

[a] W. Kabsch and C. Sander, Biopolymers (1983), 22, 2577-2637.

[b] M. Levitt and C. Chothia, Nature (London) (1976), 261, 552-558.

[c] F. Richards and C. Kundrot, Proteins, (1988), 3, 71-84.

[d] M.J. Rooman, J. Rodiguez and J.W. Shoshana, J.Mol.Biol. (1990), 212, 403-428.

[e] A. Sali and T.L. Blundell, J.Mol.Biol. (1990), 212, 403-428.

All the latest methods for classifying structural elements in proteins, irrespectively how different they might be in their approaches have one common feature, they are based on the analysis of experimentally determined X-ray structures thus they are all a-posteori.

The new method, reported here, as well as in our J.Mol.Struct. paper (Ref. 29), differs from all previous approaches in that it is based on the topological theory of peptides molecular conformations. The X-ray structures of proteins are used only as a control data set. Thus, the current method is a-priori supported by a posteriori statistics and therefore meets all the requirements for an objective method aiming at the description of protein secondary structures.

COMMENTS OF PROF. PONS

Three dimensional structure is dominated by the arrangement of secondary structure units (α-helices, β-sheets, etc.). As you have mentioned "long range" interactions (of hydrogen bands to stabilize the helix) are the ones that determine the secondary structure rather than the "local" interaction that shows up in the dipeptide picture.

Therefore, the information one may want to obtain from the dipeptide picture is more on the local dynamics than a building block for the static three dimensional structure.

LECTURER: I.G. CSIZMADIA

The α-helix and β-pleated sheets are like internal pillars or poles from which the architecture of the protein cannot be judged. See also answer to Professor Maggiora's question on β-turns (Comment No.2).

COMMENTS OF PROF. JUAN J. PEREZ

Two technical questions:

1) Have you carried out a "grid search" for the Ramachandran plot or just have localized the minima?

2) Have you checked it you get right periodicity when you recalculate a point of the potential surface after the -180 +180 range?

LECTURER: I.G. CSIZMADIA

Yes, we did carry out grid search both by ECEPP/2 and ab initio method (see Ref. 2).

Any conformational problem must be periodic after 360° rotation so we did not check the periodicity. I can't think of any reason why ab initio SCF might cheat us in this respect.

DYNAMIC SHAPE ANALYSIS OF BIOMOLECULES USING TOPOLOGICAL SHAPE CODES

Paul G. MEZEY

Mathematical Chemistry Research Unit,

Department of Chemistry and Department of Mathematics,

University of Saskatchewan, Saskatoon, Canada, S7N 0W0

ABSTRACT. An important task of the *molecular topology program* (a research program for the reformulation and description of molecular properties and chemical reactions within a differential and algebraic topological framework) is the development of methods and computer codes for the algorithmic characterization of shapes of biomolecules. Some of the new developments of biomolecular shape analysis incorporating the dynamic aspects of conformational changes and protein folding processes are reviewed.

1. Introduction

The study of the three-dimensional shapes of biomolecules is one of the fundamental steps in the quest for understanding and predicting biochemical behavior (for a sample of references, see, e.g., [1-30]). Shape analysis is also of major importance in new strategies toward the development of molecular engineering and in the eventual construction of "custom made" molecules. The *molecular topology program* (MTP, [31,32]) is a general framework for a family of concerted research projects for a differential and algebraic topological description of molecules (*Molecular Topology,* MT, [33-35]) and chemical reactions (*Reaction Topology,* RT, [36,37]).

The molecular topology program is based on the following principle: *molecules are*

83

J. Bertrán (ed.), Molecular Aspects of Biotechnology: Computational Models and Theories, 83–104.

not geometrical but topological objects [34]. Molecular topology (not to be confused with molecular graph theory) is the primary aspect of this program, which has been reviewed extensively from a quantum mechanical viewpoint in [31]. Other recent reviews have focused on a more applied context of three-dimensional (3D) shape analysis of formal molecular bodies [34] and surfaces [35]. Molecular topology provides a general framework for such a shape analysis, and incorportes the *dynamic aspects of molecular shape*. In a conformational rearrangement or in a chemical reaction, many shape features may change but some topological shape properties can remain invariant. Dynamic shape analysis deals with both the changing and the preserved aspects of shape. In fact, both aspects can be formulated in terms of topological shape features that stay invariant under minor conformational changes and the configuration space limits of such shape-feature-preserving molecular rearrangements. These invariant shape features, as well as the corresponding conformational domains, can be characterized topologically and algebraically, leading to computer programs for nonvisual dynamic shape analysis. Some of the recent results and applications of the new, dynamic molecular shape groups and nonvisual, computer-based methods for the quantification of dynamic molecular similarity will be discussed. In particular, we shall focus on the following subjects:

(i)   The *GSTE Principle* of "Geometrical Similarity as Topological Equivalence",

(ii)   Visualization vs. "Nonvisual Algebraic Shape Analysis Methods" *(NASAM)* by computer,

(iii)   The "Dynamic Shape Group" *(DSG)* and "Dynamic Shape Code" *(DSC)* methods,

(iv)   The technique of "Shape Globe Invariance Maps" *(SGIM)*,

(v)   The *ARMM Method*, "Algebraic Representation of Molecular Mobility".

2.   The GSTE Principle: Geometrical Similarity as Topological Equivalence

Topology is a powerful branch of modern mathematics suitable for the analysis of the essential features of complicated structures. In such analyses one may exploit the inherent flexibility of choosing various topologies. The infinite variety of possible topologies ranges from the simple choice of the usual (metric) topology in the three-dimensional Euclidean space (where all simply connected molecular bodies are topologically equivalent), to more intricate topologies such as those based on local geometrical properties of three-dimensional molecular functions. For example, topologies can be defined by the geometrical criteria of molecular contours such as the locally convex and concave domains on formal molecular electron density contour surfaces. Topologies can also be defined by the mutual interpenetration properties of electronic charge density and electrostatic potential contours [38-41]. For these choices, only those contour surfaces which have domain patterns that can be transformed into one another by some continuous transformations (homeomorphisms) are topologically equivalent. Note that while it is not necessary that the domain patterns agree geometrically, if such transformation exists, this certainly implies some geometrical similarity. If the pattern of the locally convex and concave domains are indeed the essential shape properties, then this topology is suitable for focusing on these essential shape features, while disregarding the "incidental", detailed geometrical properties. Geometrical similarity of these essential shape features is reflected in a topological equivalence within the framework of the given topology. Of course, topologies can be constructed by many other methods, for example, by considering the pattern of ranges of electrostatic potential on a fused spheres Van der Waals surface [42-44]. A topological equivalence of these patterns indicates a different type of similarity between two molecules or between two conformations of the same molecule.

The above considerations apply in the general case. Topology provides a rather universal approach toward the study of similarity simply by focussing on the essential. What is to be regarded essential may be chosen rather freely and this choice can be used to define the actual topology considered. Then, if the essential features of two objects agree, this similarity will appear as a topological equivalence. In this sense, the choice of topology reflects the initial decision as to which features will be considered essential. As in the case of formal molecular bodies and contour surfaces, local geometrical conditions on physical properties, such as ranges of electronic charge densities or of the values of electrostatic potentials, are often used to define domains in the 3D space. Subsequently, the pattern of these domains is analysed topologically. Hence, the similarities of the *geometrical* features and mutual arrangements of these domains will be detected as *topological equivalence*. Consequently, the general strategy for topological shape analysis can be formulated as the *GSTE principle:* Geometrical Similarity is treated as Topological Equivalence [34].

3. Visualization and Nonvisual Algebraic Shape Analysis Methods *(NASAM)* by Computer

The last decade has witnessed a spectacular development of molecular graphics methods in combination with advances in theoretical and computational chemistry leading to the almost routine computability of electronic charge densities, electrostatic potentials of medium size molecules, and realistic force fields of biopolymers. Consequently, visual inspection of molecular models on a computer screen has become a widely used, powerful (although somewhat subjective) scientific tool. Visual comparisons have become accepted as valid bases for scientific conclusions in the evaluation of molecular shape similarity, the similarity of protein folding patterns, or in dynamic model studies of folding and unfolding processes. Undoubtedly, human visual perception is a very highly developed and reliable guide in the day-to-day activities of our lives. However, there appears to be a need for a less subjective and somewhat more reproducible method when comparing molecular shapes and dynamic processes.

The topological shape analysis techniques provide a family of such nonvisual, algorithmic methods suitable for direct shape comparisons by the computer. Some of these techniques, in particular the family of shape group methods, have been reviewed recently [32,34,35] and descriptions of a variety of alternative topological shape analysis techniques are to be found in the recent literature [35,45-51]. These alternative methods include the shape matrix method [35,45], the shape graph methods [46], various algebraic characterizations of Van der Waals surfaces [47,48], and knot theoretical representations of chirality of molecules and folding patterns of chain biomolecules [49-51]. The underlying, common feature of all these topological methods is the generation of *families* of geometrical arrangements which fall within *topological equivalence classes* and the subsequent characterization of these topological equivalence classes by *algebraic* means.

All these topological techniques are based on a common principle referred to as (P,W)-*similarity* [34]. When applying this principle, one has to decide on the context of similarity, that is, one has to define a shape representation and a shape descriptor. This involves the specification of the two main components of the actual (P,W)-similarity:

(a) Specifying the *shape representation* P, that is, the physical or geometrical property or model P chosen to represent molecular shape.

The shape representation, property or model P, may be chosen in a variety of ways: as an electronic charge isodensity contour surface G(a) for a specified density value a; as a whole family of such isodensity surfaces for some range (a_1, a_2) of density values; as a fused sphere Van der Wals surface for a specified set of formal atomic radii or for a whole family of such fused sphere surfaces for a range of radii; as a geometrical description of the backbone of a protein structure as a ribbon or as a space curve; or as a sequence of simple geometrical objects such as cylinders, sheets, and rectangular blocks.

(b) Specifying a suitable topological tool W to be employed as the actual *shape descriptor* of P.

For example, if the shape representation P is an isodensity contour with a pattern of locally convex, concave and saddle type domains *relative* to an oriented ellipsoid [52], then the shape descriptor W (topological tool W) can be chosen as the shape groups [38-41] (homology groups of the truncated contour surface, obtained by eliminating domains of specified curvature properties), or shape matrices [35,45] (defined by the neighbor relations, types, and relative sizes of the domains).

The above general scheme is suitable for nonvisual, algebraic shape characterization. For a given (P,W) choice, a whole family of possible geometrical arrangements of different molecules may have a common actual realization of the shape descriptor W (e.g., a common shape group, or a common shape matrix), where each realization is called a (P,W)-*shape type*, denoted by $\tau_{(P,W)}$, or simply by τ if the (P,W) pair is implied from context. For the typical case, there are only a finite number of different shape types τ_i. Having a common shape type τ_i is a topological equivalence, representing geometrical shape similarity. Furthermore, since each shape type τ_i is specified by algebraic methods (e.g., by a group or a matrix), the technique provides a nonvisual, algebraic, algorithmic shape description, suitable for automatic, computer characterization of shapes and for the evaluation of shape similarity.

4. The "Dynamic Shape Group" *(DSG)* and "Dynamic Shape Code" *(DSC)* Methods.

Molecular shape depends on the arrangement of the nuclei. For a given overall stoichiometry, that is, for a given family of nuclei and for a specified number of electrons, all possible molecular arrangements are represented within a formal nuclear configuration space M [31]. This space M is a metric space, that is, a space wherein a suitable distance function can be defined that corresponds to the dissimilarity of nuclear configurations. Clearly, the electron distribution is dependent on the nuclear configuration, that is, the molecular shape is dependent on the nuclear arrangement. Nuclear configuration, however, is not the same as molecular shape, as molecules are three-dimensional fuzzy bodies of electron distributions with nuclei buried within the electronic clouds. In fact, two different nuclear arrangements may have very similar, or in the extreme case, identical shapes for various electronic isodensity surfaces, even if the actual nuclear arrangements differ in the two cases, e.g., in the two conformations of a given molecule, or in two different molecules. Interrelations between nuclear arrangements, their energy contents, and the associated shape properties of electron distributions can be studied within the nuclear configuration space model.

Another related aspect of molecular shape is based on the fact that molecules are not rigid, static entities but dynamic species, undergoing small and large amplitude motions and conformational changes. Furthermore, the quantum mechanical uncertainty relation introduces a fuzziness of formal nuclear positions. All these aspects certainly influence molecular shape on a fundamental level and in a rigorous model they cannot be ignored. Minor geometrical variations of nuclear arrangements may leave the topological shape properties of electron distributions invariant, and this invariance can be represented by shape invariance domains within the nuclear configuration space M. In general, such an invariance domain belongs to each (P,W)-*shape type* for the chosen shape representation P and topological shape descriptor W. The shape representation P itself may involve some parameters. For example, in the case of the shape group method as applied to isodensity contours G(a) of molecules, the shape representation P=P(a,b) is dependent on two parameters. The first of these is the isodensity value a and the second is the reference curvature value b, used to define relative convexity domains on the contour surface G(a) (the quantity 1/b may be thought of as the radius of a tangent test sphere against which the local curvature properties of the isodensity contour G(a) are compared [39]). In the general case, if the shape representation P is parametrically dependent on the components of a k-dimensional vector **p**, where **p** is regarded as an element of some vector space **P**,

$$P=P(\mathbf{p}), \tag{1}$$

then the dynamic shape space D can be defined [41] as the product space

$$D = M \otimes P. \tag{2}$$

Here the dimension of the internal nuclear configuration space M is 3N-6 for an N nucleus system, whereas the dimension of the parameter space **P** is k. Hence, the dimension of the dynamic shape space D is 3N-6+k . A given (P,W) shape type is invariant within some domain D_i of the dynamic shape space D. The extent of this invariance domain D_i in the dynamic shape space D, that is, the size of its projection within configuration space M and that within the parameter space **P**, characterize the limits of dynamic changes which preserve the essential topological shape characteristics of the molecule, as expressed by the shape representation P and shape descriptor W. The special case of P taken as the two-dimensional parameter space (a,b) of the relative convexity (b-convexity) representation of isodensity contours G(a) has been described in detail elsewhere [41,53].

One may define a distance function (a metric) d_D for the dynamic shape space D as

$$d_D = [d_p^2 + d_M^2]^{1/2} . \tag{3}$$

Here d_M is the metric in the nuclear configuration space M, whereas d_p is a suitable metric within the parameter space P, e.g., the usual metric defined as

$$d_D = [\Sigma_i \, p_i^2]^{1/2} , \tag{4}$$

given in terms of the elements p_i (i=1,2,...k) of parameter vector p.

The following techniques for dynamic shape group and dynamic shape code generation is general for real or abstract objects embedded in spaces of arbitrary dimensions, however, the examples we shall consider here are low-dimensional, chemical examples. Consider a rather general case of a family $\{D_{\omega,i}\}$ of 3D domains of a formal 3D molecular body or 2D domains of a molecular contour surface G(a), where the $D_{\omega,i}$ domains are defined by some physical or geometrical criteria. In the notation $D_{\omega,i}$, the index ω is the *domain type*, whereas i is a *serial index* for each maximum connected component of the given domain type ω. For example, we may think of the 3D electronic density function partitioned into domains $D_{\omega,i}$ by the eigenvalue distribution of the local 3D Hessian matrices of the second derivatives of the density, in a local coordinate system where one axis lies along the density gradient vector. One such partitioning can be given by taking ω as

$$\omega = \lambda, \tag{5}$$

the number of negative eigenvalues of the Hessian matrix. When taking this choice, the domain type ω represents various classes of curvature properties. The resulting family $\{D_{\omega,i}\}$ partitions the 3D space into domains according to the local curvature properties of the electronic density function. When the family $\{D_{\omega,i}\}$ is the collection of 2D domains of a molecular contour surface $G(a)$, where each domain $D_{\omega,i}$ is defined by local relative convexity as domain type, one obtains a topological decomposition of the molecular contour surface $G(a)$ suitable for the introduction of the 2D shape groups.

The following construction of the *shape groups* (SG) and *dynamic shape groups* (DSG) is general for any dimensions and all partitionings of the actual space embedding the object(s) for which a shape analysis is required. Consider the union

$$G = \bigcup_{\omega,i} D_{\omega,i} \tag{6}$$

of all domains and generate a truncated object $G(-\omega')$ by eliminating all domains of a given domain type ω':

$$G = \bigcup_{\omega,(\omega \neq \omega'),i} D_{\omega,i} . \tag{7}$$

The *shape groups* $g(G, \omega',h)$ are the homology groups of the truncated object G. In the above notation, h is the dimension of the homology group. The rank of shape group $g(G, \omega',h)$ is the h-dimensional Betti number of the homology group, that can be used for a concise characterization. (The details and various developments of shape group generation for the most common case of electronic isodensity surfaces $G=G(a)$ and homology group dimension h=1 have been reviewed extensively in references [32,34,35,38,39]).

The shape groups may be selected as shape descriptors, $W = g(G, \omega',h)$. It is important to realize that in the typical (nondegenerate) cases there are only a finite number of different shape groups within the entire dynamic shape space D for each dimension h. Hence, the continuum problem of infinitely many, geometrically different shapes can be reduced to a finite set of different $\tau_{i,(P,W)}$ shape types,

$$\tau_{i,(P,W)} = \tau_{i,(P, g(G, \omega',h))} . \tag{8}$$

The abstract group realized by all groups $g(G, \omega',h)$ with a given shape type $\tau_{i,(P, g(G, \omega',h))}$ is denoted by $g_i(G, \omega',h)$, and is referred to as the the i-th reference group. Their invariance domains $A_{i,j}(g_i(G, \omega',h))$ within the dynamic shape space D are defined as :

$$A_{i,j}(g_i(G, \omega',h)) = \{d: d \in D, \ g(G(d), \omega',h) = g_i(G, \omega',h) \} . \tag{9}$$

Here index i is the serial index of the i-th reference group $g_i(G, \omega',h)$, and where index j refers to the j-th maximum connected component $A_{i,j}(g_i(G, \omega',h))$ of the invariance set $A_i(g_i(G, \omega',h))$ of shape type $\tau_{i,}(P, g(G, \omega',h))$ within the dynamic shape space D:

$$A_i(g_i(G, \omega',h)) = \bigcup_j A_{i,j}(g_i(G, \omega',h)) . \tag{10}$$

The pair composed from a given shape group together with its invariance set within the dynamic shape space D is regarded as the *dynamic shape group:*

$$(g_i, A_i)(G, \omega',h) = (g_i(G, \omega',h), A_i(g_i(G, \omega',h))) . \tag{11}$$

The above invariance sets refer to a single shape group $g_i(G, \omega',h)$. A stronger restriction results if we insist on the *simultaneous invariance* of several shape groups. An intermediate level is obtained if for all points d of an invariance set of D the truncation type is fixed at ω' but the shape groups of all dimensions h are invariant. An even stronger restriction is obtained if invariance is required for all truncation types ω' and all dimensions h. All these more detailed dynamic shape characterizations and the associated partitionings of the dynamic shape space D can be obtained by considering intersections of the $A_{i,j}(g_i(G, \omega',h))$ invariance domains for various truncation types ω' and shape group dimensions h.

Note that the following method for the construction of numerical shape codes (*dynamic shape codes,* DSC) is also general for all possible choices of shape domains $\{D_{\omega,i}\}$, such as the choices discussed above, or, for example, for 2D domains defined by ranges of electrostatic potential values on fused sphere Van der Waals surfaces. A common feature of all the choices is the fact that all can be taken as the *shape representation* P of the general model. In order to obtain a *shape descriptor* W suitable for the generation of a simple *dynamic shape code* (DSC), first we define an order for the shape domains $D_{\omega,i}$ and then we construct a shape code in the form of a *shape matrix* that reflects some of the topological relations as well as the ordering of the shape domains.

We shall assume that the domains $\{D_{\omega,i}\}$ are ordered according to their *decreasing size*:

$$D_1, D_2, \ldots, D_s, \ldots, D_z , \tag{12}$$

where the single size index s is used to compress the information contained in the corresponding index pair (ω, i),

$$D_s = D_{\omega,i} .$$
(13)

By analogy with the "symmetric strong neighbor" relation of catchment regions of potential surfaces within the framework of reaction topology [31], we define the following N-neighbor relation for the $D_s = D_{\omega,i}$ shape domains:

$$N(D_s, D_{s'}) = \begin{cases} 1 & \text{if } (D^c_s \cap D_{s'}) \cup (D_s \cap D^c_{s'}) \neq \emptyset \\ 0 & \text{otherwise,} \end{cases}$$
(14)

where $D_s = D_{\omega,i}$ and $D_{s'} = D_{\omega',i'}$. In the notation D^c_s the superscript c stands for the set-theoretical closure within the Euclidean metric of the ordinary 3D space.

The *shape code* is given in the form of a *shape matrix* $M(K,p)$, defined as follows:

$$M(K,p)_{s,s'} = \mu_s \delta_{s,s'} + N(D_s, D_{s'})(1 - \delta_{s,s'}) ,$$
(15)

where K is a nuclear configuration in M, p is the parameter vector of shape representation P, and $\delta_{s,s'}$ is the Kronecker δ symbol. (Note that the special case of $p = (a,b)$ of shape groups of molecular isodensity surfaces has been discussed elsewhere [53]). Matrix $M(K,p)$ is chosen as the shape descriptor W.

This completes the construction of the basis for a mathematically rigorous (P,W) - similarity concept.

Note that the dependence of the shape matrix $M(K,p)$ on the nuclear configuration K allows one to consider dynamic shape properties. In particular, the $\tau_i = \tau_i (P,W)$ shape types are invariant within various subsets of the dynamic shape space D, and these invariance domains of D provide a dynamic shape description of the molecule. Hence the shape information encoded in matrix $M(K,p)$, being invariant within a $\tau_i (P,W)$ invariance domain of D, is, indeed, a dynamic shape code of the molecule. For example, the dependence of the shape matrices $M(K,p) = M(K,(a,b))$ on the two parameters a and b, where a is the isodensity value for some molecular isodensity contour G(a) and b is a reference curvature leading to a relative convexity domain partitioning of G(a), can be represented by the distribution of shape matrices within the nuclear configuration space M and along the a,b parameter plane P, that is, within the dynamic shape space D.

The shape matrices $M(K,p)$ are invariant within certain domains of the dynamic shape space D, and these invariance domains generate a complete partitioning of space D. The specification of the boundaries of these shape matrix invariance domains within the dynamic shape space D is equivalent to a complete description of the dynamic shape features of all molecular species containing precisely the given set of N nuclei. It is an important feature of this topological approach that in the typical (nondegenerate) cases there are only a *finite number* of different shape matrices called *reference matrices* :

$$M_i = M_i(K,p), \qquad i = 1,2,3,..., k. \tag{16}$$

This is the very feature that allows one to describe a *continuum of possible geometrical shapes* in terms of a *discrete topological model*. The subsets of invariance of shape matrices

$$M(d) = M(K,p) \tag{17}$$

within the dynamic shape space D are denoted by $A_i(M(d))$:

$$A_i(M_i) = \{d: d\in D, M(d) = M_i \} , \tag{18}$$

where index i is the serial index of the i-th reference matrix M_i .

A shape invariance set $A_i(M_i)$ is not necessarily connected. Each set $A_i(M_i)$ may be regarded as the union of its maximum connected components $A_{i,j}(M_i)$

$$A_i(M_i) = \bigcup_j A_{ij}(M_i). \tag{19}$$

A shape matrix $M(d)$ is a numerical code describing shape information. If this information is combined with information on the extent of shape invariance within the dynamic shape space D, then this shape code will belong to an entire range of possible nuclear configurations and choices of parameters p of the shape representation P. In the above context, the shape matrix $M(d)$ is regarded as a *dynamic shape code* of the corresponding molecular species.

Both the dynamic shape groups and the dynamic shape codes of shape matrices are means for the analysis of dynamic shape similarity of molecules. The approach is topological both in terms of the actual tools involved and in the representation obtained, thereby providing a discrete characterization of the shapes of continua, such as 3D molecular electronic densities and molecular electrostatic potential functions.

5. Shape Codes *(SC)* and Dynamic Shape Codes *(DSC)* from Shape Globe Invariance Maps *(SGIM)*.

The general method described in this section is applicable to a wide variety of primary shape representations P such as the relative convexity domain partitioning of an isodensity contour, the pattern of interpenetration of two or more molecular isoproperty surfaces (e.g., electrostatic potential ranges on an isodensity contour surface), space curves representing the backbone of biopolymers [50,51,54 and references therein], the pattern of protein structural motifs [55,56], or polyhedral models of helical domains of proteins [57-59], among others.

When devising models for shape analysis, it is natural to view molecular properties (shape descriptors P) as they would appear to an observer moving about a sphere enclosing the molecule. One simple approach that may be considered is to project a molecular image (or a shape descriptor P) onto a spherical surface, assuming a light source in the center of a sphere and regarding the sphere as a screen. This approach leads to a two-dimensional representation of the molecule on a spherical surface. However, the projected image is not suitable for distinguishing features that are present in multiply folded patterns where the light beam passes through several of these folds before reaching the spherical screen. Furthermore, each point of the spherical image contains only local shape information and the method is somewhat inefficient.

A better approach is the following whereby a global shape property of the molecule is assigned to each point of the spehere. As before, the molecule is enclosed within a sphere S. For example, we may consider the smallest possible sphere S that contains the entire shape representation P, placing the center of mass of the molecule so that it coincides with the center of the sphere. Instead of projecting the molecular descriptor P onto the spherical surface, project P onto a tangent plane R(s) at each point s of the sphere. The projection P'(s) of the shape representation P in each tangent plane R(s) can be characterized topologically, leading to a family of topological descriptors

$$F(s) = \{I(i), i=1,...k\}. \tag{20}$$

These descriptors remain invariant within some domains C_j of points r of the sphere. These projected shape invariance domains are analogous to countries on a global map, a feature that motivates the terminology *shape globe invariance maps* or simply *shape globe maps*.

An important feature of these shape globe maps is that the topological descriptors $F(s) = \{I(i), i=1,...k\}$ assigned to each point s of the sphere S provide information on a *global property* of the enclosed molecule.

For example, if P is chosen as the local relative convexity domains of an isodensity contour surface $G(a)$ of the molecule, then to each point s of S the 2D image of the corresponding curvature domain partitioning of $G(a)$ is assigned, as it is projected to the tangent plane $R(s)$ of S at point s. Truncations of certain domain types of the projected planar image define various shape groups of the 2D image, and these shape groups may be chosen as the topological descriptors ultimately assigned to points s. In this case, an invariance domain of the shape globe map is a maximum connected component of the collection of all points s of S where the shape groups of the images of local relative convexity domains projected on the tangent planes $R(s)$ are the same. Alternatively, neighbor relations of the projected local relative convexity domains define shape matrices of the 2D images, and these matrices may be regarded as the topological descriptors assigned to points s. In this case, the same shape matrix is assigned to every point of an invariance domain of the shape globe. A special case is discussed in some detail elsewhere [50,51,54], where P chosen as a space curve representing a protein backbone and the topological descriptors on the local tangent plane projections are either graphs or knots defined by the overcrossing pattern on the planar projection at each tangent plane of a sphere.

In turn, the shape globe maps can also be characterized topologically, for example, by their shape groups as defined by a specified truncation pattern (e.g., by eliminating invariance domains of certain types) or by the neighbor relations of the invariance domains on the global map. The last method leads to a treatment analogous to the shape matrix method, and the information on the size of invariance domains on the global map may be encoded by the ordering of the domains, just as it is done for "ordinary" shape matrices, discussed above. This provides an alternative shape code, based on the shape globe map approach.

Information on the dynamic shape properties of the molecule can also be included within the above framework. Clearly, if conformational rearrangements of the nuclei occur, then a given point s on the sphere S may be re-assigned to a different invariance region after the conformational change is completed. One approach involves assigning new labels to points contested by different invariance domains and considering these "no man's land" areas on the shape globe as new, separate domains. Subsequently, the characterization of the shape globe may follow the steps described above, recognizing that some of the invariance regions of the static case may become reduced in size and that new, contested regions may appear. The shape matrices of the resulting dynamic shape globe maps are shape codes including some information of the dynamic shape properties of the molecule.

This approach leads to a particular realization of the (P,W)-similarity concept [34]. Here the shape representation P (e.g., an isodensity contour with relative local convexity domains or a space curve representing a backbone of a chain molecule) is topologically characterized by a shape descriptor W, where W is the shape globe S with a topological descriptor of the pattern of spherical domains of a selected invariant (e.g., the shape matrix of the pattern projected to tangent planes of S, or the knots derived from the overcrossing pattern of chain molecule images projected to the tangent planes R(s)).

6. Mobility and the *ARMM* Method: "Algebraic Representation of Molecular Mobility".

Mobility is an energy dependent property. If an entity, such as a molecule, is provided with some energy, it may respond in two fundamentally different ways: either it accomodates the energy increase while preserving the essential structural properties or, it loses its original identity by undergoing fundamental structural changes, for example, by breaking into two or more pieces. If only a small amount of energy is available, even a "soft" object may show only little or negligible deformation. Hence, within the given energy range its low mobility is analogous to the mobility of a more rigid object subject to a much higher energy input.

Here we shall restrict the discussion to energies lower than the lowest bond dissociation energy, that is, to cases where only conformational changes and no chemical reactions are expected in the classical sense. The available molecular deformations are confined to a level set F(A) of the corresponding potential energy hypersurface E(K) [31], where the energy bound is A, and K represents a nuclear configuration. A topological characterization of the entire family of possible deformations, in fact, *the mobility of the molecule at the given energy bound* is characterized by the fundamental group $\Pi_1(A)$ of reaction mechanisms [31], that is, by the one-dimensional homology group of the level set F(A). In this context, reaction mechanisms are simply the mechanisms of conformational rearrangements, with barriers below the given energy bound A. These fundamantal groups $\Pi_1(A)$ of conformational rearrangements provide an algebraic characterization of molecular mobility. This characterization is clearly energy dependent since the group $\Pi_1(A)$ is dependent on the energy bound A.

An alternative to the above algebraic approach is also based on level sets $F(A)$ of potential energy hypersurfaces. The possible shape descriptors available for a molecule of some mobility constrained by an energy bound A are those found within the *cylindrical extension*

$$D(A) = F(A) \otimes P \tag{21}$$

of the $F(A)$ subset of the nuclear configuration space M into the actual dynamic shape space D. Here the elements **p** of the parameter space **P** are k-dimensional vectors having the p_q parameters in the shape representation $P=P(p)$ as their elements. The $\tau_{i,(P,W)}$ shape type invariance domains of cylinder $D(A)$ are

$$A_{i,j}(\tau_{i,(P,W)}) =\{d: \, d \in D(A), \, \tau_{(P,W)}(d) = \tau_{i,(P,W)} \}, \tag{22}$$

where index j refers to the j-th maximum connected component. The neighbor relations of these domains within the cylinder $D(A)$ generate a D-space shape matrix $M(A)$ (where the index ordering follows the size ordering of domains). An energy change A to A' corresponds to a transformation $T(A,A')$ of matrix $M(A)$ to $M(A')$,

$$M(A) = T(A,A') \, M(A'). \tag{23}$$

The above relation provides an algebraic representation for the shape changes induced by an increase in molecular mobility as the available energy is increased.

7. Summary

Some applications of the principle of "Geometrical Similarity as Topological Equivalence" *(GSTE)* are reviewed, relevant to the computer-based analysis of shape and similarity of both small and large scale features of molecules within the Molecular Topology Program *(MTP)*. The advantages and some of the problems of "Nonvisual

Algebraic Shape Analysis Methods" *(NASAM)* by computer are discussed, within the context of the shape group methods *(SGM)*. For the representation of molecular shape features preserved in certain dynamic processes, the "Dynamic Shape Group" *(DSG)* and "Dynamic Shape Code" *(DSC)* methods are described, with emphasis on "Shape Globe Invariance Maps" *(SGIM)* and shape codes derived from them. Alternative approaches to an "Algebraic Representation of Molecular Mobility", the *ARMM Method*, are discussed in some detail.

8. Comments and replies

8.1 *Comments of Prof. Bertran:* The topological analysis is a nice description of fundamental aspects of reality. However, I would like to see an example of how can it be applied to a real chemical problem. In other words, I would like that the research you carry out, which is basic for a mathematician, becomes basic from a chemist's point of view, that is, applied for a mathematician.

Reply of Prof. Mezey: When I want to characterize the research carried out within the molecular topology program, I also have difficulty categorizing it whether it is basic or applied research. I really think that it is both basic and applied. I see our primary goal as providing efficient tools for the chemists for the extraction of the essential information from both experimental and theoretical data, and to generate computer methods for doing this. I view topology as the mathematics of the essential. Topology is a very natural tool especially suited for chemistry, both for basic, quantum mechanical reasons (as a consequence of the inadequacy of geometrical models due to the uncertainty relation), and also for applied, practical reasons (e.g., in the computer analysis of molecular shape similarity), for example in computer-aided drug design. This latter is the area where current applications are found.

8.2 *Comments of Prof. Scheraga:* I see two possible applications of your topological approach:
  1. Use it to compare, say, a calculated conformation with an experimental one, or
  2. use it in the same way to help you understand the nature of, or the pathway of, protein folding.

If the former, what is the advantage of your method over, say, a differential geometry one to compare two conformations?

Reply of Prof. Mezey: Yes, both of the mentioned problems are treatable by the topological methods. Concerning comparisons between differential geometrical and topological methods characterizing protein folding, the former describes a formal geometrical object in detail, for example, the backbone of a protein and a particular pathway for its change. By contrast, the methods developed within the molecular topology program ignore the incidental geometrical features of individual geometrical models, and they describe the essential, common, topological features for a full range of "slightly different" geometrical models. Note that none of the individual geometrical models is fully justified on physical grounds, for example, the backbone of a protein chain is definitely not an infinitely thin line with pointlike intersections on its image projected to a plane. Furthermore, the very fact that proteins are nonrigid, dynamic objects indicates that a whole range of geometries should be considered. So a topologist might ask: why to choose one, or even several individual geometrical models and worry about their fine geometrical details, when topology can describe the essential features of a whole range, a whole continuum of formal geometries, thereby giving a truer picture of reality? On a technical level, the proposed topological descriptions can be extracted from the infinite family of differential geometrical descriptions of all models falling within a given range of geometries, so in this sense topology does not provide "additional" information. What it does well, is eliminating the incidental features, while focusing on the essential. Again, I regard topology as the mathematics of the essential.

8.3 *Comments of Prof. Csizmadia:* Can you establish topological equivalence between the active sites of proteins?
Reply of Prof. Mezey: If the shapes of two active sites are similar, this must appear as a topological equivalence at an appropriate level of resolution, that is, by an appropriate choice of the shape representation P and shape descriptor W in a (P,W)-equivalence.

8.4 *Comments of Prof. Weinstein:* I agree with Prof. Bertran's insistence on a good demonstration of the power of this method with an illustrative example. As you know, Jane Richardson has identified folding motifs in proteins in a graphic manner (e.g., greek keys, barrels, etc.) that allowed interesting insights into relations between folding patterns and the functions they subserve. The computibility of your method represents a clear advantage over such a representation, if it also carries the same ability to characterize the structures and building motifs. If you provided evidence for such an ability by providing a computationally adapted representation at a resolution that enables a characterization of folding motifs similar to that offered by Jane Richardson, your method would be widely and enthusiastically received.

Reply of Prof. Mezey: Up to date there are approximately ten papers published or in press where various applications are described; a partial list of these references can be found in this report. In my opinion, they provide some evidence of the applicability of the topological methods, however, I fully agree that more examples are needed. The level of resolution can be controlled by making appropriate choices for the shape representation P and shape descriptor W in a (P,W)-equivalence. If the mutual arrangements of the major motifs is of interest, then one may, in fact, use the very objects proposed by Jane Richardson for shape representation P and carry out a topological analysis, for example, by the shape globe model as shape descriptor W.

8.5 *Comments of Prof. Pons:* Shape analysis may be useful to describe a class of objects for which we do not have a unified representation: flexible molecules are being represented by a collection of static pictures and energies. Of course, the difficulty is likely to be to try to obtain a relationship between energy and topology of some sort, like the one existing between energy and geometry. The use of the lowest geometry within a topological "catchment" does not seem to be appropriate if there are sensible barriers for interconversion within this "topological catchment".

Reply of Prof. Mezey: One such topological energy relation is provided by the level set approach, described, e.g., in [31]. Topological constraints on energy (on the distribution of the critical points of potential surfaces) also follow from molecular symmetry; these constraints have been derived recently in the form of catchment region point symmetry theorems and vertical point symmetry theorems [60]. Note that these theorems apply to transition structures, that is, to the very barriers of interconversions you have mentioned.

8.6 *Comments of Prof. Kollman:* I agree that abstraction at the level of α-helices, β-sheets and loop representations of proteins are enormously valuable, but the key thing is that they retain a geometrical relationship between fragments: I don't understand how topological descriptions can give this same physical insight and challange you (now or in the future) to show that.

Reply of Prof. Mezey: According to the topological view of molecules, the actual relationships between protein fragments such as α-helices, β-sheets and loop representations are not geometrical but topological. For example, there is no precise, rigid angle between any two axes of formal helical fragments. Precise angles are properties of the geometrical model only; in reality, in any given protein there is a *range* of infinitely many angles between two helical fragments, where none of these angles is a correct description of reality by itself. Topology treats the entire range and pays attention only to features that are invariant. The impression of insight one may receive from a single geometrical model may be very pleasing and easily comprehensible, but it tells only a part of the full story, and may even mislead one if taken at face value. Usually, chemists take care of this intuitively, by regarding these models as that of the "most stable geometry", or that of a "time averaged structure", etc., that is, by qualifying their reliance on such geometrical models. What I suggest when using topology is to let the actual topological model take care of this qualification by eliminating the geometrical ballast, that is, by clearly distinguishing the essential, invariant (topological) features from the merely geometrical, incidental properties of individual models. One might say that the topological model of molecules can be distilled from a mathematical brew of the geometrical models of molecules and the chemical intuition of chemists. I find it advantageous that even those aspect of molecules which in geometrical models are relegated to be taken care of by chemical intuition can now be treated with some mathematical rigor within the topological framework.

8.7 *Comments of Prof. Maggiora:* It remains to be shown whether or not the topological approach proposed by Professor Mezey for characterizing protein chain folds will have sufficient discriminatory power to to provide a means for studying classes of proteins such as those described by Jane Richardson (e.g., all-α, α/β, etc.).

Reply of Prof. Mezey: As a special choice, the structural motifs of helices and sheets themselves may be chosen as elements of the topological model and in such a case the same discriminatory power is certainly available. I agree, however, that much work remains to be done. I expect this work to be rewarding, perhaps by leading to the recognition of general trends and correlations between shape and function.

Acknowledgment

Research work leading to this report has been supported by both operating and strategic research grants from the Natural Sciences and Engineering Research Council of Canada.

References

[1]   Y.C. Martin, Y.C. *Quantitative Drug Design: A Critical Introduction;* Dekker: New York, 1978.

[2]   Le Bret, M. *Biopolymers* 1979, **18**, 1709.

102

[3]   Cantor, C.R., and Schimmel, P.R. *The Conformation of Biological Macromolecules, Biophysical Chemistry, Part I;* Freeman: San Francisco, 1980.

[4]   Lesk, A.M., and Hardman, K.D. *Science* 1982, **216**, 539.

[5]   Karplus, M., and McCammon, J.A. *Annu. Rev. Biochem.* 1983, **53**, 263.

[6]   De Santis, P., Morosetti, S., and Palleschi, A. *Biopolymers* 1983, **22**, 37.

[7]   Richards, W.G. *Quantum Pharmacology;* Butterworths: London, 1983.

[8]   Franke, R. *Theoretical Drug Design Methods;* Elsevier: Amsterdam, 1984.

[9]   G.E. Schulz, G.E., and Schirmer, R.H. *Principles of Protein Structure;* Springer-Verlag: New York, 1979.

[10]   Richardson, J.S. *Methods in Enzymol.* 1985, **115**, 359.

[11]   Liebman, M.N., Venanzi, C.A., Weinstein, H. *Biopolymers* 1985, **24**,   1721.

[12]   Rawlings, C.J., Taylor, W.R., Nyakairu, J., Fox, J., and Sternberg, M.J.E. *J. Mol. Graph.* 1985, **3**, 151.

[13]   Lesk, A.M., and Hardman, K.D. *Methods in Enzymol.* 1985, **115**, 381.

[14]   Kikuchi, T., Némethy, G., and Scheraga, H.A. *J. Comput. Chem.* 1986, **7**, 67.

[15]   Carson, M., and Bugg, C.E. *J. Mol. Graph.* 1986, **4**, 121.

[16]   Dean, P.M. *Molecular Foundations of Drug-Receptor Interaction;* Cambridge University Press: New York, 1987.

[17]   Carson, M. *J. Mol. Graph.* 1987, **5**, 103.

[18]   Connolly, M.L. *Visual Comput.* 1987, **3**, 72.

[19]   Jaenicke, R. *Prog. Biophys. Molec. Biol.* 1987, **49**, 117.

[20]   Tapia, O., Eklund, H., and Brändén, C.I. Molecular, Electronic, and Structural Aspects of the Catalytic Mechanism of Alcohol Dehydrogenase. In *Steric Aspects of Biomolecular Interactions.*, Náray-Szabó,G., and Simon, K., Eds., CRC Press: West Palm Beach, 1987.

[21]   Åqvist, J., and Tapia, O. *J. Mol. Graph.* 1987, **5**, 30.

[22]   Leicester, S.E., Finney, J.L., and Bywater, R.P. *J. Mol. Graph.* 1988, **6**, 104.

[23]   Richards, F.M., and Kundot, C.E. *Protein Struct. Funct. Genet.* 1988, **3**, 71.

[24]   Abagyan, R.A., and Maiorov, V.N. *J. Biomol. Struct. Dynam.* 1988, **5**,   1267.

[25]   Dearden, T. *J. Comput. Chem.* 1989, **10**, 529.

[26]   Hao, M.-H., and Olson, W.K. *Biopolymers,* 1989, **28**, 873.

[27]   Mitchell E.M., Artymiuk, P.J., Rice, D.W., and Willett, P. *J. Mol. Biol.* 1990, **212**, 151.

[28]   Fisher, C.L., Tainer, J.A., Pique, M.E., and Getzoff, E.D., *J. Mol. Graph.* 1990, **8**, 125.

[29] Colloc'h, N., Mornon, J.P. *J. Mol. Graph.* 1990, **8**, 133.

[30] Wang, D., Driessen, H.P.C., and Tickle, I.J. *J. Mol. Graph.* 1991, **9**, 50.

[31] Mezey, P.G. *Potential Energy Hyperfsurfaces;* Elsevier: Amsterdam, 1987.

[32] Mezey, P.G. Topological Quantum Chemistry. In *Reports in Molecular Theory,* Náray-Szabó, G., and Weinstein, H., Eds., CRC Press: Boca Raton, 1990.

[33] Mezey, P.G. Topological Theory of Molecular Conformations. In *Structure and Dynamics of Molecular Systems,* Daudel, R., Korb, J.-P., Lemaistre, J.-P., and Maruani, J., Eds., Reidel: Dordrecht, 1985.

[34] Mezey, P.G. Three-Dimensional Topological Aspects of Molecular Similarity. In *Concepts and Applications of Molecular Similarity,* M. A. Johnson and G.M. Maggiora, Eds., Wiley: New York, 1990.

[35] Mezey, P.G. Molecular Surfaces. In *Reviews in Computational Chemistry,* Lipkowitz, K.B., and Boyd, D.B., Eds., VCH Publ.: New York, 1990.

[36] Mezey, P.G. Topological Model of Reaction Mechanisms. In *Structure and Dynamics of Molecular Systems,* Daudel, R., Korb, J.-P., Lemaistre, J.-P., and Maruani, J., Eds., Reidel: Dordrecht, 1985.

[37] Mezey, P.G. Reaction Topology. In *Applied Quantum Chemistry,* Smith, V.H., Schaefer III, H.F., and Morokuma, K., Eds., Reidel: Dordrecht, 1986.

[38] Mezey, P.G. *Internat. J. Quantum Chem. Quant. Biol. Symp.* 1986, **12**, 113.

[39] Mezey, P.G. *J. Comput. Chem.* 1987, **8**, 462.

[40] Mezey, P.G. *Internat. J. Quantum Chem. Quant. Biol. Symp.* 1987, **14**, 127.

[41] Mezey, P.G. *J. Math. Chem.* 1988, **2**, 299.

[42] Arteca, G.A., and Mezey, P.G. *Internat. J. Quantum Chem. Quant. Biol. Symp.* 1987, **14**, 133.

[43] Arteca, G.A., Jammal, V.B., Mezey, P.G.,Yadav, J.S., Hermsmeier, M.A., and Gund, T.M. *J. Mol. Graph.* 1988, **6**, 45.

[44] Arteca, G.A., Jammal, V.B., Mezey, P.G. *J. Comput. Chem.* 1988, **9**, 608.

[45] Mezey, P.G. *IEEE Eng. in Med. & Bio. Soc. 11th Annual Int. Conf.,* 1989, **11**, 1907.

[46] Mezey, P.G. The Topology of Molecular Surfaces and Shape Graphs. In *Computational Chemical Graph Theory",* Rouvray, D.H., Ed., Nova: New York, 1990.

[47] Arteca, G.A., and Mezey, P.G. *Internat. J. Quantum Chem.* 1988, **34**, 517.

[48] Arteca, G.A., and Mezey, P.G. *J. Comput. Chem.* 1988, **9**, 554.

[49] Mezey, P.G. *J. Amer. Chem. Soc.* 1986, **108**, 3976.

104

[50] Arteca, G.A., and Mezey, P.G. *J. Mol. Graphics,* 1990, **8**, 66.

[51] Arteca, G.A., Tapia, O., and Mezey, P.G. *J. Mol. Graphics,* 1991, **9**, 148.

[52] Mezey, P.G. *J. Math. Chem.* 1988, **2**, 325.

[53] Mezey, P.G. Non-Visual Molecular Shape Analysis: Shape Changes in Electronic Excitations and Chemical Reactions. In *Computational Advances in Organic Chemistry (Molecular Structure and Reactivity),* Ogretir, C., and Csizmadia, I., Eds., Kluwer: Dordrecht, 1991.

[54] Arteca, G.A., and Mezey, P.G. Algebraic Approaches to the Shape Analysis of Biological Macromolecules. In *Theoretical Chemistry, vol. 2,: Structure and Interactions,* S. Fraga, S., Ed., Elsevier: Amsterdam, 1991.

[55] Richardson, J.S. Adv. Protein Chem. 1981, **34**, 167.

[56] Richardson, J.S. Methods in Enzymol. 1985, **115**, 359.

[57] Murzin, A.G., and Finkelstein, A.V. *J. Mol. Biol.* 1988, **204**, 749.

[58] Chothia, C. *Nature* 1989, **337**, 204.

[59] Maggiora, G.M., Mezey, P.G., Mao, B., Chou, K.C. *Biopolymers,* 1990, **30**, 211.

[60] Mezey, P.G. *J. Amer. Chem. Soc.* 1990, **112**, 3791.

COMPUTER SIMULATION OF BIOMOLECULES:
COMPARISON WITH EXPERIMENTAL DATA

W.F. VAN GUNSTEREN, R.M. BRUNNE, A.E. MARK
Laboratory of Physical Chemistry
Swiss Federal Institute of Technology Zurich
ETH Zentrum
8092 Zürich
Switzerland

and

S.P. VAN HELDEN
Department of Pharmacology
University of Utrecht
P.O. Box 80.082
3508 TB Utrecht
The Netherlands

ABSTRACT. The computer simulation technique of molecular dynamics is briefly reviewed. The validity of the force field and simulation protocol that is used, is demonstrated by a comparison of simulated and experimental results. Firstly, the population of different conformers of proline residues in the cyclic decapeptide antamanide in solution is analyzed. Good agreement with populations and average J-coupling values derived from NMR spectroscopy is observed. Secondly, the relative free enthalpies of complex formation between four para-substituted phenols and α-cyclodextrin in aqueous solution are computed. Here, the agreement with experiment is about 1.8 kJ mol^{-1}.

105

J. Bertrán (ed.), Molecular Aspects of Biotechnology: Computational Models and Theories, 105–122.
© 1992 Kluwer Academic Publishers.

1. Introduction

Over the past twenty years simulation of molecular dynamics on a computer has developed from simulation of a few hundred small molecules such as H_2O or N_2 over a few picoseconds (Rahman and Stillinger, 1971) to simulation of a protein-DNA complex in aqueous solution over a hundred picoseconds (de Vlieg et al, 1989). Thereby, it has become possible to study molecular systems of practical interest at the atomic level. It has to be kept in mind, however, that the quality and reliability of a computer simulation depends on (1) the quality of the atomic interaction function or force field used, and (2) the suitability of the chosen simulation procedure. The quality of the simulated properties can be judged by comparison with those measured experimentally. If a force field and simulation technique are sufficiently accurate, computer simulation can be used to predict molecular properties for which experimental determination is too costly, too time-consuming or even impossible.

For molecules in solution few atomic properties can be measured directly. Nuclear magnetic resonance (NMR) techniques yield atom-atom distances derived from nuclear Overhauser (NOE) experiments, and so-called J-coupling constants (Ernst et al, 1987). Other spectroscopic techniques like fluorescence depolarisation or quenching measurements for a chromophore yield only information on special groups of atoms in a molecule. Thermodynamic quantities, such as the free enthalpy of solvation or of complex formation, also allow for a comparison of simulated and experimental values for solutions of molecules.

Since the most widely used force fields for biomolecular systems have been applied to a variety of molecules, the detailed comparison of simulated and measured properties is spread over the literature. A number of examples of such a comparison of various atomic and system properties for different compounds simulated using the GROMOS force field (van Gunsteren and Berendsen, 1987) has been given in (van Gunsteren and Berendsen, 1990). In this review, however, neither J-coupling constants nor free enthalpies of complex formation were considered, since data were not yet available.

Here, a comparison of simulated and measured values of these quantities is presented and discussed. J-coupling constants are calculated from stochastic dynamics simulations of the cyclic decapeptide antamanide. Free enthalpies of complexation are calculated from molecular dynamics simulations of α-cyclodextrin complexed with differently para-substituted phenols.

2. Methodology

2.1. MOLECULAR DYNAMICS SIMULATION

In the molecular dynamics (MD) simulation method a trajectory (molecular configurations as a function of time) is generated by simultaneous integration of Newton's equations of motion

$$dr_i(t)/dt = v_i(t) \tag{1}$$

$$dv_i(t)/dt = m_i^{-1} F_i(t) \tag{2}$$

for all atoms ($i = 1.. N$) of the molecular system. The Cartesian position vector r_i and the velocity v_i of atom i, with atomic mass m_i, is a function of time t. The force F_i exerted on atom i by the other atoms of the molecular system is given by the negative gradient of the atomic interaction function V which depends on the coordinates of all N atoms in the system:

$$F_i(t) = -\partial V(r_1(t), r_2(t), ..., r_N(t))/\partial r_i(t) \tag{3}$$

For small time steps Δt, eq. (2) may be approximated by

$$v_i(t+\Delta t/2) = v_i(t-\Delta t/2) + m_i^{-1} F_i (r_1(t), r_2(t), ...,r_N(t)) \Delta t \tag{4}$$

and eq. (1) likewise by

$$r_i(t+\Delta t) = r_i(t) + v_i(t+\Delta t/2) \Delta t \ . \tag{5}$$

Eqs. (4) and (5) form the so-called leap-frog scheme by which eqs. (1) and (2) can be integrated in small time steps Δt, typically 1-10 fsec for molecular systems.

2.2. STOCHASTIC DYNAMICS SIMULATION

The method of stochastic dynamics (SD) is an extension of the molecular dynamics technique. A trajectory of a molecular system is generated by integration of the stochastic Langevin equation of motion, that is, eq. (2) is replaced by

$$dv_i(t)/dt = m_i^{-1} F_i(t) + m_i^{-1} R_i(t) - \gamma_i v_i(t) \ . \tag{6}$$

Two terms have been added to the right-hand-side of eq. (2), a stochastic force $R_i(t)$ and a frictional force proportional to a friction coefficient γ_i. The stochastic term introduces energy, the frictional term removes energy from the system, the condition for zero energy loss being

$$<R_i^2(t)> = 6m_i\gamma_i k_B T_{ref} \ , \tag{7}$$

where k_B is Boltzmann's constant, T_{ref} the reference temperature of the system and the brackets $<...>$ denote an average over time. The leap-frog scheme to integrate eqs. (1) and (6) using condition (7) is given in (van Gunsteren and Berendsen, 1988).

Stochastic dynamics simulation can be used to mimic the solvent effect. The solvent degrees of freedom are omitted from the equations of motion, and their effect on the solute atoms (degrees of freedom) is approximated by the stochastic and frictional forces: the stochastic force $R_i(t)$ represents collisions of solute atoms with solvent molecules, and the frictional term $-\gamma_i v_i(t)$ represents the drag exerted by the solvent on the solute atom motion. By the omission of the solvent degrees of freedom generally a factor of 10-100 in computing time is saved, at the expense of a loss of accuracy due to the approximate treatment of solvent effects.

2.3.   THE ATOMIC INTERACTION FUNCTION OR FORCE FIELD

A typical atomic interaction function for biomolecular systems has the form (van Gunsteren and Berendsen, 1987):

$$V(r_1, r_2, ..., r_N) =$$

$$= \sum_{all\,bonds} \frac{1}{2} K_b[b-b_o]^2 + \sum_{all\,bond\,angles} \frac{1}{2} K_\theta[\theta-\theta_o]^2$$

$$+ \sum_{\substack{improper \\ dihedrals}} \frac{1}{2} K_\xi\,[\xi-\xi_o]^2 + \sum_{dihedrals} K_\varphi\,[1+\cos(n\varphi-\delta)]$$

$$+ \sum_{all\,pairs\,(ij)} \left[C_{12}(i,j)/r_{ij}^{12} - C_6(i,j)/r_{ij}^6 + q_i q_j/(4\pi\varepsilon_o\varepsilon_r r_{ij}) \right] \tag{8}$$

The first term represents the covalent bond-stretching interaction along bond b. It is a harmonic potential in which the minimum energy bond length b_o and the force constant K_b vary with the particular type of bond. The second term describes the bond-angle bending interaction in similar form. Two forms are used for the dihedral-angle interactions: a harmonic term for dihedral angles ξ that are not allowed to make transitions, e.g. dihedral angles within aromatic rings, and a sinusoidal term for the other dihedral angles φ, which may make 360 degree turns. The last term is a sum over all pairs of atoms and represents the effective non-bonded interaction, composed of the van der Waals and the Coulomb interactions between atoms i and j with charges q_i and q_j at a distance r_{ij}.

A particular force field is characterized by its functional form, e.g. (8), by the values of its parameters, K_b, b_o, K_θ, θ_o, K_ξ, ξ_o, K_φ, n, δ, $C_{12}(i,j)$, $C_6(i,j)$, q_i, q_j and ε_r, and by the choice of dihedral angles occurring in the summations of the third and fourth terms of (8). The GROMOS force field used in this study is specified in (van Gunsteren and Berendsen, 1987).

2.4. USE OF CONSTRAINTS

The length of a time step Δt in a molecular dynamics or stochastic dynamics simulation is bounded by the highest frequency motions occurring in the system: $\Delta t << v^{-1}_{max}$. By freezing or rigidly constraining the generally uninteresting high-frequency bond-length vibrations, v^{-1}_{max} is increased, which allows for a longer time step Δt. The application of the SHAKE algorithm (Ryckaert et al, 1977) to constrain bond lengths saves about a factor of 3 in computing effort (van Gunsteren and Berendsen, 1977). In the simulations presented here all bond lengths have been constrained using the SHAKE method.

2.5. SPATIAL BOUNDARY CONDITIONS

When simulating a system of finite size ($N << Avogadro's\ number$) some thought must be given to the way the boundary of the system is treated. The best way to minimize edge effects in a finite system is to use periodic boundary conditions. The solute is surrounded by solvent molecules such that they fill a periodic space-filling box, which is treated as if it is surrounded by identical, translated images of itself. Since the α-cyclodextrin complexes are rather spherical, truncated octahedron periodic boundary conditions were used in the simulations of these complexes in water.

When the solvent degrees of freedom are omitted from the simulation, like in the stochastic dynamics simulations on antamanide, it makes little sense to use periodic boundary conditions. The vacuum boundary condition is used instead.

2.6. THE COMPUTATION OF RELATIVE FREE ENTHALPIES

Molecular dynamics simulations can be performed at constant temperature and pressure by a weak coupling of the system to a heat bath and to a pressure bath (Berendsen et al, 1984). Such simulations allow for a determination of the relative free enthalpy of complexation of molecules in solution by using the concept of a thermodynamic cycle (Tembe and McCammon, 1984). A thermodynamic cycle is based on the fact that the free enthalpy, G, is a thermodynamic state function. This means that as long as the system is changed in a reversible way, the change in free enthalpy, ΔG, will be independent of the path of change. Therefore, $\Delta G=0$ along a closed path or cycle. For example, the cycle for the complex formation of two differently para-substituted phenols, $P(A)$ and $P(B)$ with α-cyclodextrin (αCD) is

$$
\begin{array}{ccc}
P(A) + \alpha CD & \xrightarrow{\ 1(exp.)\ } & [\ P(A){:}\alpha CD\ ]_{complex} \\
\Big\downarrow 3(sim) & & \Big\downarrow 4(sim) \\
P(B) + \alpha CD & \xrightarrow[\ 2(exp.)\ ]{} & [\ P(B){:}\alpha CD\ ]_{complex} \ .
\end{array} \tag{9}
$$

Since complex formation involves the removal of solvent molecules from the binding area of the phenol inside α-cyclodextrin, a simulation of processes 1 and 2 in a reversible manner is impossible. Since (9) is a thermodynamic cycle we have

$$\Delta G_2 - \Delta G_1 = \Delta G_4 - \Delta G_3 \tag{10}$$

and the desired result can be obtained by simulating the non-chemical processes 3 and 4, in which only a few atoms making up the differences between the para-substituted phenols $P(A)$ and $P(B)$ are to be changed.

In order to compute the free enthalpy change $\Delta G_{BA} = G(B) - G(A)$ between two systems with atomic interaction functions indicated by V_B and V_A, the potential energy function V (8) is made a function of a coupling parameter λ, such that $V(\lambda_A) = V_A$ and $V(\lambda_B) = V_B$. Since the force field $V(\lambda)$ has become dependent on λ, the free enthalpy $G(\lambda)$ is also dependent on λ. Using statistical mechanics the free enthalpy change ΔG_{BA} can be expressed as an integral, see e.g. (van Gunsteren, 1988):

$$\Delta G_{BA} = \int_{\lambda_A}^{\lambda_B} <\partial V / \partial \lambda>_\lambda \, d\lambda \tag{11}$$

where the brackets $<...>_\lambda$ denote averaging over an equilibrium ensemble generated with the potential energy function $V(\lambda)$. Using molecular dynamics simulation to generate an ensemble of molecular configurations, the integral (11) can be evaluated in different ways. The coupling parameter λ can be made a function of time $\lambda(t)$ such that it slowly changes from λ_A to λ_B during a MD simulation:

$$\Delta G_{BA} = \sum_{t_i=t_A}^{t_{B-1}} \partial V / \partial \lambda \big|_{\lambda(t_i)} [\lambda(t_{i+1}) - \lambda(t_i)] \tag{12}$$

The time points in the simulation are indicated by t_i, where $\lambda=\lambda_A$ at $t=t_A$ and $\lambda=\lambda_B$ at $t=t_B$. This procedure is called continuous change or slow growth. Alternatively, the ensemble average in (11) may be evaluated at a number of discrete λ-values by performing a separate MD simulation at each chosen λ-value λ_i. These ensemble averages are then used to obtain the integral by numerical quadrature:

$$\Delta G_{BA} = \sum_{\lambda_i=\lambda_1}^{\lambda_M} <\partial V / \partial \lambda>_{\lambda_i} \Delta \lambda_i \tag{13}$$

The points λ_i $(i=1,2,..,M)$ and weights $\Delta \lambda_i$ can be chosen according to Gauss-Legendre quadrature, or any other numerical quadrature method. This procedure will be referred to as thermodynamic integration by numerical quadrature. Both formulae (12) and (13) were used to compute relative free enthalpies of α-cyclodextrin complexes.

3. Comparison of simulated and experimental results

3.1. J-COUPLING VALUES FOR ANTAMANIDE

The cyclic peptide antamanide is shown in Figure 1. It contains four proline residues, the dynamics of which has been analysed recently using NMR experiments by Mádi et al (1990). It was found that Pro^2 and Pro^7 interconvert between two energetically similar

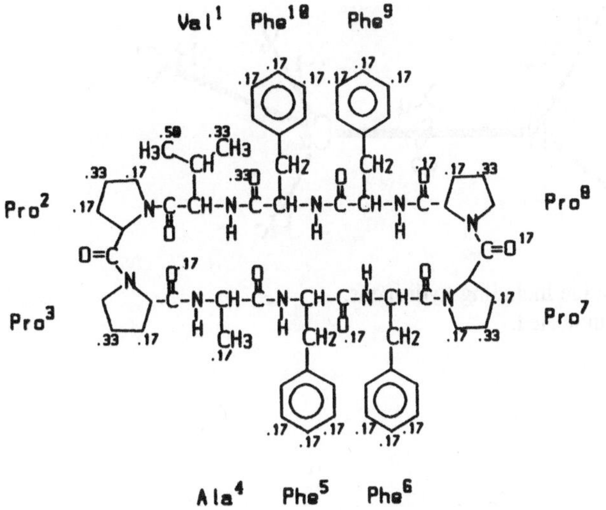

Figure 1. Primary structure of antamanide. Non-zero atomic accessible area weight factors w_i for a representative conformation are listed with the atoms (Brunne et al, 1992).

conformations. The populations of Pro^3 and Pro^8 are inverted with respect to Pro^2 and Pro^7 and much more asymmetric, see Table 1. The vicinal J-coupling constants for the 10 pairs of protons attached to covalently bonded carbon atoms (C_α : 1 proton; C_β, C_γ, C_δ : 2 protons, denoted as *cis* and *trans*; see Figure 2) as measured by NMR are given in Table 2. It is this data that was to be reproduced by a simulation using the GROMOS force field.

The NMR measurements concern antamanide in chloroform, a rather non-polar solvent. The solvation properties of such a solvent can be effectively approximated by a stochastic dynamics treatment in which the atomic friction coefficients γ_i are taken proportional to the atomic accessible surface area w_i (which depends on the solute conformations) with the proportionality constant γ being representative for the solvent friction (Shi Yun-yu et al, 1988):

$$\gamma_i = \gamma w_i \tag{14}$$

Therefore, the antamanide conformational dynamics was simulated using the stochastic dynamics technique. A value of $\gamma = 19 \, psec^{-1}$ was used. Typical w_i values are shown in

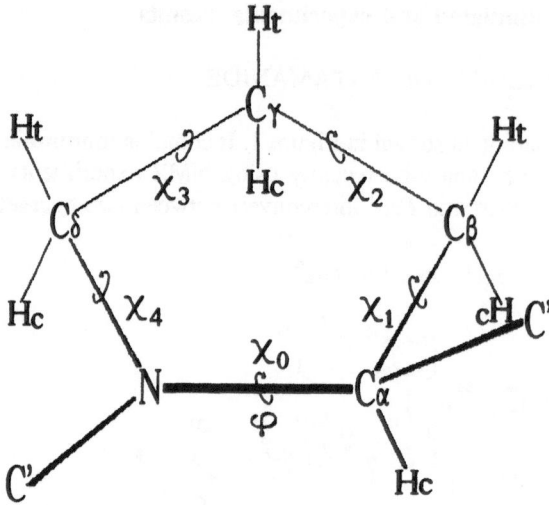

Figure 2. Proline residue including hydrogens.
Torsional angles are indicated.

φ : C-N-C_α-C

χ_0 : C_δ-N-C_α-C_β

χ_1 : N-C_α-C_β-C_γ

χ_2 : C_α-C_β-C_γ-C_δ

χ_3 : C_β-C_γ-C_δ-N

χ_4 : C_γ-C_δ-N-C_α

Fig. 1. The SD simulation covered a nanosecond, the last half of which was used for analysis. Details of the simulation protocol will be given elsewhere (Brunne et al, 1992).

Each proline residue indeed populates two conformers, which can be distinguished by the value of the χ_2-angle being larger or smaller than zero. From Table 1 it can be concluded that the experimental NMR data are well reproduced in the simulation. The X-ray crystallographic data clearly represent a mixture of two or more conformers.

The J-coupling constants in Table 2 are obtained as an average over the SD trajectory, and using a modified Karplus relation. There appears to be good qualitative agreement with the experimentally measured values.

We conclude that a 1 nsec stochastic dynamics simulation, that mimics a solution of antamanide in chloroform, and that uses the standard GROMOS force field, can reproduce the conformationally different properties of the 4 proline residues as measured by J-coupling constants and conformer populations. A more complete analysis of the simulation will be given elsewhere (Brunne et al, 1992).

Table 1. Populations p_i and mean torsional angles (φ, χ_0 - χ_4, in degrees) of the observed proline conformations in antamanide, as obtained from stochastic dynamics (SD) simulation, NMR spectroscopy (Mádi et al, 1990) and X-ray (XR) crystallography (Karle et al, 1979).

	Pro^2					Pro^7				
	XR	NMR		SD		XR	NMR		SD	
p_i		.65	.35	.75	.25		.55	.45	.71	.29
φ	-64			-59	-62	-62			-58	-62
χ_0	-1	+4	-5	+8	-2	-10	-6	-8	+9	-2
χ_1	-15	-30	+23	-26	+21	+11	-22	+26	-27	+20
χ_2	+26	+45	-32	+34	-31	-11	+42	-34	+34	-31
χ_3	-27	-42	+29	-29	+30	+6	-45	+29	-29	+30
χ_4	+15	+24	-15	+13	-18	+3	+31	-13	+12	-18

	Pro^3					Pro^8				
	XR	NMR		SD		XR	NMR	SD		
p_i		.90	.10	.84	.16			.83	.17	
φ	-80			-90	-85	-92		-88	-83	
χ_0	-9	-15	-24	-20	-13	-16	-14	-19	-11	
χ_1	+25	+35	-4	+32	-9	+29	+34	+32	-11	
χ_2	-33	-42	+30	-33	+27	-31	-40	-33	+27	
χ_3	+27	+33	-45	+21	-34	+23	+31	+22	-33	
χ_4	-10	-11	+42	0	+29	-5	-10	-2	+28	

3.2. FREE ENTHALPIES OF BINDING FOR α-CYCLODEXTRIN WITH DIFFERENT INCLUSION COMPOUNDS

Cyclodextrins (CD) are a family of cyclically closed, torus-shaped oligosaccharides consisting of six (α), seven (β), and eight (γ) glucose units covalently linked by α(1-4) bonds. Their most remarkable property, the inclusion of guest molecules in the annular cone-shaped cavities with 5 to 7 Å diameter, have been studied in great detail with spectroscopic and crystallographic methods (Cramer et al, 1967, Saenger, 1980). The guest molecules can vary from hydrophilic to hydrophobic in character, the only condition for inclusion being that they are small enough to fit spatially into the hydrophobic cavity, the rims of which are lined with hydrophilic O-H groups.

TABLE 2. Vicinal $^3J_{HH}$ coupling constants (in Hz) for four proline residues in antamanide, as obtained from stochastic dynamics (SD) simulation and from NMR spectroscopy (Mádi et al, 1990).

^3J	Pro^2		Pro^7		Pro^3		Pro^8	
	NMR	SD	NMR	SD	NMR	SD	NMR	SD
$\alpha\beta_c$	7.97	7.80	8.81	7.82	8.48	7.90	8.08	7.96
$\alpha\beta_t$	7.40	7.76	5.78	7.51	1.12	2.26	0.88	2.36
$\beta_c\gamma_c$	7.08	8.28	7.39	8.32	6.76	8.73	6.83	8.67
$\beta_c\gamma_t$	4.80	3.21	5.93	3.63	12.00	9.75	12.97	9.74
$\beta_t\gamma_c$	8.96	8.99	7.52	8.54	2.36	2.19	1.38	2.25
$\beta_t\gamma_t$	7.08	8.41	7.04	8.43	6.47	8.56	6.42	8.50
$\gamma_c\delta_c$	7.61	8.22	7.34	8.24	7.61	8.74	7.34	8.64
$\gamma_c\delta_t$	8.50	7.96	7.36	7.55	2.11	3.44	1.46	3.36
$\gamma_t\delta_c$	4.60	3.55	5.35	3.93	10.29	7.56	10.88	7.71
$\gamma_t\delta_t$	7.02	8.03	6.99	8.08	8.53	8.92	8.81	8.83

Molecular dynamics simulations of crystalline α-cyclodextrin using the GROMOS force field yield very good agreement with X-ray and neutron diffraction data (Koehler et al, 1987). The experimental positions of the α-cyclodextrin atoms are reproduced to within 0.25 Å (Figure 3). Features, such as hydrogen bond flip-flops (Koehler et al, 1988a) and three-center hydrogen bonding (Koehler et al, 1988b) are well reproduced in the simulations. Structural and dynamical differences between α-cyclodextrin in the crystal and in aqueous solution were also studied by these authors (Koehler et al, 1988c). The conclusion of these studies was that the GROMOS force field is capable of describing the structural properties and conformational equilibria of cyclodextrins in detail. On this basis it is worthwhile to investigate whether experimentally measured

Figure 3. Time and space averaged structure of crystalline α-cyclodextrin at room temperature.
Thick lines: molecular dynamics simulation
Thin lines: model based on neutron diffraction data.
Root mean square difference between the two structures equals 0.25 Å (Koehler et al, 1987).

binding constants of guest molecules may be correctly calculated from simulations based on this force field. Here, we present some preliminary results with respect to this topic. A detailed account of the work will be given elsewhere (van Helden, 1992).

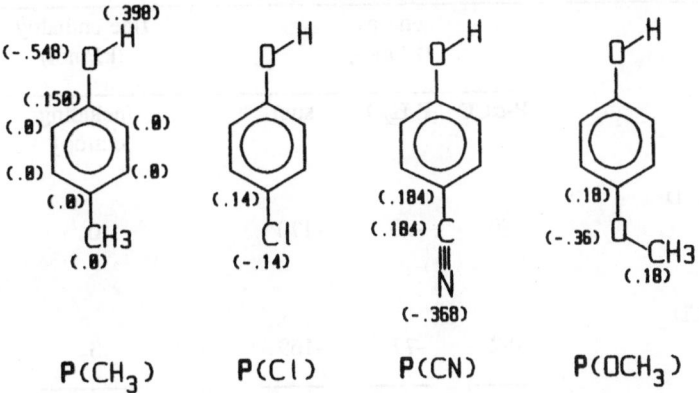

Figure 4. Para-methyl-phenol and three para-substituted inclusion compounds of α-cyclodextrin. The numbers in parentheses are the atomic partial charges (in e) of the GROMOS force field (van Helden, 1992).

The four guest molecules for which the relative free enthalpy of binding to α-cyclodextrin in aqueous solution was calculated, are shown in Figure 4, para-methyl-phenol and three other para-substituted compounds. The crystal structure of the complex of para-hydroxy-benzoic acid with α-cyclodextrin (Harata, 1977) was used to obtain atomic coordinates for the different complexes. In order to simulate process 4 in scheme (9) the complex was put into a truncated octahedron filled with 508 water molecules. An octahedron of identical size was used in the simulation of process 3 in scheme (9): 544 water molecules were needed to solvate the smaller solute. The simulations were carried out at constant temperature and pressure, using the weak coupling technique (Berendsen et al, 1984). All bond lengths were constrained using the SHAKE method (Ryckaert et al, 1977). Water molecules were modelled by the SPC water model (Berendsen et al, 1981). The GROMOS force field was used, with extensions derived similarly for non-standard moieties. The cut-off radius for non-bonded interactions was taken equal to 10 Å, and periodic boundary conditions were applied. Every molecule or complex was equilibrated for at least 10 psec before initiating a free enthalpy calculation.

The first question to be addressed concerns the binding mode of the guest molecule: with the CH_3 group inside (OH group in solution) or with the OH group inside (CH_3 group in solution). A preliminary free enthalpy calculation based on formula (12), in which the methyl group was changed to a hydroxyl group and concurrently the hydroxyl group to a methyl group in the complex of para-methyl-phenol with α-cyclodextrin, favours the binding mode with the methyl group inside, see Table 3. This binding mode was assumed in subsequent simulations. Table 3 also illustrates the importance of solvent and entropic effects when computing binding constants of cyclodextrin host-guest complexes. If only the interaction energy between guest molecule

TABLE 3. Different contributions to the relative free enthalpy of binding of para-methyl-phenol (P) to α-cyclodextrin (αCD) in aqueous solution (H_2O) for two different binding modes (CH_3 inside αCD, OH inside αCD), as obtained from MD simulations.

Binding mode	energy (kJ/mol)			free enthalpy (kJ/mol)
	$P\text{-}\alpha CD$	$P\text{-}H_2O$	sum	including entropy
- CH_3 inside αCD OH in water	-80	-90	-170	0
- OH inside αCD CH_3 in water	-92	-77	-169	3
- Difference	+12	-13	-1	-3

and α-cyclodextrin is considered, the energy difference between the two binding modes is *12 kJ mol⁻¹* ($\approx 5\ k_BT$): hydroxyl to cyclodextrin hydrogen bonding favours the "hydroxyl inside" binding mode. When the interaction with the water molecules is also considered, the picture is inverted: the hydroxyl to water hydrogen bonding slightly favours the "hydroxyl outside" binding mode by *1 kJ mol⁻¹*. When entropic effects are included, this trend is increased: the free enthalpy difference becomes *3 kJ mol⁻¹*. The entropic contribution is twice as large as the net enthalpic one.

Formula (11) to determine the free enthalpy difference ΔG_{BA} between two states of a system is only strictly valid if the system always remains close to equilibrium for each λ-value, and when a representative ensemble is sampled to obtain $<...>_\lambda$ at each λ-value in going from the initial to the final state. When applying formula (12) these conditions will be better satisfied the longer the period τ_{MD} over which the system is changed, is chosen. Generally one requires (van Gunsteren, 1988)

$$\tau_{MD} > \tau_{system} \quad , \tag{15}$$

the simulation must be much longer than the relaxation time of the molecular system, τ_{system}. Partial information on the reversibility and on the adequacy of the sampling can be obtained by comparing the calculated free enthalpy for the forward process, changing λ from λ_A to λ_B, to that of the reverse process, changing λ from λ_B to λ_A. If the change is carried out reversibly, the hysteresis will be zero:

$$\Delta G_{BA} + \Delta G_{AB} = 0 \tag{16}$$

Yet this is only a necessary, not a sufficient condition for obtaining an accurate estimate of ΔG_{BA} using formula (11). If the opposite of condition (15) is true, that is, if

$$\tau_{MD} \ll \tau_{system} \tag{17}$$

equation (16) can also be approximately satisfied, since the change of λ is proceeding so fast that the system cannot adapt to the change in the potential energy function $V(\lambda)$. Only if

$$\tau_{MD} \approx \tau_{system} \tag{18}$$

the irreversibility of the change will appear as significant hysteresis:

$$\Delta G_{BA} + \Delta G_{AB} \neq 0 \tag{19}$$

TABLE 4. Change in free enthalpy (in kJ/mol) upon changing the guest molecule from $P(Cl)$ to $P(CH_3)$ when complexed to α-cyclodextrin in aqueous solution (process 4) and when unbound in solution (process 3), for different lengths of the MD simulation (τ_{MD} in psecs) during which the change Cl (state A) to CH_3 (state B) is made. The slow growth or continuous change formula (12) was used.

- τ_{MD}	25	50	100	300
- process 4 (complex)				
$[\Delta G_{BA}-\Delta G_{AB}]/2$	3.68	4.16	4.18	3.42
$\Delta G_{BA}+\Delta G_{AB}$ (hysteresis)	0.51	0.77	1.48	0.51
- process 3 (unbound)				
$[\Delta G_{BA}-\Delta G_{AB}]/2$	-4.33	-4.90	-4.50	
$\Delta G_{BA}+\Delta G_{AB}$ (hysteresis)	0.25	1.22	0.21	
- difference				
$\Delta\Delta G=\Delta G_4-\Delta G_3$	8.0	9.1	8.7	7.9 [a]
hysteresis	0.8	2.0	1.7	0.7 [a]

[a] For process 3 the $\tau_{MD} = 100$ psec values were used.

These considerations are illustrated in Table 4. For the host-guest complex the hysteresis becomes larger when the simulation time is lengthened from 25 psec to 50 psec to 100 psec. It is only when τ_{MD} reaches 300 psec that the hysteresis is again reduced. For the unbound guest molecule the maximum hysteresis is observed for changing λ from λ_A to λ_B over a shorter period of 50 psec, as is to be expected: the relaxation time of the unbound guest molecule is shorter than that of the complex. When applying the slow growth or continuous change formula (12) the value of λ is changed by a very small amount at each time step. The number of λ-values sampled is, therefore, equal to the number of time steps (10^4-10^5) in the simulation. Thus, the method is characterized by a very large number of λ-values but poor sampling at each point. In order to avoid significant hysteresis the change in λ must be sufficiently small to allow the system to adapt continuously. In this way the ensemble average $<\partial V/\partial \lambda>_{\lambda_i}$ is approximated by the mean of $\partial V/\partial \lambda$ over a set of different λ-values close to λ_i.

An alternative to the use of formula (12) is the use of formula (13). The used formula is characertized by a much smaller number of λ_i-values, about 3 to 10. The ensemble average $<\partial V/\partial \lambda>_{\lambda_i}$ is, however, obtained by thorough sampling using a separate MD simulation for each λ_i-value. If the function $<\partial V/\partial \lambda>_{\lambda}$ is a smooth function of λ, the integral in (11) can be well approximated by numerical methods. In Table 5 results are shown for the same host-guest complex as discussed above, but calculated using formula (13) in combination with 3-point or 5-point Gauss-Legendre numerical quadrature. Ideally, the length of the MD simulation at each λ_i point should be chosen such that the ensemble average $<\partial V/\partial \lambda>_{\lambda_i}$ does not change upon extension of the simulation. For the complex the change of $<\partial V/\partial \lambda>_{\lambda_i}$ is less than 0.5 kJ/mol for τ_{MD} longer than 30 psec. In the simulation of the unbound guest molecule the corresponding time is 20 psec. We observe that both the 3-point and the 5-point integration yield $\Delta\Delta G$ values which agree well with those obtained using the slow growth technique.

TABLE 5. Change in free enthalpy (in kJ/mol) upon changing the guest molecule from $P(Cl)$ to $P(CH_3)$ when complexed to α-cyclodextrin in aqueous solution (process 4) and when unbound in solution (process 3), using formula (13) in combination with 3-point or 5-point Gauss-Legendre numerical integration. Equilibration time τ_{equil} and sampling time τ_{MD} per λ_i-value are given in psecs.

	3-point	5-point
- process 4 (complex)		
τ_{equil}	20	20
τ_{MD}	60	60
ΔG_{BA} (Cl to CH_3)	3.68	3.51
- process 3 (unbound)		
τ_{equil}	10	10
τ_{MD}	50	50
ΔG_{BA} (Cl to CH_3)	-4.64	-4.58
- difference		
$\Delta\Delta G = \Delta G_4 - \Delta G_3$	8.3	8.1

Formula (13) and Gauss-Legendre numerical quadrature was used to compute the relative free enthalpies of complexation with α-cyclodextrin in aqueous solution for the four guest molecules shown in Figure 4. When the number of non-hydrogen atoms is equal in two guest molecules, the 3-point formula is used, otherwise the 5-point formula is applied. The simulations of the host-guest complex at each λ_i-value involved 20 psec of equilibration (τ_{equil}) followed by 40 psec of sampling (τ_{MD}). For the simulations of the unbound guest molecules it was sufficient to use shorter values, $\tau_{equil} = 10\ psec$ and $\tau_{MD} = 20\ psec$. The results are summarized in Figure 5. Ideally, within a closed thermodynamic cycle, ΔG must be zero. We find, however, for the four possible cycles

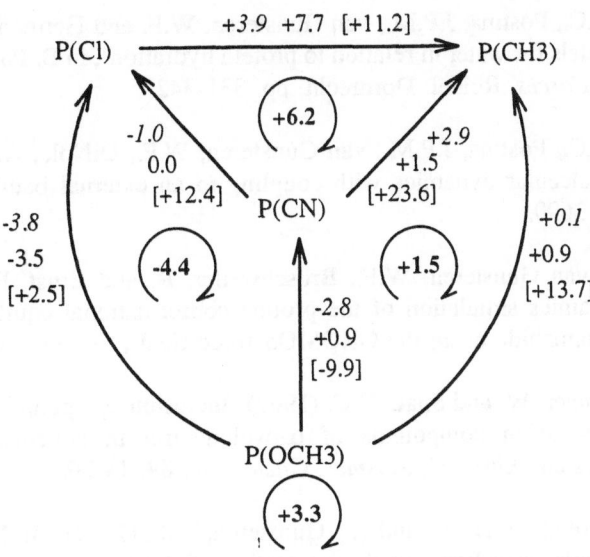

Figure 5. Relative free enthalpy differences ($\Delta\Delta G = \Delta G_4 - \Delta G_3$) for complex formation of four different inclusion compounds with α-cyclodextrin. All values in kJ mol^{-1}. The experimental values are given in italics. Relative host-guest energies are given in square brackets. The fat numbers are simulated free enthalpy changes for the indicated closed thermodynamic cycles. Data from (van Helden, 1992).

values ranging from 1.5 kJ/mol to 6.2 kJ/mol. The values are non-zero due to finite equilibration and sampling times at each λ_i-value and possibly due to the use of too few λ_i points for the numerical quadrature. The average value of this error per leg or per individual ΔG_{BA} value in Figure 5 is 1.3 kJ/mol ($\approx \frac{1}{2}k_BT$). The agreement between the calculated and the experimental (in italics) free enthalpies of complexation ($\Delta\Delta G$ values) is very good. The average absolute difference equals 1.8 kJ/mol, the largest deviation being 3.8 kJ/mol, the smallest 0.3 kJ/mol. The GROMOS force field reproduces for these α-cyclodextrin complexes free enthalpies of binding within $\frac{3}{4}\ k_BT$. In Figure 5, the values within square brackets are the host-guest energy changes, that is, the relative complexation enthalpies in the absence of water and without entropic contributions. The average absolute difference between these calculated and the experimental values equals 11.4 kJ/mol, an error which is 6 times larger than the one obtained from a full treatment of solvent and entropic effects. This implies that simple host-guest modelling and the determination of interaction energies will not yield reliable binding constants for flexible complexes in aqueous solution. To obtain accurate predictions of binding enthalpies free energy calculations, which properly include solvation and entropic effects, are required.

Stimulating discussions with R. Brüschweiler, R.R. Ernst and L.H.M. Janssen are gratefully acknowledged.

References

- Berendsen, H.J.C., Postma, J.P.M., van Gunsteren, W.F. and Hermans, J. (1981) 'Interaction models for water in relation to protein hydration', in B. Pullman (ed.), *Intermolecular Forces*, Reidel, Dordrecht, pp. 331-342.

- Berendsen, H.J.C., Postma, J.P.M., van Gunsteren, W.F., DiNola, A. and Haak, J.R. (1984) 'Molecular dynamics with coupling to an external bath', *J. Chem. Phys.* **81**, 3684-3690.

- Brunne, R.M., van Gunsteren, W.F., Brüschweiler, R. and Ernst, R.R. (1992) 'Molecular dynamics simulation of the proline conformational equilibrium and dynamics in antamanide using the GROMOS force field', to be published.

- Cramer, F., Saenger, W. and Spatz H.C. (1967) 'Inclusion compounds, XIX. The formation of inclusion compounds of α-cyclodextrin in aqueous solutions. Thermodynamics and kinetics', *J. Amer. Chem. Soc.* **89**, 14-20.

- De Vlieg, J., Berendsen, H.J.C. and van Gunsteren, W.F. (1989) 'An NMR based molecular dynamics simulation of the interaction of the *lac* repressor headpiece and its operator in aqueous solution', *Proteins* **6**, 104-127.

- Ernst, R.R., Bodenhausen, G. and Wokaun, A. (1987) Principles of Nuclear Magnetic Resonance in One and Two Dimensions, Clarendon, Oxford.

- Harata, K. (1977) 'The structure of the cyclodextrin complex. V. Crystal structures of α-cyclodextrin complexes with p-nitrophenol and p-hydroxybenzoic acid', *Bull. Chem. Soc. Japan* **50**, 1416-1424.

- Karle, I.L., Wieland, T., Schermer, D. and Ottenheym, H.C.J. (1979) 'Conformation of uncomplexed natural antamanide crystallized from CH_3CN/H_2O', *Proc. Natl. Acad. Sci. U.S.A.* **76**, 1532-1536.

- Koehler, J., Saenger, W. and van Gunsteren, W.F. (1987) 'A molecular dynamics simulation of crystalline α-cyclodextrin hexahydrate', *Eur. Biophys. J.* **15**, 197-210.

- Koehler, J., Saenger, W. and van Gunsteren, W.F. (1988a) 'The flip-flop hydrogen bonding phenomenon. Molecular dynamics simulation of crystalline β-cyclodextrin', *Eur. Biophys. J.* **16**, 153-168.

- Koehler, J., Saenger, W. and van Gunsteren, W.F. (1988b) 'On the Occurrence of Three-Center Hydrogen Bonds in Cyclodextrins in Crystalline Form and in Aqueous Solution: Comparison of Neutron Diffraction and Molecular Dynamics Results', *J. Biomol. Struct. Dyn.* **6**, 181-198.

- Koehler, J., Saenger, W. and van Gunsteren, W.F. (1988c) 'Conformational Differences Between α-Cyclodextrin in Aqueous Solution and in Crystalline Form. A Molecular Dynamics Study', *J. Mol. Biol.* **203**, 241-250.

- Mádi, Z.L., Griesinger, C., Ernst, R.R. (1990) 'Conformational dynamics of proline residues in antamanide. J-coupling analysis of strongly coupled spin systems based on E-COSY spectra', *J. Amer. Chem. Soc.* **112**, 2908-2914.

- Rahman, A. and Stillinger, F.H. (1971) 'Molecular Dynamics Study of Liquid Water', *J. Chem. Phys.* **55**, 3336-3359.

- Ryckaert, J.-P., Ciccotti, G. and Berendsen, H.J.C. (1977) 'Numerical integration of the cartesian equations of motion of a system with constraints: molecular dynamics of *n*-alkanes', *J. Comput. Phys.* **23**, 327-341.

- Saenger, W. (1980) 'Cyclodextrin inclusion compounds in research and industry', *Angew. Chem. Int. Ed. Engl.* **19**, 344-362.

- Shi Yun-yu, Wang Lu and van Gunsteren, W.F. (1988) 'On the approximation of solvent effects on the conformation and dynamics of cyclosporin A by stochastic dynamics simulation techniques', *Mol. Simulation* **1**, 369-388.

- Tembe, B.L. and McCammon, J.A. (1984) 'Ligand-Receptor Interactions', *Computers & Chemistry*, **8**, 281-283.

- van Gunsteren, W.F. (1988) 'The role of computer simulation techniques in protein engineering', *Protein Eng.* **2**, 5-13.

- van Gunsteren, W.F. and Berendsen, H.J.C. (1977) 'Algorithms for macromolecular dynamics and constraint dynamics', *Mol. Phys.* **34**, 1311-1327.

- van Gunsteren, W.F. and Berendsen, H.J.C. (1987) 'Groningen Molecular Simulation (GROMOS) Library Manual, Biomos, Nijenborgh 16, Groningen, The Netherlands.

- van Gunsteren, W.F. and Berendsen, H.J.C. (1988) 'A leap-frog algorithm for stochastic dynamics', *Mol. Simulation* **1**, 173-185.

- van Gunsteren, W.F. and Berendsen, H.J.C. (1990) 'Computer Simulation of Molecular Dynamics: Methodology, Applications and Perspectives in Chemistry', *Angew. Chem. Int. Ed. Engl.* **29**, 992-1023.

- van Helden, S.P. (1992) Structure and Stability of Cyclodextrin Complexes. A Molecular Modelling Study, thesis, University of Utrecht, Utrecht, The Netherlands.

MOLECULAR DYNAMICS COMPUTER MODELLING AND PROTEIN ENGINEERING

O. TAPIA and O. NILSSON
Department of Physical Chemistry
University of Uppsala
Box 532
S-751 21 Uppsala
Sweden

ABSTRACT. A set of molecular dynamics computer simulation studies applied in protein design is reviewed. The picture begins to emerge that with present day force fields and molecular dynamics simulation techniques the essentials of proteins' structural and dynamical features around their native states may be obtained. The perspective of computer simulation as a tool in molecular engineering is explored. We discuss modelling of collective motion in proteins' secondary structural elements, thermal stability of protein structures, differential stability in protein folds and surface plasticity properties of proteins. Properties of the following proteins are touched: the carboxy terminal fragment of the L7/L12 ribosomal protein from *Escherichia coli*, the potato carboxypeptidase A protein inhibitor, bacteriophage T4 glutaredoxin and the retinol binding protein. Appended are discussions of the chemical mechanism of hydride transfer in horse liver alcohol dehydrogenase, the fundamentals of interactive dynamical computer graphics analysis and a survey of the theoretical framework of molecular dynamics simulations.

1. INTRODUCTION

Molecular dynamics (MD) computer simulation techniques are invaluable assistants in refining structures of biomacromolecules extracted from X-ray diffraction or nuclear magnetic resonance (NMR) measurements (van Gunsteren, 1988; van Gunsteren & Berendsen, 1990; McCammon & Harvey, 1987; Karplus & Petsko, 1990; Brooks III et al., 1988; Brünger & Karplus, 1991). In molecular biotechnology, and protein engineering in particular, MD techniques are evolving as auxiliary tools in the characterization of mutant structures. Here, we overview a number of theoretical studies of structural and dynamical properties of protein molecules. They provide examples of how these methods have been used to obtain potentially useful information in this field.

The success of molecular dynamics simulations partly resides in their power to explore significant regions of the molecular conformational space around the folded state. Aside an average structure, MD simulations provide atomic fluctuations which can be

J. Bertrán (ed.), *Molecular Aspects of Biotechnology: Computational Models and Theories*, 123–152.
© 1992 *Kluwer Academic Publishers*.

compared with the temperature factors derived from X-ray diffraction measurements. Most of the force fields used in simulations render the main features of a folded structure adequately (van Gunsteren, 1988; van Gunsteren & Berendsen, 1990; McCammon & Harvey, 1987; Karplus & Petsko, 1990; Brooks III et al., 1988; Weiner & Kollman, 1981; Weiner et al., 1984; Åqvist et al., 1985; Åqvist et al., 1986; Tapia & Åqvist, 1989; Nilsson et al., 1990).

In our work, the GROMOS force field has been used consistently. A number of MD simulations employing the GROMOS parameter set yield a qualitatively adequate structural and dynamical description of water-soluble and fairly globular protein molecules, as exemplified by the carboxy-terminal fragment (CTF) of the ribosomal L7/L12 protein from *Escherichia coli*, bacteriophage T4 glutaredoxin (T4-glx) and the potato carboxypepatidase A inhibitor protein (PCI). The first two molecules have a high density of charged residues residing on their surfaces. For practical reasons most of the molecular dynamics simulations are carried out with the protein framework only. Solvent and counter ions, warranting electroneutrality, are treated with an especial model. This one is presented in the following section. In recent work, we refer to this as the non-inertial solvent simulation scheme.

The relationships between a molecule's structure and fluctuation pattern with its function are at present not fully understood. Part of the work summarized here attempts to explore aspects of this problem.

For the monomeric CTF protein extensive simulations have been performed with the non-inertial solvent scheme and with an explicit solvent representation. Here, for the first time, a collective motion in subdomains of a protein was clearly identified. In the dimeric form of CTF – with a structure suggested by X-ray crystallography – monomer-monomer motions emerge which appear to originate in the simple monomer subdomain motions. As discussed in section 2, these movements may be of relevance for the biological function of CTF in the L7/L12 protein.

Although the speed with which three-dimensional protein structures can be determined by X-ray crystallography is steadily increasing, the number of known protein sequences and genes that can be expressed is far too large for any one to expect that structural determinations can be made for all of them. Therefore structural model-building has become an important tool. In this respect, molecular dynamics computer simulations can provide information to validate static model-built structures. Such an approach was used for the CTF protein from spinach chloroplasts which was model-built into the parent structure of the *E. coli* CTF protein. This was a first step towards integrating molecular dynamics simulations in protein design. In section 3 we report on two examples: the CTF and PCI cases (Nilsson, 1990; Oliva et al., 1991b; Oliva et al., 1991a; Horjales et al., 1987).

Side-chains undoubtedly play an essential role in determining the actual shape of protein folds. However, once the protein has attained its native fold, questions could be raised regarding the contribution to the protein stability from the main-chain constituent. The issue may be explored in a computer simulation experiment by construction of a polyglycine mutamer of a given protein fold – where all residues along the protein's wild-type sequence have been changed into glycines – and endowing that mutamer with the conformation of the native protein's fold. Since MD simulations of the full protein render the global fold quite accurately, the computer experiment may yield information on the properties of the main-chain. Theses ideas have been tested for the T4-glx and the CTF folds. Results are discussed in section 4.

Protein surfaces are essential for molecular recognition processes. The surfaces can be changed locally by site-specific mutations, or more globally by loop displacements. In T4-glx, extensive molecular dynamics simulations with a strong coupling to a thermal bath facilitated detection of a particular loop which is responsible for large changes on the surface around the T4-glx active site. The results correlate with experimental data, and a model for protein surface plasticity as a response to ligand binding could be formulated; these results are discussed in section 5.

The theoretical framework for the MD equations, the analytical molecular graphics techniques and some aspects of enzyme reactions necessary to understand the floor discussion are summarized in the Appendix.

2. MOLECULAR DYNAMICS SIMULATIONS OF PROTEIN STRUCTURES

Simulations using the non-inertial solvent representation (Åqvist et al., 1985; Nilsson et al., 1990) have been successful in rendering protein molecules with good agreement to X-ray crystallographic structure (Åqvist et al., 1985; Åqvist et al., 1986; Tapia & Åqvist, 1989; Nilsson et al., 1990; Makinen et al., 1989; Tapia et al., 1990). The root mean square (rms) deviation for alpha carbon atoms between average MD structures and the one derived by X-ray diffraction typically ranges between 0.9-2.5Å ($1Å = 10^{-10}$m); inclusion of side-chain atoms in the rms calculation yields deviations between 2-3Å. These differences between the experimental and modelled average structures can be rationalized by considering the absence of explicit solvent molecules, the unequal molecular environments – a protein crystal versus the solvated protein – and the differences in averaging conditions. In the experimental case the order of a μmol of molecules are observed in the seconds time-scale, whereas in model studies one or a few molecules are observed only a few hundred picoseconds ($1ps = 10^{-12}$s). Exclusion of explicit solvent in the model leads to a slight implosive effect, which is impeded if the solvent molecules are explicitly included. This is an artificial limitation of non-inertial solvent models. However, for all cases studied so far, the actual fold is respected in the simulation. Having the above-mentioned limitation in mind, non-inertial solvent schemes can be used to avoid a resource-consuming detailed treatment of water molecules in the simulation model. Instead of the explicit introduction of counter ions (these keep the system electroneutral) the globally charged groups are electrically neutralized, but partial atomic charges are kept to facilitate hydrogen bond interactions. Formally, this approach corresponds to a functional inclusion of the (electro-static) molecular field effects due to the environmental solvent and counter ions surround-ing the solute. Inertial collisional effects are absent.

2.1 CTF: THE C-TERMINAL FRAGMENT OF THE RIBOSOMAL L7/L12 PROTEIN

Efficient polypeptide synthesis in bacteria requires the L7/L12 protein (L=large; the protein is a constituent of the large, 50S, ribosomal subunit); (Liljas, 1982). Experiments including electron microscopy show that the 50S ribosomal subunit contains four copies of the L7/L12 molecule. The protein is found in a thin protuberance called the "stalk" where it interacts via its N-terminal fragment (NTF) with the L10 protein located at the base (Liljas, 1982). Ribosomes from a number of distantly related bacteria all have a stalk protuberances in their 50S subunits (Marquis et al., 1981); similarly is found for the *B. stearothermophilus* bacterium (Marquis & Fahnestock, 1978; Marquis & Fahnestock, 1980). The L7/L12 proteins are independently fairly flexible (Gudkov et al., 1982).

A detailed molecular mechanism for the interaction of the L7/L12 protein with the elongation factors is not available, but some clues exist. The functionally significant unit seems to be the dimeric association of L7/L12 molecules in the stalk (Liljas, 1982). The high flexibility of this part of the protein is likely due to the residues 38-51 connecting the two highly structured fragments: NTF (residues 2-37 in *E. coli*) and CTF (residues 52-120 in *E. coli*) (Bushuev et al., 1989). Ribosomal GTPase activity is probably triggered by a conformational change in the L7/L12 dimer (Bushuev et al., 1989). The fact is that significant GTPase activity can be achieved by EF-Tu in association with the L7/L12 protein without any other components of the protein synthesis machinery (Donner et al., 1978). The L7/L12 protein – and its structural homologs from other species – seems to be essential for high GTPase activity, and important for elongation factor binding (Liljas, 1982). The importance of the C-terminal fragment has been elicited by its removal from the ribosome: certain functions are hampered dramatically. The binding of EF-Tu, EF-G is reduced and the factor-dependent GTP hydrolysis vanishes almost (for references see Leijonmarck & Liljas, 1987).

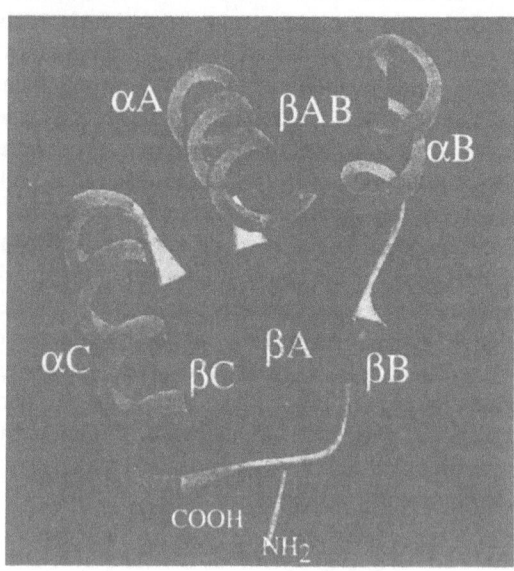

Figure 1. The monomeric *E. coli* CTF protein in a ribbons representation; α-helices are denoted by αA, αB and αC, β-strands by βA, βAB, βB and βC. The βAB strand in the $\alpha\alpha$-corner continues the β-sheet of the adjacent CTF molecule in the dimer.

The structure determination of the *E.coli* molecule at 1.7Å resolution and with a 17% R-value opened up possibilities for a more detailed analysis. These dimers may have functional significance (Leijonmarck & Liljas, 1987). In the protein crystal, CTF forms dimers where the individual monomeric subunits are related by a two-fold symmetry axis. A patch of invariant amino acid residues, probably involved in functional interactions with elongation factors, was suggested by the structural data.

The available high resolution X-ray CTF structure facilitated model studies of the monomeric and dimeric CTF forms, exploration of the molecule's dynamical properties, and investigation of the possible relationships between the protein's dynamics and its structure.

The CTF protein from *E.coli* folds into an β-α-α-β-α-β pattern (Leijonmarck & Liljas, 1987). Two domains are clearly distinguishable: an $\alpha\alpha$-corner (Efimov, 1984) with helices αA and αB, and a twisted β-sheet. The $\alpha\alpha$-corner is articulated to the βA strand via a type II' turn involving residues 61-64, while the turn connecting to βB is shorter and contains *cis* Pro-91. The αC helix appears to be tightly bound to the β-sheet since the two turns connecting to it are both short. The α-helices and the β-sheet are arranged roughly on two layers, assembled one on top of the other. The interior protein core between the two layers is entirely hydrophobic. In Figure 1 a ribbons picture (Carson, 1987) of the CTF monomer from *E.coli* is shown.

The CTF molecule has a higher percentage of charged residues than what is average for proteins at the expense of polar but uncharged ones (Liljas, 1982). At the surface there are 15 acidic and 11 basic residues rendering it highly charged (Leijonmarck & Liljas, 1987). Not surprisingly, the simulation of the CTF structure without explicit solvent, i.e. in a pseudo vacuum, presented some difficulties. Uncompensated charges and their inhomogeneous distribution over the protein surface produce large forces. In aqueous solution, such charges are balanced by counter ions and attenuated by solvent. The forces are also screened by polarization effects when interacting with other proteins in the ribosome. Due to these problems, the non-inertial solvent model was developed.

Several molecular dynamics simulations of CTF have been performed: the monomer was modelled both in non-inertial solvent conditions and in water with counter ions; the dimeric form was modelled with the crystallographically detected water molecules included in the non-inertial solvent scheme (Åqvist et al., 1985). The agreement between the experimental and modelled structures and between the atomic fluctuation patterns show the quality of the employed calculation scheme (Åqvist et al., 1985; Åqvist et al., 1986).

Studies of the CTF MD trajectories with techniques derived from signal analysis procedures show varied patterns of low frequency fluctuations involving inter- and intra-secondary structural motions. The folding pattern and the frequency distribution for collective fluctuations correlate nicely. The β-sheet fluctuates with frequencies that are much higher than those associated to helical fluctuations in the $\alpha\alpha$-domain. Helix αC, which is tightly bound to the sheet, presents a fluctuation spectrum with frequencies larger than $10 cm^{-1}$. The $\alpha\alpha$-domain via its αB helix is endowed with a relatively large flexibility. This feature is present also in solution conditions (Tapia et al., 1990).

In the dimer, the segment connecting the flexible helices αA and αB forms a β-strand which is incorporated into the β-sheet of the partner as indicated in the Figure 2. One would expect that such an arrangement would damp the flexibility of the dimer. The computer simulation result is contrary to this expectation. The correlation function and power spectra for inter-domain secondary structures show that a collective motion in the $5 cm^{-1}$ is clearly present (Åqvist et al., 1989). This result relates to the physics of the resonance-mediated energy transfer between two coupled harmonic oscillators of equal masses and force constants.

The assembly of CTF monomers in the dimer appears as a dynamically stable structure. The flexibility present in the αα-domain in the dimer is communicated to it via the tight contacts made by the loop connecting the constituent helices. When the coordinates are fitted to one monomer, the other monomer looks as being involved in a large amplitude motion. The twofold symmetry enforced in the crystallographic data is lost. These model results show that the CTF dimer may exist in at least two conformational states. One corresponds to the crystallographic arrangement, which is unstable under non-inertial solvent conditions, the other is the MD average structure. This finding may be of pertinence for the shape of the dimer in the ribosome.

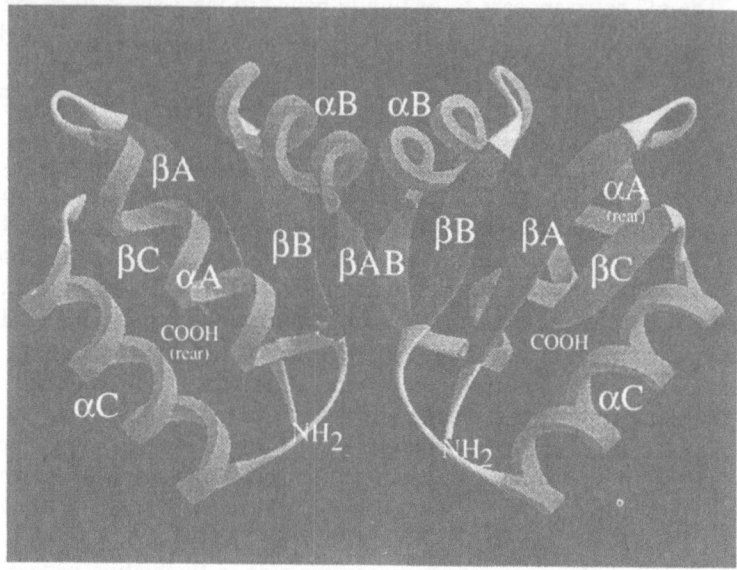

Figure 2. The dimeric form of *E.coli* CTF in a ribbons representation.

From the ensemble of simulation results on the CTF system one can conclude with a mild speculation: the CTF fold seems to be designed for achieving a differential fluctuation pattern in its subdomains which leads to the functional properties of the dimer; in this manner it may participate as an effector when bound to elongation factors. Of course, experimental investigations will eventually close the issue.

3. PROTEIN DESIGN

The spectacular progress in contemporary genetics, molecular biology and microbiology, has resulted in useful and rapid procedures to produce mutant protein sequences. Genetic engineering can be employed to tailor proteins with specific desirable properties. This enterprise, however, has its problems. Point mutations introduce non-local changes in the three-dimensional (3D) structure, these may affect the proteins' functionality in a barely predictable manner. Thus, the use of quasi-static molecular modelling, which is uniquely based on energy optimization procedures, is not sufficient to detect changes occurring in regions distant from the mutated residue. The structures of proteins, their thermal stability and functional behavior can be modulated – or even be drastically modified – by altering their amino acid sequences. Molecular dynamics simulation may overcome this problem and consequently contribute with information to assist the design of mutant proteins. The quality of theoretical predictions depends on careful studies of wild-type structures and appropriate gauging of theoretical information against experimental data.

One of the problems experimentalists may confront is that of selection of which mutants to construct out of a large number of possible candidates in order to obtain a desired effect. The trial and error procedure is not necessarily the best choice. Today, if the structure is known, it is possible to construct a mutation to change specific hydrogen bonds or modify the pattern of electrostatic interactions. But, it is not possible to predict a mutation's effect on the the accessible conformational space. Such changes (if any) may be important in fine-tuning protein's functionality or, in other cases, they may even modify the fluctuation pattern to such an extent that the molecule's function may be drastically changed. Cases where these phenomena are highly important include the cooperative allosteric effect in proteins such as haemoglobin. Dynamical conformational alterations in one subunit of the tetrameric protein modifies the conformations of the active sites in the other haemoglobin subunits, and thereby also the oxygen binding affinity.

3.1 MODEL BUILDING OF HOMOLOGOUS STRUCTURES

Model-built structures (MBS) provide useful information for computer-assisted modelling of protein-ligand interactions and drug design. A MBS is usually energy-optimized with standard protein force fields. Such a process eliminates bad contacts between side-chains, while only small changes in the fold are achieved. The method was used by Leijonmark to construct the structure of chloroplast CTF (cCTF) from spinach using as a template the structure of CTF from *E. coli*. The cCTF has 60% sequence identity with the *E. coli* homolog; 27 amino acids are substituted and one is inserted in the cCTF sequence. As discussed above, MD simulations of CTF with the non-inertial solvent force field have shown excellent agreement with the experimental data. Starting from the MBS of cCTF, a more than two nanoseconds long molecular dynamics trajectory at 293K has shown a dynamically stable structure with similar atom fluctuation patterns and analogous collective motions in its secondary structures as those found in the CTF from *E. coli* (Tapia et al., 1990; Nilsson, 1990; Åqvist et al., 1985). The MBS cCTF model structure may be considered now as representing a realistic structure endowed with substantial thermal stability. Its ability to form dimers with functionally significant capabilities follows form the study made on the *E. coli* CTF.

3.2 SITE-SPECIFIC MUTANTS OF PROTEASE INHIBITORS

Protease inhibitors are frequently subjected to protein engineering studies. These molecules are involved in important biological processes, such as hormone and neuropeptide processing, defense mechanisms, fertilization and virus replication. It is then desirable to develop target-oriented inhibitors to control these processes (Schnebli & Braun, 1986; Kotler et al., 1988; Navia et al., 1989a; Navia et al., 1989b). The potato carboxypeptidase A inhibitor protein, shown in Figure 3, is well-suited for molecular dynamics simulation studies because it is small (Hass & Ryan, 1982), its structure is known in aqueous solution (Clore et al., 1987), and in a crystal complex with the receptor protein carboxypeptidase A (CPA); (Rees & Lipscomb, 1982). The present view on PCI's inhibitory mechanism is that it strongly binds to CPA in a competitive fashion (its K_i is in the nM range); (Hass & Ryan, 1982; Rees & Lipscomb, 1982; Hass et al., 1976). The functional importance of the primary contact site to the CPA enzyme – the C-terminal tail consisting of residues 35-38 – and the short stretch of residues located around Trp-28 – i.e. the secondary binding site which covers residues 28-31 – have been probed experimentally by chemical modification studies (Hass et al., 1976).

The quality of the employed force field was gauged by simulation of the isolated wild-type PCI. Results show that the PCI fold is intrinsically stable. The primary and the secondary binding sites to CPA were fairly well reproduced in the average MD structure, whereas the only exceptional conformational change occurred in the N-terminal tail. The technical aspects of the simulations have been reported elsewhere (Oliva et al., 1991b).

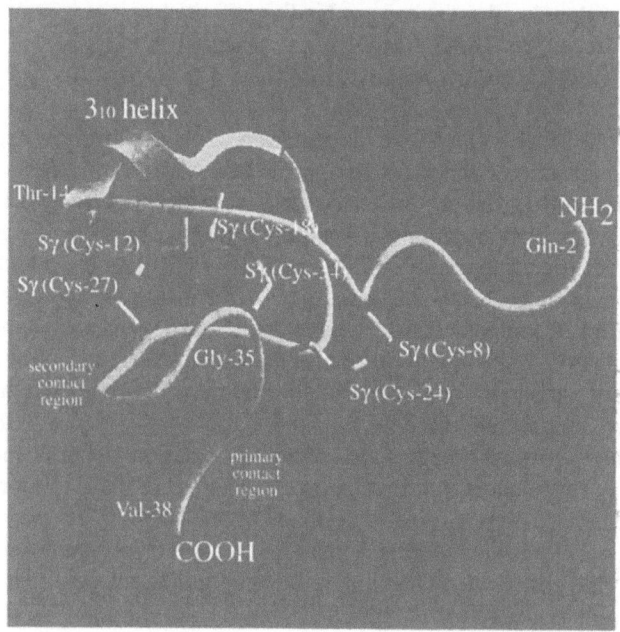

Figure 3. The PCI in its X-ray conformation when bound to the carboxypeptidase A receptor molecule. Ribbons representation.

Molecular graphics analysis of the wild-type PCI average MD structure docked onto CPA was used to identify possible inhibitory structural determinants. It was shown that all residues except Gly-35 in the C-terminal tail make contacts with CPA. As first noticed by Lipscomb, Pro-36 contacts Tyr-198, which is part of the binding site S_3 in CPA. This position in CPA has been proposed to be a recognition site; there, a sliding mechanism for transport of the substrate into the active site pocket initiates. Interactive three-dimensional computer graphical animation (Nilsson, 1990) of the PCI trajectories superposed onto the location of PCI in the PCI-CPA complex shows that Gly-35 does not approach residues in the CPA active site while Pro-36 always fluctuates in the neighborhood of Tyr-198. A mutant was designed following two simple criteria: (1) Only a minimal change should be introduced in the inhibitor-receptor interaction zone; (2) A maximal change in the fluctuation pattern should be achieved. A Pro-36-Gly mutant of the PCI X-ray structure as it is found in complex with CPA was selected. Note that a Pro→Gly mutation is not uncommon; for example in cCTF at the position 62 there is a proline while in *E.coli* a glycine is found, and Gly-62 is conserved in all bacterial species reported by Leijonmark and Liljas (Leijonmarck & Liljas, 1987).

The dynamically equilibrated structure of the mutant PCI presents structural differences with respect to the wild-type X-ray and the average MD structures of PCI (Oliva et al., 1991a; Tapia et al., 1991). The overall fold is basically retained. In both simulations the N-terminal tail bends toward the PCI core, and it has higher fluctuations in the mutant PCI model than in the wild-type one. The C-terminal tail, essential for interaction with CPA, displays lower fluctuations in the mutant PCI than in the wild-type; in fact, it packs to the core of the protein and changes its direction with respect to secondary binding site. The mutation of Pro-36, located in the center of the C-terminal tail, produces a change in the accessible conformational space of the mutant compared to the wild-type model.

The C-terminal tail's orientation with respect to the body of the protein was quantitated with an angle measuring its position with respect to the core. For wild-type PCI, the simulation showed an angular distribution function with a peak near the value PCI has in its complex with the enzyme, while for the mutant PCI the distribution is clearly shifted away from this value. A very simple hypothesis concerning functionality was suggested by this result: It is assumed that a good inhibitor will have an angle distribution function peaking as near as possible to the value corresponding to PCI in its complex with CPA. Therefore, it is suggested that the potency of the Pro-36→Gly mutant PCI will be lower, and that complex-formation kinetics will be slower, than for wild-type PCI.

Experimental work is in progress to evaluate the accuracy of this prediction.

4. STABILITY IN PROTEIN FOLDS

The three-dimensional structures of proteins are functions of their amino acid sequences (Anfinsen, 1973; Skolnick & Kolinski, 1990; Eklund et al., 1984; Rossman & Moras, 1974; Lesk & Chothia, 1980; Alber, 1989; Alber et al., 1987; Alber et al., 1988; Reidhaar-Olsen & Sauer, 1990; Creighton, 1983). There are numerous examples of proteins with different amino acid sequences, but with very similar three-dimensional folds (Chothia & Lesk, 1986) and sometimes similar functions (Bowie et al., 1990). As a result families of topologically related proteins can be constructed. For example, although there is little amino acid sequence homology between bacteriophage T4 glutaredoxin and *E. coli* thioredoxin they share the same fold, apart from an additional α-helix and β-strand in the N-terminal region of the latter (Eklund et al., 1984). The nucleotide binding domain (the

Rossman fold) is a common structural feature of nicotinamide dependent dehydrogenases (Rossman & Moras, 1974); globins share the same topological traits (Lesk & Chothia, 1980). Furthermore, amino acid substitutions at many positions in different proteins have been engineered without affecting the three-dimensional structure or the stability (Alber, 1989; Alber et al., 1987; Alber et al., 1988; Reidhaar-Olsen & Sauer, 1990).

4.1 THE POLYGLYCINE MODEL OF PROTEIN FOLDS

The observed degeneracy of the primary sequence to tertiary structure coding in the amino acid alphabet raises a question concerning the role assigned to side-chains as sources to the stability of a fold. One may wonder whether the main-chain conformation can contribute to the fold stability beyond what is obviously provided by the hydrogen bonding factor. That is, the main-chain can be locked in a particular fold thereby contributing to the stability of the overall structure and conferring additional robustness to the folded structural entity. The hypothesis may be investigated by mutating all residues of a protein into glycines in the folded state. The marginal stability rendered by the folded state itself would thereby be probed. Glycine has no statistical preference to obtain either α-helical or β-strand conformations (Creighton, 1983; Levitt, 1978). Consequently, if segments of an initially folded polyglycine chain maintain secondary- and tertiary structures in a fairly long time-scale, fresh information on the properties of the folded state can be obtained.

These ideas have been numerically tested with encouraging results. Molecular dynamics computer simulation at room temperature (293K) with the polyglycine model is thus used as an exploratory tool to sense possibly relevant stability features of protein folds. Up till now the T4-glx (Nilsson & Tapia, 1991) and the CTF (Soares et al., 1991) folds have been analyzed. Note that presence of explicit solvent is not desired in the model, since the focus is placed on the intrinsic differential structural stability of molecules.

4.2 THE BACTERIOPHAGE T4 GLUTAREDOXIN FOLD

The computer model of the T4-glx fold possesses denoted differential stability, as evidenced by a 300ps non-inertial solvent molecular dynamics simulation of polyglycine folded into the wild-type T4-glx conformation. Figure 4 shows a ribbon model of the canonical view of the wild-type T4-glx X-ray conformation. Note the well-formed elements of secondary structure; a four-stranded central β-sheet in the protein core is surrounded by two almost parallel α-helices on one side of the sheet and one perpendicularly oriented α-helix on the sheet's other side. Separated by an almost vertical pseudo-symmetry plane in the canonical view, two folding units can be distinguished: one $\beta\alpha\beta$ and one $\beta\beta\alpha$ (on the left and right sides in the canonical view in Figure 4). The units are interconnected by the perpendicularly oriented helix $\alpha 2$ and two long loops, L2 and L3 (Söderberg et al., 1978; Eklund et al., 1991). The elements of secondary structure are interconnected with three long loops on the active site face (the front face in the canonical view) of the molecule.

The most pronounced feature in the polyglycine model of the fold is that the two folding units have a differential stability behavior, the first unit appears more robust than the second (Figure 5); (Nilsson & Tapia, 1991). This could relate to properties of the wild-type T4-glx molecule; atomic fluctuations appear larger in the second folding unit than in the first in X-ray diffraction data (Nilsson et al., 1990), and denaturation-renaturation experiments have indicated presence of domains with different folding characteristics (Borden & Richards, 1990).

Furthermore, the four-residue segment containing the redox-active cysteines in the wild-type T4-glx molecule is located in a very stable part of the main-chain. Structural shifts and atomic fluctuations in this region are low in experimental data and in all simulation studies. In the polyglycine model the conformation of this segment is remarkably well retained. This stability may be due to the actual conformation of this zone, and in part by it being located close to the robust helix α1. Such features may be important in providing functional reversibility and the exact stereochemistry for the protein's redox-reaction. The robustness of the active site segment has been evidenced experimentally through site-directed mutagenesis: it has been possible to substitute the two residues between Cys-14 and Cys-17 by several other, without impairing the protein's structure (Eklund et al., 1991), or its function (Joelson et al., 1990). The computer simulation results show that a disulfide bridge is not necessary to maintain the global shape of the active site segment in the time-scale corresponding to the model, and that this region possesses a marginal stability partly endowed by the folded state.

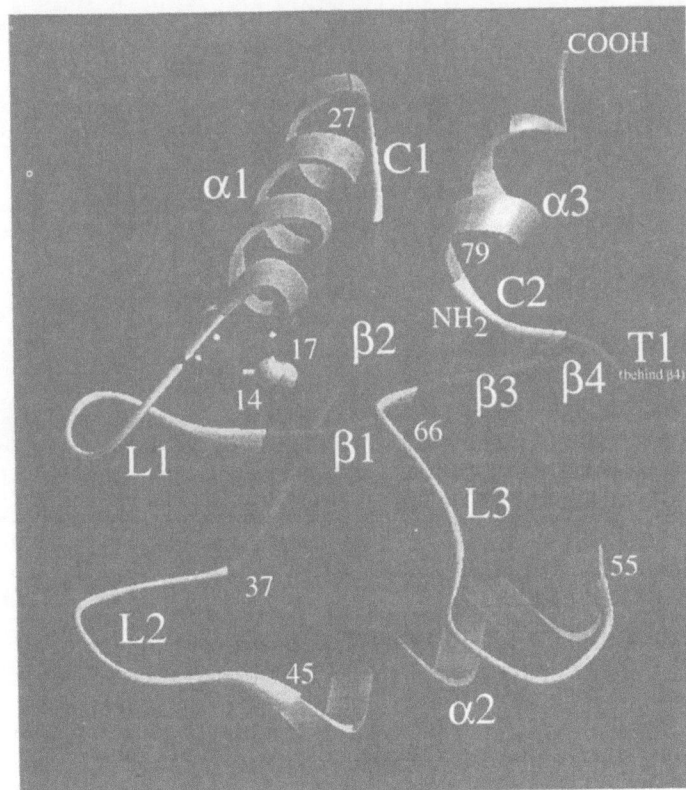

Figure 4. The X-ray conformation of wild-type T4-glx in a ribbons representation; α-helices are denoted by α1-α3; β-strands by β1-β4; loops by L1-L3; connecting regions by C1-C2; a turn by T. The active site contains Cys-14 and Cys-17. Two folding units can be appreciated if the molecule is partitioned vertically between β1 and β3 and through the center of α2.

From a global topological perspective, the T4-glx fold is well retained in the 300ps simulation, however. The relative locations of elements of secondary structure are not altered, although some of these elements are deformed.

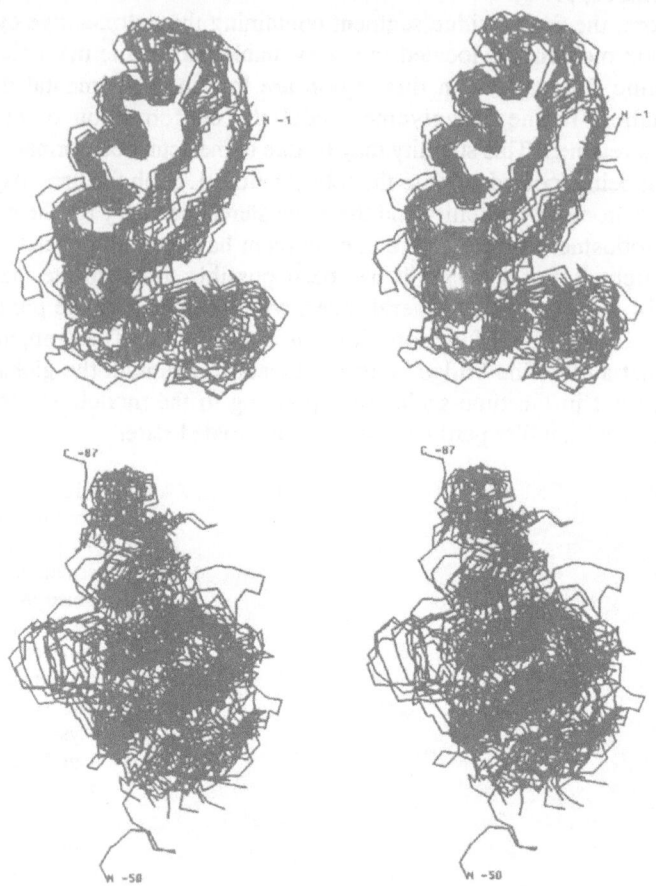

Figure 5. Differential stability behavior in T4-glx' two folding units. (a; top) The first folding unit is well preserved after 300ps of non-inertial solvent molecular dynamics simulation, whereas (b; bottom) the second unit is rapidly deformed. These are stereoviews seen from the center of the T4-glx molecule.

4.3 THE CTF FOLD

In the polyglycine model of the CTF fold, the first part of helix αA and most of αC are well preserved in the time-course of 100ps of simulation. Also, parts of β-strands A and B maintain their conformations, whereas the short β-strand AB in the αα-corner (cf. Figure 1) and the parts of αA and αB connecting to it, as well as the C-terminal section of βB and the whole of βC become deformed (Soares et al., 1991).

The residues in the αα-corner are highly conserved among homologous CTF sequences (see the Figure 11 in Leijonmarck & Liljas, 1987 for the amino acid sequences of

L12 proteins from bacterial species and spinach). Five residues in each helix are involved in the packing of these helices. These residues are fully conserved. Val-68 and -72 are also involved in packing interactions between αA and αC. In the polyglycine model these interactions are suppressed and, quite naturally, the corresponding regions become deformed. The folded state of this protein class has a dependence on the invariant residues.

4.4 THE BACTERIOPHAGE T4 GLUTAREDOXIN AND THE CTF FOLDS

From the molecular dynamics simulations on the bacteriophage T4 glutaredoxin and the CTF folds the following picture emerges: In the T4-glx fold large sections, which basically correspond to elements of secondary structure, are structurally robust. This protein shows a clear two-domain behavior. The protein is a redox system acting in both directions: the active site main-chain framework appears to be an extremely stable structure. The CTF fold, on the other hand, presents basically a one-domain behavior. The stability of its functionally important domain appears to depend crucially on the side-chains. The extremely high degree of residue conservation in this region is then quite suggestive.

In both cases folded α-helical sections appear more robust than folded β-sheets. Globally, the topology of the T4-glx fold appears better preserved than that of the CTF fold, although using a measure of local stability distribution the differences between the folds are small. This could be due to the fact that T4-glx has longer loops than CTF – in particular those which connect the two folding units in T4-glx. These can dissipate tension forces and act as buffer regions.

4.5 ELECTROSTATIC MIMICRY OF PROTEIN DENATURATION

As an initial attempt to model deformation in protein structures and protein folds a scheme which employs intra-molecular electrostatic attenuation of polyglycine models of protein folds was used to reach putative denatured conformers. This setup mimicks a situation that can be achieved experimentally; for instance, it is possible to denature protein molecules by altering the ionic strength, or by adding compounds which shift the relative dielectric constant of a protein solution (Goto et al., 1990a; Goto et al., 1990b; Tanford, 1968).

In the case of the T4-glx fold it is observed that its local features are lost very rapidly in a cooperative fashion. All hydrogen bonds which render the shape of secondary structural elements are lost; these are highly electrostatic in nature. For the helices, it appears as if their middle sections survive slightly longer than their ends, and helix α2, which is connected only loosely to the rest of the T4-glx framework, seems to survive longer than the other helices in the model. This may be due to the energy-dissipating features of the connecting loops. The overall topology of the T4-glx fold is not drastically altered. The final "denatured" object is "molten" and rather globular, but without any of the secondary structural features characteristic of the native folded state. The reader is referred to the work by Nilsson & Tapia, 1991 for further details.

Actually, the model is such that only a marginal decrease from normal values in the electric field (~20%) is required to achieve "unfolded" conformers. Thus, the electrostatic contribution to the non-inertial solvent model's force field appears to be optimal in rendering secondary structure features.

5. BIOMOLECULAR SURFACES

Characterization of biomolecular surfaces is an important issue in contemporary molecular biology. The complex and elaborate, yet organized, networks of physical and chemical processes that take place mediated by these surfaces provide a lower level foundation for the processes of life. Accurate determinations of (bio)surfaces may lead to an understanding of the relationships between molecules' structures and functions (Meyer, 1986). The specific intra-molecular associations in multimeric proteins, polypeptide hormone-receptor interactions and the epitopes (antigenic regions) on protein surfaces illustrate some of the important roles that (bio)surfaces play.

In the early seventies, Lee and Richards proposed a solvent-accessible molecular surface model (SAS); (Lee & Richards, 1971; Richards, 1977). In the eighties a number of algorithms were reported for estimation of molecules' areas and volumes (Lee & Richards, 1971; Richards, 1977; Hermann, 1972; Meyer, 1986; Drummond, 1988). Among these, Connolly's molecular surface (MS) approach has been widely used. All the proposed methods have not yet provided a fully satisfactory solution for (bio)surface characterization, and in particular, Connolly's scheme has been criticized for its oscillatory behavior with respect to the internal parameters (Meyer, 1986). Recently, an algorithm based on sphere tessellation was designed to provide surfaces and volumes of simple and complex molecules (Pascual-Ahuir et al., 1987; Pascual-Ahuir & Silla, 1989); the scheme has been extended and endowed with analytical computer graphical features (Silla et al., 1990).

The concept of surface fractality has also been explored and applied to study the retinol binding protein in its interaction with prealbumin (Åqvist & Tapia, 1987). For each residue on the protein's surface a fractality index is computed; this value contains information regarding the 'roughness' of the surface. It appears plausible that molecular surface regions involved in inter-molecular contacts suitable for complex-formation should present a 'rough' surface to prevent sliding and allowing optimization of favorable interactions, and that active sites, which should support possibilities for substrate diffusion, should be characterized by a fractally more 'smooth' surface. For the case of retinol-prealbumin system it was possible to model-build a docking mode of the two molecules where regions of high surface fractality coincided with regions of topographic complementarity (Åqvist & Tapia, 1987).

5.1 SURFACE PLASTICITY IN BACTERIOPHAGE T4 GLUTAREDOXIN

The bacteriophage T4 glutaredoxin protein presents interesting properties that can be analyzed in terms of molecular surface changes. In this redox-active 87 amino acid residue molecule all the charged side-chains are located on the surface and a cluster of hydrophobic residues forms the molecule's central core. Adjacent to the T4-glx active site there is a rather extensive crevice in the structure derived by X-ray diffraction beset by three flexible loops (Figure 6; cf. Figure 4). *In vivo* and *in vitro* T4-glx has the ability to interact with several ligands of varying size, shape and polarity, ranging from a tripeptide (glutathione) to a tetrameric protein (bacteriophage T4 ribonucleotide reductase; ~2000 residues).

Model studies of T4-glx employing molecular dynamics computer simulation suggest that the charges on its surface are prone to alter their distribution quite significantly under the influence of thermal energy and collisions with solvent molecules. After 150ps of non-inertial solvent MD simulation at room temperature the longest T4-glx loop (loop L3; cf. Figure 4) makes a spectacular cooperative conformational transition; the atoms in the loop which move the most traverse a span of ~18Å (Figures 7; cf. Figure 6). The simulation was carried out at constant temperature with a rather tight coupling to a thermal bath favouring energy exchange between the solute and its thermal bath. No simulated annealing scheme was employed. The analyses with atom fluctuation ellipsoids show that the loops and the turn enter into concerted, large amplitude oscillations that help finding paths towards other regions of the conformational space. Thus, the tight coupling to the thermal bath and the nature of the structure impart high amplitude and apparently low frequency motions to the atoms constituting the loop system. Flexible regions in the structure can explore larger fractions of the conformational space. In the present case the L3 loop may alter its position with seeming ease. No high barrier is detected in the potential energy for the detected conformational change.

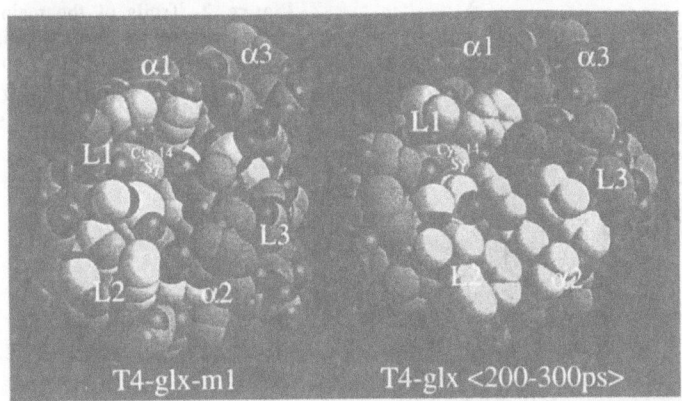

Figure 6. Van der Waals surfaces of T4-glx in the X-ray conformation (left side) and of the 200-300ps average MD structure after the transition of loop L3 (right side).

Post-transitionally, a new surface appears adjacent to the active site. This one exposes more hydrophobic amino acids than the initial surface (in the MD model there is an ~100Å² gain in exposure of hydrophobic residues in loop L3), and it is slightly concave.

In view of the MD model results, it could be conceived that T4-glx is capable of switching active-site surfaces depending on the global or local environmental conditions. In a polar solvent most T4-glx molecules in a population should present a polar surface; here interactions with glutathione would be favoured. A more hydrophobic T4-glx surface may be prompted forth if T4-glx' surrounding environment becomes increasingly hydrophobic. It might also be thought that loop L3 could be displaced when another large protein molecule approaches the active site face of the T4-glx. In any case, the characteristic feature of the T4-glx MD model surface is its structural heterogeneity. One could then expect some

138

difficulties when carrying out experimental determinations of T4-glx structures – and related properties – in solution.

The modelled dynamical possibilities for T4-glx to largely alter its envelope surface's shape and polarity distribution provides one possible explanation for the observed substrate versatility of this molecule.

Interestingly, there is experimental data that point in the same direction as the molecular dynamics simulation study. The existence of three T4-glx average structures derived by X-ray diffraction (Söderberg et al., 1978; Eklund et al., 1991) allows for an estimation of the molecule's surface plasticity and potential dynamical properties. This view could be elaborated further as the number of X-ray structures increases. In particular for T4-glx, it is found that charged residues on the surface may occupy different positions also in the expectedly constrained crystal environment. Loop L3 is the most flexible region of the T4-glx molecule. For this loop, there are atom positional differences ranging up to 5Å between the three X-ray structures. The experimental pattern of differential flexibility and deformability among the T4-glx active site surface loops L1<L2<L3 coincides with the model results. Thus, it is substantiated that the molecular dynamics computer simulation model captures the essential trends of the global dynamical properties of bacteriophage T4 glutaredoxin.

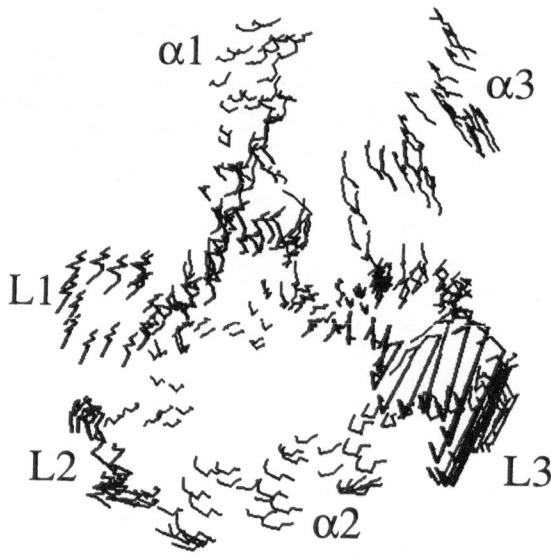

Figure 7. Trails of the main-chain atoms of T4-glx during the 120-200ps section of the non-inertial solvent MD trajectory. Loop L3 (down right) makes a spectacular transition.

6. DISCUSSION

Molecular biotechnology is a new area of development at the interface of molecular biology, molecular genetics, biochemistry, biomolecular crystallography, protein engineering and modern (computational) theoretical chemistry. The list is not exhaustive. In the present paper, some examples of applications of molecular dynamics computer simulations and molecular graphics methods have been presented.

Although not explicitly mentioned, perturbation simulation techniques were used to create protein mutants. This procedure is at the base of computer assisted free energy difference calculations. The success of such methods depend on the invariance of the conformational space accessible to the initial and final states. In practice, this is not always granted. It is in this respect that simulations as reported for the PCI represent a first step towards examining its accessible conformational space. While exploring the conformational space for different mutamers, qualitative indications concerning possible structural and dynamical effects generated by mutations can be gathered.

Stationarity of a given computed trajectory is a necessary condition for validation of methods based upon statistical mechanics. In the two nanoseconds long trajectory for cCTF this objective was achieved. The auto- and cross-correlation functions do not depend on the time origin, signaling the system's stationarity. It is obvious that such stationarity is not achieved for short trajectories – a few hundred picoseconds – where conformational changes may take place. The molecular dynamics simulations must then be carefully interpreted. A time-drift in the total potential energy nearly always signals structural changes. The conformational effects should then be put in a perspective using experimental data and scientific judgment. This point is illustrated by the study on surface plasticity of bacteriophage T4 glutaredoxin.

The folded polyglycine models represent the main-chains of generic proteins. The results found until now suggest that it may be a useful sensor for differential structural stability of a given fold. These trajectories are typically of non-equilibrium character. By monitoring persistence times new information on the properties of the fold is gathered. If the image of "protein quakes" introduced by Frauenfelder is employed, our results suggest that there are more or less seismic zones in protein folds. Thus, regions "melting" rapidly may sense seismic segments in the basic fold: mutations in the parent molecule affecting those zones may produce changes in the structure's properties. Or conversely, robust main-chain zones may absorb mutations to a better extent than seismic regions. Further work along this line is necessary before a clear picture emerges.

We acknowledge financial support of the Swedish Natural Science Research Council (NFR) and the Swedish National Board for Technical Development (STU). We are indebted to Dr. H. Eklund for making X-ray diffraction data of bacteriophage T4 glutaredoxin available to us prior to publication. Furthermore we are most grateful to our colleagues for their contributions to this work: J. Andrés, X. Avilés, M. Campillo, R. Cardenas, F. Colonna, B. Oliva, E. Querol, C. Soares, M. Wästlund and J. Åqvist.

7. REFERENCES

Alber, T. (1989) 'Mutational effects on protein stability', Annu. Rev. Biochem., 58, 765-798.

Alber, T., Bell, J. A., Dao-Pin, S., Nicholson, H., Wozniak, J. A., Cook, S. & Matthews, B. W. (1988) 'Replacements of Pro-86 in phage T4 lysozyme extend an α-helix but do not alter protein stability', Science, 239, 631-635.

Alber, T., Dao-Pin, S., Wilson, K., Wozniak, J. A., Cook, S. P. & Matthews, B. W. (1987) 'Contributions of hydrogen bonds of Thr-157 to the thermodynamic stability of phage T4 lysozyme', Nature, 330, 41-46.

Anfinsen, C. B. (1973) 'Principles that govern the folding of protein chains', Science, 181, 223-230.

Borden, K. L. B. & Richards, F. M. (1990) 'Folding kinetics of phage T4 thioredoxin', Biochemistry, 29, 3071-3077.

Bowie, J. V., Reidhaar-Olson, J. F., Lim, W. A. & Sauer, R. T. (1990) 'Deciphering the message in

140

protein sequences: Tolerance to amino acid substitutions', Science, 247, 1306-1310.

Brooks III, C. L., Karplus, M. & Pettit, B. M. (1988) 'Proteins: A theoretical perspective of dynamics, structure, and thermodynamics', Adv. Chem. Phys., 71, 1-259.

Brünger, A. T. & Karplus, M. (1991) 'Molecular dynamics simulations with experimental restraints', Accounts of Chemical Research, 24, 54-61.

Bushuev, V. N., Gudkov, A. T., Liljas, A. & Sepetov, N. F. (1989) 'The flexible region of protein L12 from bacterial ribosomes studied by proton nuclear magnetic resonance', J. Biol. Chem., 264, 4498-4505.

Carson, M. (1987) 'Ribbon models of macromolecules', J. Mol. Graph., 5, 103-106.

Chothia, C. & Lesk, A. (1986) 'The relation between the divergence of sequence and structure in proteins', EMBO J., 5, 823-826.

Clore, G. M., Gronenborn, A. M., Nilges, M. & Ryan, C. A. (1987) 'Three-dimensional structure of potato carboxypeptidase inhibitor in solution. A study using nuclear magnetic resonance, distance geometry, and restrained molecular dynamics.', Biochemistry, 26, 8012-8023.

Creighton, T. E. (1983) 'Secondary structure' in Proteins. Structures and molecular properties, W.H. Freeman and Company, New York, USA., pp. 235.

Donner, D., Villems, R., Liljas, A. & Kurland, C. G. (1978) 'Guanosinetriphosphatase activity dependent on elongation factor Tu and ribosomal protein L7/L12', Proc. Natl. Acad. Sci. USA, 75, 3192-3195.

Drummond, M. L. J. (1988) 'A supertensor formalism for solute-continuum solvent interactions with arbitrarily shaped cavity. II. Preliminary apllication to model systems', J. Chem. Phys., 88, 5021-5026.

Efimov, A. V. (1984) 'A novel super-secondary structure of proteins and the relation between the structure and the amino acid sequence', FEBS Lett., 166, 33-38.

Eklund, H., Cambillau, C., Sjöberg, B.-M., Holmgren, A., Jörnvall, H., Höög, J.-O. & Brändén, C.-I. (1984) 'Conformational and functional similarities between glutaredoxin and thioredoxins', EMBO J., 3, 1443-1449.

Eklund, H., Ingelman, M., Söderberg, B.-O., Ulin, T., Nordlund, P., Nikkola, M. & Joelsson, T. (1991) 'The structure of oxidized bacteriophage T4 glutaredoxin (thioredoxin)', submitted.

Goto, Y., Calcaiano, L. J. & Fink, A. L. (1990a) 'Acid-induced folding of proteins', Proc. Natl. Acad. Sci. USA, 87, 573-577.

Goto, Y., Takahashi, N. & Fink, A. L. (1990b) 'Mechanism of acid-induced folding of proteins', Biochemistry, 29, 3480-3488.

Gudkov, A. T., Gongadze, G. M., Bushuev, V. N. & Okon, M. S. (1982) 'Proton nuclear magnetic resonance study of the ribosomal protein L7/L12 in situ', FEBS Lett., 138, 229-232.

Hass, G. M., Ako, H., Grahn, D. T. & Neurath, H. (1976) 'Carboxypeptidase inhibitor from potatoes. The effects of chemical modifications on inhibitory activity', Biochemistry, 15, 93-100.

Hass, G. M. & Ryan, C. A. (1982) 'Carboxypeptidase inhibitor from potatoes', Meth. Enzymol., 80, 778-791.

Hermann, R. B. (1972) 'Theory of hydrophobic bonding. II. The correlation of hydrocarbon solubility in water with solvent cavity surface area', J. Phys. Chem., 76, 2754-2759.

Horjales, E., Åqvist, J., Leijonmarck, M. & Tapia, O. (1987) 'Aspects of model building applied to the C-terminal domain of the L7/L12 protein from chloroplast ribosomes: A molecular dynamics study', Biochem. Biophys. Res. Commun., 148, 954-961.

Joelson, T., Sjöberg, B., -M. & Eklund, H. (1990) 'Modifications of the active center of T4 thioredoxin by site-directed mutagenesis', J. Biol. Chem., 265, 3183-3188.

Karplus, M. & Petsko, G. A. (1990) 'Molecular dynamics simulations in biology', Nature, 347, 631-639.

Knowles, J. R. (1987) 'Tinkering with enzymes: What are we learning?', Science, 236, 1252-1258.

Kotler, M., Katz, R. A., Danho, W., Leis, J. & Skalka, A. M. (1988) 'Synthetic peptides as substrates and inhibitors of a retroviral protease', Proc. Natl. Acad. Sci. USA, 85, 4185-4189.

Kraut, J. (1988) 'How do enzyme work?', Science, 242, 533-540.

Kubo, R. (1959) 'Lectures in theoretical physics' in Brittin, W. E. & Dunham, L. G. (eds.), Interscience, London, UK, pp. 120.

Lee, B. & Richards, F. M. (1971) 'The interpretation of protein structures: Estimation of static accessibility', J. Mol. Biol., 55, 379-400.

Leijonmarck, M. & Liljas, A. (1987) 'Structure of the C-terminal domain of the ribosomal protein L7/L12 from *Escherichia coli* at 1.7Å', J. Mol. Biol., 195, 555-580.

Lerner, R. L. & Tramontano, A. (1987) 'Antibodies as enzymes', TIBS, 12, 427-430.

Lesk, A. M. & Chothia, C. (1980) 'How different amino acid sequences determine similar protein structures: The structure and evolutionary dynamics of the globins', J. Mol. Biol., 136, 225-270.

Levitt, M. (1978) 'Conformational preferences of amino acids in globular proteins', Biochemistry, 17, 4277-4285.

Liljas, A. (1982) 'Structural studies of ribosomes', Prog. Biophys. Molec. Biol., 40, 161-228.

Makinen, M. W., Troyer, J. M., van der Werff, H., Berendsen, H. J. C. & van Gunsteren, W. F. (1989) 'Dynamical structure of carboxypeptidase A', J. Mol. Biol., 207, 201-216.

Marquis, D. M. & Fahnestock, S. R. (1978) 'A complex of acidic ribosomal proteins. Evidence of a four-to-one complex of proteins in the *Bacillus stearothermophilus* ribosome', J. Mol. Biol., 119, 557-567.

Marquis, D. M. & Fahnestock, S. R. (1980) 'Stoichiometry and structure of a complex of acidic ribosomal proteins', J. Mol. Biol., 142, 161-179.

Marquis, D. M., Fahnestock, S. R., Henderson, E., Woo, D., Schwinge, S., Clark, M. W. & Lake, J. A. (1981) 'The L7/L12 stalk, a conserved feature of the prokariotic ribosome, is attached to the large subunit through its N-terminus', J. Mol. Biol., 150, 121-132.

McCammon, J. A. & Harvey, S. C. (1987) Dynamics of proteins and nucleic acids, Cambridge University Press, Cambridge, UK.

Meyer, A. (1986) 'The size of molecules', Chem. Soc. Rev., 15, 449-474.

Navia, M. A., Fitzgerald, P. M. D., McKeever, B. M., Leu, C.-T., Heimbach, J. C., Herber, W. K., Sigal, I. S., Darke, P. L. & Springer, J. P. (1989a) 'Three-dimensional structure of aspartyl protease from human immunodeficiency virus HIV-1', Nature, 337, 615-620.

Navia, M. A., McKeever, B. M., Springer, J. P., Lin, T. Y., Williams, H. R., Fluder, E. M., Dorn, C. P. & Hoogsteen, K. (1989b) 'Structure of human neutrophil elastase in complex with peptide chloromethyl ketone inhibitor at 1.84Å resolution', Proc. Natl. Acad. Sci. USA, 86, 7-11.

Nilsson, O. (1990) 'Molecular conformational space analysis using computer graphics: Going beyond FRODO', J. Mol. Graph., 8, 192-200.

Nilsson, O. & Tapia, O. (1991) 'Electrostatic forces and the structural stability of a modelled bacteriophage T4 glutaredoxin fold: Molecular dynamics simulations of polyglycine 87-mers', J. Mol. Struct. (Theochem), *in press*.

Nilsson, O., Tapia, O. & van Gunsteren, W. F. (1990) 'Structure and fluctuations of bacteriophage T4 glutaredoxin modelled by molecular dynamics', Biochem. Biophys. Res. Commun., 171, 581-588.

Oliva, B., Nilsson, O., Wästlund, M., Cardenas, R., Querol, E., Avilés, F. X. & Tapia, O. (1991a) 'A molecular dynamics study of a model built Pro-36-Gly mutant derived from the potato carboxypeptidase A inhibitor protein', Biochem. Biophys. Res. Commun., 176, 627-632.

Oliva, B., Wästlund, M., Nilsson, O., Cardenas, R., Querol, E., Avilés, F. X. & Tapia, O. (1991b) 'Stability and fluctuations of the carboxypeptidase A protein inhibitor fold: A molecular dynamics study', Biochem. Biophys. Res. Commun., 176, 616-621.

Pascual-Ahuir, J. L. & Silla, E. (1989) 'Gepol: A method to calculate the envelope surface. Computations of changes in conformational area and volume of n-octanol' in Carbó, R. (ed.), Quantum chemistry - Basic aspects, actual trends, Elsevier Scientific Publishers B.V., Amsterdam, The Netherlands, pp. 597-603.

Pascual-Ahuir, J. L., Silla, E., Tomasi, J. & Bonacorsi, R. (1987) 'Electrostatic interaction of solute with a continuum. Improved description of the cavity and of the surface cavity bound charge distribution', J. Comput. Chem., 8, 778-787.

Rees, D. C. & Lipscomb, W. N. (1982) 'Refined crystal structure of the potato inhibitor complex of carboxypeptidase A at 2.5Å resolution', J. Mol. Biol., 160, 475-498.

Reidhaar-Olsen, J. F. & Sauer, R. T. (1990) 'Functionally acceptable substitutions in two α-helical regions of λ repressor', Proteins, 7, 306-316.

Richards, F. M. (1977) 'Areas, volumes, packing and protein structure', Annu. Rev. Biophys. Bioeng., 6,

142

151.

Rossman, M. G. & Moras, D. (1974) 'Chemical and biological evolution of a nucleotide-binding protein', Nature, 250, 194-199.

Schnebli, H. P. & Braun, N. J. (1986) 'Proteinase inhibitors as drugs' in Salvesen & Barret (eds.), Proteinase Inhibitors, Elsevier Science Publishers BV, Amsterdam, pp. 613-627.

Silla, E., Villar, F., Nilsson, O., Pascual-Ahuir, J. L. & Tapia, O. (1990) 'Molecular volumes and surfaces of biomacromolecules via GEPOL: A fast and efficient algorithm', J. Mol. Graph., 8, 168-172.

Skolnick, J. & Kolinski, A. (1990) 'Simulations of the folding of a globular protein', Science, 250, 1121-1125.

Soares, C., Nilsson, O. & Tapia, O., *unpublished results*.

Söderberg, B.-O., Sjöberg, B.-M., Sonnerstam, U. & Brändén, C.-I. (1978) 'Three-dimensional structure of thioredoxin induced by bacteriophage T4', Proc. Natl. Acad. Sci. USA, 75, 5827-5830.

Tanford, C. (1968) 'Protein denaturation', Adv. Prot. Chem., 23, 121-282.

Tapia, O. (1991) 'Theory of solvent effects and chemical reactions', Rep. Mol. Th., *in press*.

Tapia, O., Cardenas, R., Andres, J., Krechl, J., Campillo, M. & Colonna-Cesari, F. (1991) 'Electronic aspects of LADH catalytic mechanism', Int. J. Quantum. Chem., 39, 767-786.

Tapia, O., Nilsson, O., Campillo, M., Åqvist, J. & Horjales, E. (1990) 'Low frequency motions in protein's secondary structures. Molecular dynamics studies on carboxy terminal fragment of L7/L12 ribosomal protein' in Sarma, R. H. & Sarma, M. H. (eds.), Structure & Methods, DNA Protein Complexes., Adenine Press, New York, USA, pp. 147-170.

Tapia, O., Oliva, B., Nilsson, O., Querol, E. & Avilés, F. X. (1991) 'Molecular dynamics simulation as a tool to design mutant proteins: Comparative studies of the modelled potato carboxypeptidase A inhibitor and its Pro-36→Gly mutant with the receptor-docked X-ray structure' in Giralt, E. & Andreu, D. (eds.), Peptides 1990, ESCOM Science Publishers B.V., Leiden, The Netherlands, pp. 581-584.

Tapia, O. & Åqvist, J. (1989) 'Molecular dynamics as a tool for structural and functional predications: The retinol binding and chloroplast C-terminal fragment of the L7/L12 ribosomal protein', Prog. Clin. Biol. Res., 289, 55-64.

van Gunsteren, W. F. (1988) 'The role of computer simulations techniques in protein engineering', Protein Engineering, 2, 5-13.

van Gunsteren, W. F. & Berendsen, H. J. C. (1990) 'Computer simulation of molecular dynamics: Methodology, applications, and perspectives in chemistry', Angewandte Chemie, 29, 992-1023.

Weiner, P. K. & Kollman, P. A. (1981) 'AMBER: Assisted model building with energy refinement. A general program for modeling molecules and their interactions', J. Comput. Chem., 2, 287-303.

Weiner, S. J., Kollman, P. A., Case, D. A., Singh, U. C., Ghio, C., Alagona, G., Profeta, S. & Weiner, P. (1984) 'A new force field for molecular mechanical simulation of nucleic acids and proteins', J. Am. Chem. Soc., 106, 765-784.

Åqvist, J., Leijonmarck, M. & Tapia, O. (1989) 'A molecular dynamics study of the C-terminal fragment of the L7/L12 ribosomal protein', Eur. Biophys. J., 16, 327-339.

Åqvist, J., Sandbolm, P., Jones, T. A., Newcomer, M. E., van Gunsteren, W. F. & Tapia, O. (1986) 'Molecular dynamics simulation of the holo and apo forms of the retinol binding protein. Structural and dynamical changes induced by retinol removal', J. Mol. Biol., 192, 593-604.

Åqvist, J. & Tapia, O. (1987) 'Surface fractality as a guide for studying protein-protein interactions', J. Mol. Graph., 5, 30-34.

Åqvist, J., van Gunsteren, W. F., Leijonmarck, M. & Tapia, O. (1985) 'A molecular dynamics study of the C-terminal fragment of the L7/L12 ribosomal protein. Secondary structure motion in a 150 picosecond trajectory', J. Mol. Biol., 83, 461-477.

8. APPENDIX

8.1 THEORETICAL FRAMEWORK OF MOLECULAR DYNAMICS SIMULATION

Molecular dynamics simulation algorithms are based on classical statistical mechanics. The atom dynamics in the Cartesian space – the nuclear conformational space – is driven with a total hamiltonian H which is a sum of molecular Hamiltonians representing the medium surrounding (H_m), the subsystem of interest (H_s) and the intermolecular interactions (V_{ms}):

$$H(X,P) = H_m(P_m,R_m) + H_s(P_s,R_s) + V_{ms}(R_m,R_s) \qquad [1]$$

here P is the linear momentum in Cartesian coordinates of the particle ensemble; P_m and P_s identify the momenta associated to each subsystem, and R_m and R_s denote their Cartesian coordinates, $X=(R_m, R_s)$.

The equation of motion for the atoms (groups) of the subsystem under the effects of the field provided by the surrounding medium at the instant time t can be obtained by using appropriate projection techniques. If both subsystems have local thermal equilibrium where fluctuations in local temperatures are allowed, then, in the time-scale of atomic motions, energy exchange between the systems would ensure global thermal equilibrium. According to the premises used to get separability of both systems, the S-system looks like a Brownian particle. Time-dependent processes associated with a velocity dependent dragging force and the Langevin stochastic force are two sources of interactions. An approximate set of such equations is given by (Tapia, 1991):

$$\dot{P}_s = -\partial V(R_s)/\partial R_s + F_s(t) + \int dt'\Phi(t-t')\, P_s(t') \qquad [2]$$

The first term describes the force created by the atoms of the system of interest, that is, the (intra)molecular force field. It depends implicitly upon time only. The other two forces are time-dependent: one is the stochastic Langevin force $F_s(t)$ and the other, the delayed (memory) term or systematic force. These forces are the subject of particular modelling so that one solves the MD equations for the subsystem of interest only. In order to achieve this, the fluctuation-dissipation theorems are most useful to relate the surrounding medium forces. Thus, the autocorrelation matrix of the stochastic force $F_s(t)$ is related to the kernel of the non-Markovian term by:

$$\Phi(t) = (F_s(t), F^+_s(0))\, (P_s(0), P^+_s(0))^{-1} \qquad [3]$$

which is the second fluctuation-dissipation theorem. The scalar product is defined in terms of matrix elements with the help of standard correlation functions:

$$<F_i(t)F_j(0)> = \int d\Gamma_o \ f(\Gamma_o)F_i(t)F_j^*(0) = (F_i(t), \ F_j^*(0)) \qquad [4]$$

where $f(\Gamma_o)$ is the probability density at the point Γ_o in phase space, the star indicates complex conjugation. Similar definitions hold for the momenta of the particle system **P**.

The time integral of $\Phi(t)$ is a generalized friction function:

$$\gamma(t) = \int dt'\Phi(t-t') \qquad [5]$$

The last two terms in eqn. 2 are responsible for energy exchanges between the subsystems. If $V_{ms}(\mathbf{R}_m,\mathbf{R}_s)$ is small compared with H_m and H_s each subsystem can be considered as nearly in local equilibrium. The temperatures of the subsystems do not necessarily have the same value at a given time. The local temperatures will be T_m and T_s, respectively. T_o is the target (global) temperature during a particular molecular dynamics simulation. As it has been shown by Kubo (Kubo, 1959) the rate of change of the energy in time is given by:

$$-dW_s \ /dt = B(T_s - T_m)/T_o =(T_s - T_m)/T_o \int_o^{\tau_o} dt \int_o^{\beta} < \dot{H}_s(-ih\lambda/2\pi) \ \dot{H}_s(t)> \ d\lambda \qquad [7]$$

where $\beta=1/k_B T_o$; k_B and h are the Boltzman and Planck constants, respectively. In eqn. 7, the characteristic time τ_o must be long compared to the correlation time of the integrands and short on a macroscopic scale. The brackets in the last term indicates thermodynamic averaging.

In a molecular dynamics simulation, the temperature of the bath is very near the one for the global system: $T_m \approx T_o$ and the heat capacity (C_s) of the medium surrounding is assumed to be large compared to the generalized solute. Then, eqn. 7 can be written as:

$$dT_s / dt = -(B/T_o C_s) \ (T_s - T_o) = - (1/\tau_r) \ (T_s - T_o) \qquad [8]$$

where τ_r is a relaxation time, with the second equality serving as definition of it. This is the basic algorithm used to couple the system of interest to a thermal bath. Using it, and following the prescription of the fluctuation dissipation theorems, the algorithm used in the GROMOS package obtains. The molecular dynamics equations where the system of interest is coupled with a thermal bath are:

$$\dot{\mathbf{P}}_s = -\partial V(\mathbf{R}_s)/\partial\mathbf{R}_s - \mathbf{P}_s(t) \cdot \gamma(t) \ (1-T_o/T) \qquad [9]$$

the matrix γ is usually taken as a diagonal scalar matrix: $\gamma\mathbf{1}$. In this case:

$$\gamma = 1/2t_r \qquad [10]$$

that is, all atoms are subjected to the same damping coefficient.

The molecular system's potential energy function, $V(R_s)$, describes the interactions between the atoms of interest. For a protein, or any biomacromolecule, the potential is composed of terms representing covalent bond stretching; bond angle bending; quadratic dihedral bending which considers of out-of-plane and out-of-tetrahedral configuration motion; sinusoidal dihedral torsion; interatomic repulsive forces and van der Waals attractive interactions, which make up for the Lennard-Jones potential used in normal liquid simulations, and Coulomb interactions.

8.2 INTERACTIVE DYNAMICAL COMPUTER GRAPHICS ANALYSIS

The vast amount of numerical data generated by macromolecular computer simulations – and their dynamical nature – require three-dimensional computer graphical tools for their efficient analysis. To meet these needs we developed the mdFRODO software (Nilsson, 1990). The guiding principle is that the interactive three-dimensional analysis of an ensemble of molecular structures is extended to involve the time-domain. A set of graphical operators have been devised for this purpose:

The standard animation technique has the analytical drawback that it instantaneously neither contains the history of the shown process nor the relation of any instantaneous conformation to any other conformation:

$$S(t) = F(n) \qquad\qquad ; \text{where } n = [1,2,3 \ldots m] \text{ and } n \Rightarrow t \qquad [11]$$

where $S(t)$ is the screen shown on a computer display at time t in an animation and $F(n)$ a frame containing the chosen molecular representation of the n:th conformation in the analytical set of m conformations. For a molecular dynamics simulation the order of conformations in the trajectory naturally infers the corresponding time.

The lacking history in the standard animation approach may be remedied by keeping previous frames displayed on the screen along the animation in such a way that:

$$S(t) = F(1) + F(2) + F(3) + \ldots + F(n) \qquad ; n \Rightarrow t \qquad\qquad [12]$$

and the missing relation problem can be overcome by including a statically displayed conformation – for instance an X-ray average structure – together with the animation of the dynamical process through construction of:

$$S(t) = F(n) + F(k) \qquad\qquad ; n \Rightarrow t \qquad\qquad [13]$$

where k denotes a fixed conformation in the analysis set. The most complete view of the modelled dynamical process' history is rendered by:

$$S(\text{all } t's) = F(1) + F(2) + F(3) + \ldots + F(m) \qquad\qquad [14]$$

but at the expense of dynamics. The full geometrical conformational space traversed by the system during a simulation may be examined.

If only a limited amount of history is of interest, or if conformational change is to be measured, it is convenient to select a small window of frames – three frames is normally an adequate choice – from the time-series for display in each screen along the analytical animation:

$$S(t) = F(n-1) + F(n) + F(n+1) \qquad ; n \Rightarrow t \qquad [15]$$

This animation mode provides a useful compromise between display of history and rendered dynamics.

Should border cases (conformations a and b) be of interest; e.g. the initial and the final conformations ($a=1$, $b=m$), or the two most deviating conformations ($a=c_{min}$, $b=c_{max}$), these may be included in the animation described in eqn. 15:

$$S(t) = F(n-1) + F(n) + F(n+1) + F(a) + F(b) \qquad ; n \Rightarrow t \qquad [16]$$

Thereby, it becomes possible to inspect the relation of a dynamical process with its border cases.

The bonded representation of the structural data is for historical reasons commonly used; all atoms that are covalently bound to each other in a molecule are connected with a line, and thereby the topology of the molecular object is shown. This information is of use for instance when X-ray crystallographers determine the structure of a molecule by building its topology into the deduced electron density map. However, in the analysis of molecular dynamics simulation data, the topology normally does not change. The display of the topology in each analytical frame is superfluous and it adds undesirable noise to the dynamical analysis. Instead, the attention should be focused on fluctuations in atom positions rather than changes in the positions of the (artificial) constraints between atoms. This can be achieved by construction of particle traces for the individual atoms by connecting the isoatomic positions in a set of, for instance three, consecutive conformations:

$$S(t) = \sum \left\{ [r_i(n-1) - r_i(n)] + [r_i(n) - r_i(n+1)] \right\}$$
$$; i=1 \dots N_{atoms} \text{ and } n \Rightarrow t \qquad [17]$$

where $r_i(n)$ is the position of the i:th atom of the n:th conformation in the set. As the analytical animation advances, the short particle traces display the time-evolution of the system's dynamics for the selected atoms. This may be the most powerful of the outlined techniques to render the dynamics of a system, it allows an overall view of correlations in the motion of different atoms, and appreciation of plasticity, elasticity and fluidity properties of a model.

The full history of the atomic positional time-evolution can be constructed by display of the complete particle traces:

$$S(\text{any } t) = \sum \left\{ [r_i(1) - r_i(2)] + [r_i(2) - r_i(3)] + \dots + [r_i(m-1) - r_i(m)] \right\}$$

$$; i=1 \dots N_{atoms} \qquad [18]$$

This last approach is a very efficient for spotting regions that present structural differences in an otherwise rather invariant system, such as structurally limited conformational changes, which might be loop transitions, or amino acid side-chain flips.

8.3 ENZYME REACTIONS

In terms of the transition state theory, enzyme catalysis can be understood by means of relatively simple ideas. One of them states that strong binding interactions are required by enzymes to reduce the energy barriers along the reaction path between the substrates and the products. This can be accomplished by stabilizing the activated complex of the reaction catalyzed by the enzyme; the active site will provide an environment complementary in structure to the relevant transition state; this principle was stated by Linus Pauling in 1948. Another way to achieve a similar result is via differential energy stabilization of reactants (R) and products (P) for potential rate limiting steps. Thus, if R and P have a large enthalpy difference for the reaction *in vacuo*, and entropy does not determine the free energy issue, the complexes formed with the enzyme, E-R and E-P say, will bring both states to a matching of internal energies thereby ensuring that the equilibrium probability of reaching the transition state from either side becomes equalized. Both mechanisms for changing the energy profile are complementary (Knowles, 1987). A corollary follows: substances mimicking the transition state structure of a particular enzymic reaction would tightly bind to the corresponding enzymes. As a matter of fact, the transition state (TS) analog concept is now firmly established in biochemistry, and it is a fruitful scheme for designing enzyme inhibitors. This very approach has been instrumental in tailoring "abzymes" or catalytic antibodies (Lerner & Tramontano, 1987).

In the final part of our talk, electronic aspects of the catalytic mechanism of horse liver alcohol dehydrogenase (LADH) were summarily discussed by using results obtained from *ab initio* analytical gradient self-consistent field molecular orbital (SCF MO) methods. The LADH protein catalyzes the oxidation of primary and secondary alcohols and their corresponding aldehydes or ketones with the reduction/oxidation of the nicotinamide adenine dinucleotide coenzyme, NAD^+ and NADH, respectively. The SCF MO results were discussed from the vantage point suggested by the general principles described above. The focus was placed on two global aspects: (1) the geometry and the electronic structure of the transition state for the hydride transfer reaction; (2) the factors affecting the energy gap for the hydride transfer step, namely: substrate binding to zinc, reaction field and effects of Ser-48 on the corresponding potential energy profile. For details, the reader is referred to the paper by Tapia *et al.* (Tapia et al., 1991). In Figure 8 the LADH active site and the saddle point structure are displayed; a short description of the system and mechanism is given in the caption.

The hydride transfer step was examined in view of obtaining geometrical and electronic information of the transition structure and the intrinsic geometry of the reactants *in vacuo*. With such information conjectures can be obtained concerning the role played by the shape of the active site. Two model systems were examined. The first is the pyridinium cation

(Py^+)/ 1,4-dihydropyridine (PyH) coupled to the complementary redox system methanolate (CH_3O^-)/formaldehyde (CH_2O). The second, cyclopropenyl cation (CP^+)/cyclopropene (CPH) coupled to CH_3O^-/CH_2O, is constructed to study the effects of particular residues found at the walls of the active site. The structure of the saddle point obtained in the former system was compared with experimental data concerning primary and secondary kinetic isotope effects. A normal mode analysis showed that the calculated transition structure is in excellent agreement with the one derived from isotope measurements.

The spatial orientation and intermolecular distance of the saddle point structure are those corresponding to the equivalent atoms in ternary complexes of LADH. This fact illustrates Pauling's 1948 statement. The study of the cyclopropenyl cation (CP^+)/cyclopropene (CPH) coupled to CH_3O^-/CH_2O was carried out with the aim of finding a smaller model where other features of the active site could be introduced at the quantum chemical level of description. The transition state obtained therefrom mimicks the one corresponding to the more realistic model. In this sense, this system can be considered a faithful model to study hydride transfer.

For the reaction *in vacuo* the hydride transfer step has a too large energy gap between R and P. It is expected that effects due to the surrounding protein will modulate the relative energy gaps. We considered two factors: (1) the role of a reaction field; and (2) the effects of Ser-48 on the hydride transfer potential energy profile (Ser-48 is located at the active site within hydrogen bond distance of the alcoholate oxygen). As expected, the effect of the model Ser-48 on the hydride transfer step was to decrease the energy gap between the reactant and product of this step reaction. Replacing this polar residue by a non-polar group should profoundly alter the energetics of the hydride transfer step and, consequently, one would expect a change of catalytic properties for such a mutated enzyme. If the enzyme were mutated at position 48 to a residue like alanine the theoretical prediction is that it would function better during alcohol oxidation than in aldehyde reduction.

The results concerning the reaction field effects show that the ion-pair structure is strongly stabilized with respect to the neutral aldehyde-PyH complex. The transition structure is also affected by this coupling. Its position on the energy scale suggests that it follows the trend expected from the Hammond principle.

The differential binding principle is based on free energy profiles (Knowles, 1987). The numerical examples presented here refer to differences in internal energy. The entropy factor has not been taken into account. This latter element may be of secondary importance as reactants and products are tightly bound at the active site of the enzyme during the hydride transfer process. Substrate binding to zinc, alcohol deprotonation, the Ser-48 effect and the general reaction field effects taken independently from each other all contribute in filling the energy gap found for the hydride transfer step reactants *in vacuo*. Negative cooperative effects, however, cannot be discarded. The conclusion from the present and related theoretical studies is that protein binding effects will work in the same sense as is required by conventional biochemical theories.

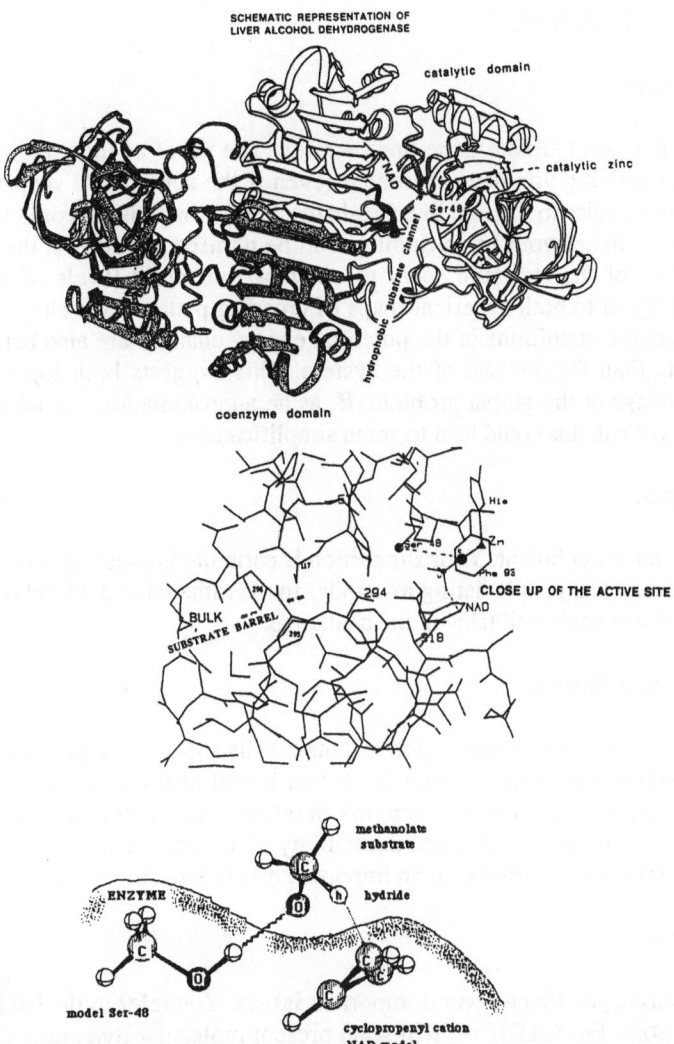

SCHEMATIC REPRESENTATION OF
LIVER ALCOHOL DEHYDROGENASE

catalytic domain

catalytic zinc

Ser48

coenzyme domain

CLOSE UP OF THE ACTIVE SITE

BULK
SUBSTRATE BARREL

methanolate
substrate

ENZYME

hydride

model Ser-48

cyclopropenyl cation
NAD model

Saddle point structure calculated with ab initio MO methods

Figure 8. The structure and chemical reaction mechanism of horse LADH. The enzyme is a dimer constructed from two identical subunits with a total molecular weight of 80000 Da. Each subunit (374 amino acid residues) has a coenzyme binding domain constituting the central core of the molecule and two catalytic domains; each subunit has two tightly bound zinc ions, one of them being essential for catalysis.

In the initial step the enzyme is in an open conformation. Coenzyme binding triggers a large conformational change which will be achieved after substrate binding. The conformational changes yields a closed structure. First, a loop containing residues 294 to 298 in the coenzyme binding domain are subject to a substantial local displacement. This motion relieves contacts with the catalytic domain producing two effects: (1) it widens the substrate barrel cross-section; (2) the catalytic domain may rotate about the molecular hinge once the substrate (inhibitor) binds.

NAD^+ binding is associated with a pK_a shift of the catalytic zinc-bound water leading to a hydroxyl group bound to zinc. Alcohol binds to the binary complex NAD^+-LADH, it exchanges a proton with the hydroxyl group and binds to the zinc as an alcoholate; the corresponding water molecule leaves the active site and the catalytic domain closes off; in fact, in the holo-structure there is no space for a water molecule at the active site. The hydride transfer step is discussed in the text.

9. FLOOR DISCUSSIONS

Prof. Paul G. Mezey:

Cervantes of the Don Quixote fame once wrote: "The proof of the pudding is in the eating". I like the pudding you have presented (even if the superposed snapshots of the protein backbone dynamics look more like spaghetti), but I wonder if we could have it with a different flavor. Some of your basic findings can be phrased as follows: the active site region and the rest of the structure seem to obey two different levels of topological invariance with respect to both spherical maps of crossing patterns and shape groups. In addition, the geometric conditions of the potential energy changes are also better defined for the active site than for the rest of the system. This suggests both topological and energetic partitionings of the global problem. If, as an approximation, we take these two partitionings as identical, this could lead to some simplifications.

Answer by O. Tapia:

Thank you for the compliment. Your suggestion is certainly interesting. The problem is to find bold and competent people daring to wander in the rather abstract field of topology. I'm sure readers of this book well take up the challenge.

Question by Prof. Juan Bertrán:

You have shown that the catalytic region is dynamically more stable than the rest of the protein in your MD calculations. Do you think that it will also be true if you take into account the coupling with the chemical system? In other words, will perturbations owing the substrate and chemical reaction change the stability of the catalytic region? On the other hand, will fluctuations in this region play an important role in inducing the reaction?

Answer by O. Tapia:

Dear Prof. Bertran, you touch several important issues. You refer to the bacteriophage T4 glutaredoxin work. For LADH we will soon present molecular dynamics simulations concerning the active site. Let me answer in order:
(1) The active site of T4-glx is an extremely stable entity. This, of course, does not mean that it is static. It is continuously fluctuating. If you construct a complex, say with glutathione, you must expect some adjustment of the atoms participating in the reaction. One can therefore expect well defined average structures. The stability of the main-chain tells you that no large conformational changes can be expected in that region. But, the fluctuations will always be there. (2) The answer to your last question is positive. They affect the rates. You can use the image of electron transfer theory in Marcus' sense.

Question by Prof. G. Naray-Szabó:

I will make a provocative statement which will look like a question at the end. We may formulate the electrostatic principle of enzymatic rate acceleration as follows: In enzymatic mechanisms known by us to date, the transition state(s) is (are) always strongly polar or even charged (proteases, metalloenzymes, dihydrofolate reductase, ribonuclease, etc.). Thus, the source of catalytic rate acceleration is the electrostatic stabilization of the transition state by the enzymatic environment – protein molecule(s) – Hydrogen bonding is considered as an electrostatic effect. Do you know any exception to this principle?

Answer by O. Tapia:

I think the answer is yes. As I mentioned in the appendix, the current understanding of enzyme catalysis is dominated by the hypothesis – proposed by Linus Pauling – about shape complementarity of the active site and the transition structure of the reaction catalyzed by a given enzyme. This factor is not directly related to the electrostatics, it is an intrinsic electronic factor. Our paper on the saddle point of the model system: pyridinium cation-methanolate illustrates this point. However, this is not the only source of catalytic activity. Your own work, and many others including the beautiful contributions by Arieh Warshel, clearly show that the electrostatic factor does play a central role.

Let me elaborate this point. If you mutate all residues participating in the catalytic triad of serine proteases, and in particular if serine is replaced by a residue that occupies the same volume, the enzyme mutamer will still possess some catalytic activity. As commented by Kraut (Kraut, 1988) when Ser-221, His-64 and Asp-32 in subtilisin are all replaced by alanines, the residual activity still produces a reaction rate 1000 times the uncatalyzed rate. In this case, it is evident that the binding energy has been used to deform the substrate into the conformation it should have in the real reaction.

Question by Prof. Arieh Warshel:

Experimentally, one finds no barrier for *gas phase* hydride transfer. How this is related to your *ab initio*? I would be happy to assume that the experiment reveals no barrier because of tunneling.

Answer by O. Tapia:

This is a complicated question. It is true that in the *gas phase* that hydride transfer reactions are extremely exothermic. Using the simple scheme provided by Hammond principle, one would expect an extremely low activation barrier, if any. In our case, we identify first the saddle point structure for a reaction involving a potential hydride transfer. This point is very high up on the energy hypersurface if we compare with the molecular associations the reactant partners acquire. In other words, for this reaction the probability to go about via out saddle point is extremely small. Since the reactants cannot approach each other following the direction of our saddle point, then you can expect tunneling to be the mechanism.

152

Color Plate 1. Van der Waals surfaces of bacteriophage T4 glutaredoxin seen from the canonical view (the active site face; cf. Figure 6). Carbon atoms are rendered grey, nitrogens blue and oxygens red. Hydrogen atoms are omitted. Some of the hydrophobic amino acid side-chains which may be important in ligand-binding have been given gold or white color. The red arrow indicates the accessible Sγ of Cys-14 in the active site. (**a**; top) The MD average structure <20-120ps> before the transition of loop L3 (see the text for details). (**b**; bottom) The MD average structure <200-300ps> after the modelled loop L3 transition. Note the largely altered topography and charge-distribution on the active-site face surface of the molecule.

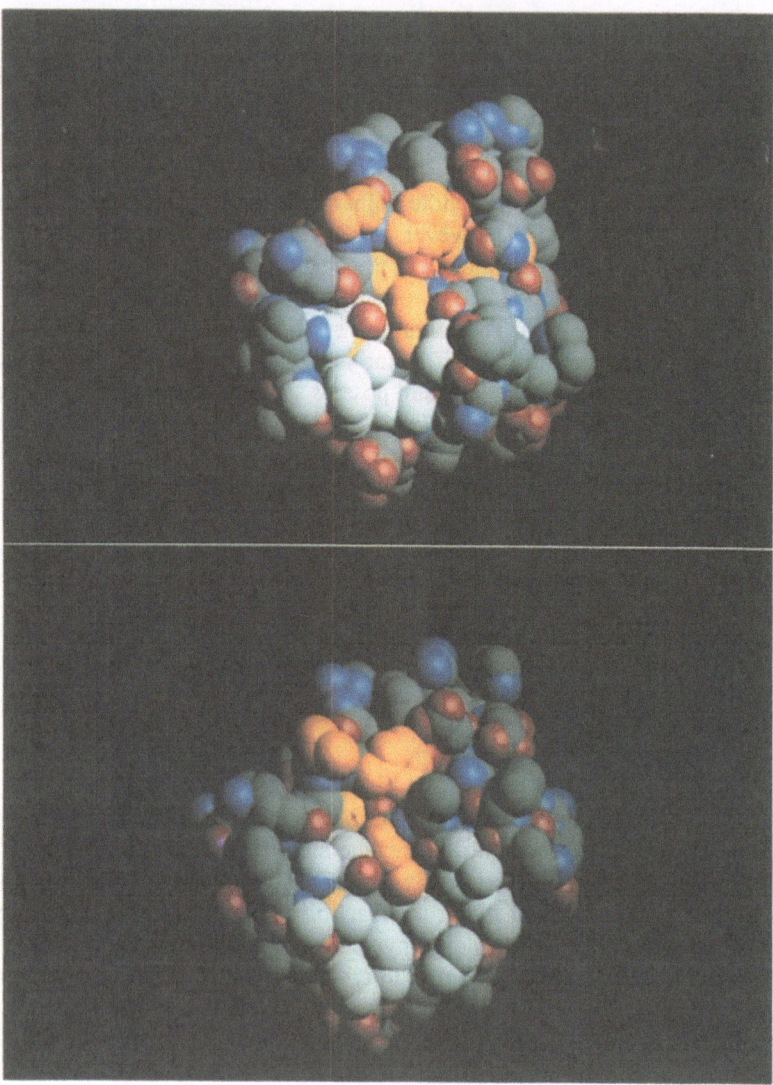

STRUCTURAL SPECIFICITY IN THE ENGINEERING OF BIOLOGICAL FUNCTION: INSIGHTS FROM THE DYNAMICS OF CALMODULIN

HAREL WEINSTEIN and ERNEST L. MEHLER
Department of Physiology and Biophysics, Box 1218
Mount Sinai School of Medicine of the City University of New York
New York, NY 10029-6745, USA

ABSTRACT. The goals of molecular biotechnology to modulate, modify, and mimic biological processes for a specific technological aim, rest on our current understanding of the structural specificity of the molecular mechanisms underlying biological function. We use the current insights we obtained from molecular dynamics simulations of Calmodulin (CAM) to illustrate how they can guide experimental engineering of the molecular structure. CAM is a modulatory protein that regulates cellular activity by binding to and activating a large number of macromolecules, and the key elements in the structure-function relations are the selectivity for calcium ions, and the mechanism of the calcium–dependent conformational changes.

1. Introduction

The rapid growth in understanding of biological systems and their mechanisms, combined with the technological capabilities of molecular biology, have engendered the promise of molecular engineering in biology. For this promise to be realized fully, and to be able to reach further along the goals of molecular biotechnology, this new field requires detailed structural information, at the atomic level of resolution, about an increasingly complex array of biomolecular systems. Such structural information, appropriately related to data on the functions of the biomolecules, is essential for the successful manipulation and mimicry of biological mechanisms through molecular biotechnology. Molecular inferences obtained from the techniques of molecular biology form a necessary solid basis for structural research, but do not provide in themselves sufficient structural detail for reliable mechanistic descriptions at the resolution required for molecular engineering to proceed. Unfortunately, structural information about biological systems continues to be a limiting factor in the development of molecular biotechnology because direct sources, such as x-ray crystallography and multidimensional NMR, remain costly in time and resources. For these reasons, computational simulations of molecular mechanisms, which rest on data and inferences from both molecular biology and structural biology, have grown in importance. As a third form of scientific inquiry, complementing theory and experimentation, computational simulations have become a major factor in modern research. When applied to the problems of biological structure and function, these computational simulations have become a pervasive approach in molecular biotechnology because they complement and help

153

J. Bertrán (ed.), Molecular Aspects of Biotechnology: Computational Models and Theories, 153–173.
© 1992 *Kluwer Academic Publishers.*

define and refine essential structural information. More importantly, these computational approaches also provide a clear way of interpreting the structural information in its relation to biological function [1-6]. It was therefore of considerable interest to review in the context of this Workshop some current insights obtained from such computational simulations, and to illustrate how they can guide experimental engineering of biomolecular structure and function. For this purpose, some recent studies of calcium-binding proteins are reviewed here, with emphasis on our own contributions to this area of research.

2. Calcium Binding and Biomolecular Function

The engineering of molecular properties of calcium-binding proteins is of enormous practical interest in view of the essential and multifaceted functions that this large family of proteins performs in biological systems. From regulation of gene expression to conduction of nerve impulses, and from muscle contraction to blood coagulation, physiological processes in animals and humans depend on the functions of calcium-binding proteins [7-9]. Many of these physiological processes are beginning to be understood at the cellular and biochemical level, so that direct manipulation and interference in the underlying molecular mechanisms becomes feasible. An example of the application of bioengineering approaches to calcium-binding systems has been demonstrated recently for the relatively simple, and yet very important cases of enzyme systems that are modulated by calcium binding. Thus, specific structural modifications of the subtilisin BPN' molecule were designed to affect Ca^{2+}-binding, and were implemented successfully to yield the expected result of increased thermal stability of the protein [10, 11]. More subtle manipulations of enzyme function through the modification of calcium binding properties were suggested for enzymes that incorporate both a structural and a modulatory calcium binding site, such as proteinase K [12] as well as subtilisin BPN' [10]. In particular, the results of an analysis of the sensitivity to Ca^{2+} binding in proteinase K revealed a mechanism of structural rearrangement reminiscent of the process proposed for the modulatory calcium-binding proteins such as calmodulin that is discussed further in a section below. Thus, comparisons of the crystal structures of proteinase K with and without Ca^{2+} at the modulatory "tight binding site" have identified a "concerted domino-like movement" caused by removal of Ca^{2+} from the site that affects enzymatic activity at a distally situated active site [13]. Such specific information provides a challenging prospect for subtle modulation of enzymatic activity based on the effects of calcium binding to specific sites.

These insights emphasize the essential role of structural information in molecular engineering efforts that would make possible the design of changes, improvements, and modifications of the calcium-binding properties of the proteins, in order to modulate their normal functions, or to endow them with new ones. Such information is available for just a few of the proteins involved in the binding of Ca^{2+}, or activated by the binding of calcium ions. Consequently, it becomes all the more important to identify the structural commonalities among them, and their relation to the functions performed by these proteins. The practical significance of such an understanding was illustrated recently in the report of an artificial calcium binding site created in human lysozyme [14] according to the template offered by the structural motif of a Ca^{2+}-binding "EF-hand" (see description below). The incorporation of such a folding motif that is common to many calcium-binding proteins [15] in the structure of the lysozyme molecule resulted in a protein with the expected enhanced structural stability caused by the binding of Ca^{2+} [14]. Similarly, the grafting of a calcium-binding loop from the known structure of the thermolysin Ca^{2+}-binding site onto the neutral protease of *Bacillus subtilis* [16] by molecular engineering, achieved the desired enhancement in the stability of the protein structure.

All the attempts to understand and mimic the effects of calcium binding described above were based on the structures and properties of protein folding motifs that are known to bind Ca^{2+} selectively and with high affinity. These insights prompted us to design a program of study, that is ongoing, with the aim of elucidating structure-function relations in calcium-binding proteins for which the structural details were available at atomic resolution. With computational simulations of molecular properties and dynamics, we were able to investigate both the structural and the dynamic elements of the relation between the structures and the functions of a number of calcium-binding proteins that are active both in the storage of Ca^{2+} and in calcium-dependent modulation of cellular functions.

3. Structure-Function Relations in Calcium-Binding Proteins

3.1 ON THE STRUCTURAL REQUIREMENTS FOR CA2+ BINDING

Various proteins recognize Ca^{2+} ions, bind them with considerable affinity, and react to the binding with structural rearrangements that prepare the proteins for participation in specific biological processes. A first goal of our inquiry was, therefore, to identify the role of secondary and tertiary structure of such proteins in their ability to bind Ca^{2+}, in order to gain an understanding of the basis for the selectivity of proteins for Ca^{2+}, and in order to account for the special role of Ca^{2+}-binding in the biological functions of these proteins. The group of proteins addressed in this study includes *troponin C* (TNC), which binds calcium to initiate the intracellular event in muscle contraction, *calmodulin* (CAM), a ubiquitous cellular protein known to activate a very large number of different target proteins when it is itself activated by binding calcium, and the vitamin D-dependent calcium binding protein *calbindin-D9k* (CABD). Their structures have been elucidated by x-ray crystallography [17-19]. These calcium-binding proteins have in common a highly specialized structural motif, known as the EF hand, where the calcium binding occurs [20]. In its usual form, the EF hand consists of a loop of 12 residues flanked by two helices positioned at roughly right angles to one another [21]. Calmodulin and troponin C each contain four such calcium-binding regions, while calbindin-D9k has two of them arranged symmetrically and connected by a short linker segment. The calcium-containing EF hands in the structures studied with x-ray crystallography were shown to have a very high degree of structural similarity, leading to near identity of the super-secondary structural motifs (composed of the helix-loop-helix group) that bind Ca^{2+} [22]. The amino acid sequences of calcium-binding loops in many different EF hands is highly conserved, leading to a consistent coordination to the calcium.

The loop sequences are rich in carboxylate side chains and the ligands of the Ca^{2+} ion form an approximately octahedral binding site [21]. The importance of identifying the ability to bind Ca^{2+} for the characterization of protein function is evident from the numerous proposals in the literature of schemes for inferring calcium-binding functionality in proteins based on such primary properties as the homology of amino acid sequences that have been shown to bind Ca^{2+}. One such example, in which calcium-binding was proposed as a biological mechanism based only on the similarity in amino acid sequence to known Ca^{2+}–binding loop sequences [23] prompted us to evaluate the role of the secondary and tertiary structure of the protein surrounding the binding loop in determining the structural parameters for calcium binding [24, 25]. Based on a comparative analysis with the structural properties of the loops in CABD, the results of our dynamics simulations showed that the conformational space available to the calcium-binding loops can be severely limited by the structure of the rest of the protein, so as to impose the conformation that is optimal for the binding of the Ca^{2+} ion [25]. If this finding is general, it limits the use of sequence similarity as the sole criterion for calcium-binding ability, and explains some of the constraints related to the rest of the protein structure. A practical conclusion is that sequence comparisons can serve as guides for the construction of Ca^{2+}-binding peptides, or for the identification of Ca^{2+} binding properties, only if they are complemented with information regarding the appropriate structural constraints imposed by the tertiary structure of the protein in which they are imbedded. Specific details about the constraints imposed by the tertiary structure, and the response of the protein structure to the binding of Ca^{2+}, obtained from our work on the structural rearrangements related to the binding of calcium in CAM and TNC, which contain several EF-hands, reinforce this conclusion. This work is summarized in section 4, below.

3.2   ON THE STRUCTURAL BASIS OF CA^{2+} SELECTIVITY

Based on the evolutionary conservation of the amino acid sequences in calcium binding loops [15], sequence similarity has also been invoked to explain the selectivity of certain proteins for calcium. The mechanistic explanations for selectivity have been based on the electronic properties of the ion expressed in its ionic radius and its charge, both of which determine the interaction with the binding loop (for a recent discussion see [26]). But such considerations are not sufficient if they ignore the properties required of the overall structure that surrounds the ion and shields it from the aqueous solvent. Our studies [27] showed that understanding the selectivity for Ca^{2+} over other ions, such as Mg^{2+} or Na^+ which are often present in much higher concentrations in the biological environment, requires an evaluation of the entire energetic process of extraction of the ions from the aqueous solvent into the protein binding site. As proposed earlier, such an evaluation for the calcium-binding systems requires a computational approach combining the methods of quantum chemistry, molecular mechanics, and molecular dynamics [25, 28]. The first model for selectivity in ion binding we studied was the binding of Ca^{2+} and Mg^{2+} to the small synthetic peptide cyclo-(-L-Pro-Gly)$_3$, termed PG3 [29]. Experimental observations concerning the binding of the divalent ions to PG3 showed that while Mg^{2+} is octahedrally coordinated to PG3 in a 1:1 complex, the Ca^{2+} cation is sandwiched between two PG3 molecules in a 1:2 complex [29]. The comparison of the binding energies of Ca^{2+} and Mg^{2+} to PG3 in their 1:1 as well as in the 1:2 complexes was carried out by means of the statistical mechanical simulation procedure called the adiabatic charging process, implemented as described in detail [27]. Application of this method to the complexes of PG3 with the calcium and magnesium ions yielded the differences in the free energies of solvation of the two ions in water and in the PG3 complexes, showing that the selectivity for calcium stems from the larger difference between the solvation energies of Ca^{2+} and Mg^{2+} in aqueous solutions than in proteins. Thus, Mg^{2+} was found to be better stabilized than Ca^{2+} by both the 1:1 and the 1:2 complexes with PG3, but its excess stabilization compared to Ca^{2+} is even greater in water [27]. Consequently, we learned that Mg^{2+} does not form a 1:2 complex because in such a complex the ion is sequestered from solvent and the interaction with the protein is not sufficiently strong to compensate for the loss of solvation energy. In contrast, Ca^{2+} does form such a sequestered complex because its solvation energy is lower than that of Mg^{2+}, and is compensated by the interaction with the protein. These quantitative results thus provide a direct explanation for the different stoichiometries in the binding of the two ions to PG3, and specific hints about the likely structural differences between high and low affinity, as well as high and low selectivity binding sites for calcium and magnesium in proteins with or without EF hand structures. These indications are very important for future studies of calcium binding systems because our results show that the degree of solvent-accessibility of the Ca^{2+} ion bound in a specialized loop is clearly determined by the nature of the ligands and the conformation of the binding loop, but the specific role of the helix–loop–helix structure of the EF–hand, and the advantage of the occurrence of such Ca–binding EF-hands in pairs, are still topics of intense investigation (e.g. see [30]).

4. Functional Consequences of Structural Changes Induced by Calcium Binding

4.1 THE COMPACTION OF CALMODULIN IN SOLUTION

CAM contains two pairs of EF-hands, each of which forms nearly globular domains at the N–terminal and the C–terminal ends of the molecule [18]. The two domains are connected by a tether helix that continues the C–terminal helix of the second EF–hand in the N-terminal lobe, and passes directly into the N–terminal helix of the third EF-hand in the C-terminal lobe. In this way, the two similar globular Ca–binding domains (lobes) of the CAM molecule are separated by a long helical stretch as shown in Figure 1 for the structure of CAM in the crystal.

Recent experimental results have suggested that the Ca-bound (but not Mg-bound) structures of CAM in solution may be more compacted than those observed in the crystal, due to a kinking or bending of the helix linking the two domains [31, 32]. We carried out molecular dynamics simulations with the CHARMM program package [33] to investigate whether such global structural alterations due to changed environmental conditions can be accounted for by observations from computational approaches. The results were used to determine the structural consequences of a compaction of CAM through a bending of the central helix tether connecting the two calcium-binding domains (see Fig. 1), and the functional implications of such a structural rearrangement. All the calculations were started from the available crystal structure. Different solvent models were tested for the simulations, including a distance dependent dielectric permittivity, ε, the inclusion of the crystallographically identified waters, and the immersion of the molecule in a water bath simulated by a large number of water molecules and periodic boundary conditions.

The first simulations were carried out with a simple, but apparently quite reliable model which includes crystallographic waters and $\varepsilon = r$ [34]. These simulations resulted in a compaction of the structure of CAM relative to the structure observed in the crystal [34]. The compaction resulted from a bending of the central helix, with preservation of the secondary and tertiary architecture in the Ca^{2+}-binding lobes, as indicated in Figure 1. The calculations leading to these conclusions [34] were carried out with the default parameters in CHARMM, i.e., a nonbonded interaction cutoff of 7.5Å used with the CHARMM shift function. The parametrization of Ca^{2+} was taken from Hori et al. [35]. Preliminary energy minimization was done to relax the coordinates of crystallographic water molecules as well as residues 1-4 and 148 of CAM which had to be placed by optimization because no crystallographic coordinates had been reported for them. Prior to any dynamics runs, full energy optimizations were carried out for all the structures to relax them in the model environments used to account for solvent effects as described below. Each structure was first heated in 5 degree steps over a period of 6ps to 300K, equilibrated (20-80ps), and the dynamics simulation was continued until the structure had stabilized or was rejected according to specific criteria for structure selection [34]. A structure was considered to be equilibrated if the atomic RMS deviations of the average coordinates from a 40ps trajectory showed less than ca. 0.8Å differences for most atoms. The structures stabilized after about 100-150 ps of dynamics. Structural RMS fluctuations, as well as those of the energy were monitored, and found to remain stable over simulation time periods of 300ps. The average structures were optimized to eliminate obvious stereochemical artifacts such as flips of carboxylate oxygens. No external constraints were used in the calculations.

Figure 1 clearly shows the somewhat **S** shaped result of the compacted structure of CAM. This structure is in good agreement with experimental observations of complexes between CAM and peptides such as melittin [36], mastoparan [37], and components of cellular proteins modulated by CAM [38, 39] for which some structural information enables a direct comparison. As shown in Figure 2, this structure presents to the environment two pockets containing several nonpolar residues, and an increased accessibility of some of the hydrophobic residues known from experiment [40-42] to be involved in the complexation of CAM with cellular proteins and inhibitors [37]. Initial modeling studies of interaction of the neuroleptic drug trifluoperazine with CAM suggest that these hydrophobic patches could also be involved in the binding of CAM inhibitors [43].

The exploration of the direct effects of solvent on the structure of CAM is continuing with our current studies of the molecular dynamics of CAM in a water bath, as depicted in Figure 3. In these simulations, CAM is surrounded by a shell of discrete water molecules reaching out to a radius of 4Å from the outermost atoms of the protein. The water shell comprising 923 molecules was built with the SOAK routine of the Insight II software package. The starting CAM structure was the same as described before [34], with all the crystallographic waters deleted before the SOAK routine was applied. The runs were carried out both with a dielectric constant of unity, $\varepsilon=1$, and with a distance dependent dielectric, $\varepsilon=r$, to account for the bulk effects of the solvent. The energy minimizations preceding the dynamics runs were done in two stages. First the water positions were minimized with the protein constrained, and then the constrains were relaxed and the entire system was optimized with standard CHARMM options. The two-stage procedure was also followed in the initiation of the dynamics runs: The water shell was first heated to 300K (in 6ps) and then equilibrated for 34ps while the protein remained constrained at the optimized structure. Relaxation of the constraints and heating to 300K (in 6ps) was undertaken in the second step only after minimization of the structure resulting from the heating of the water bath in the first 6-40ps. The structure shown in Fig. 3 is a snapshot at 129ps from the production run trajectory computed under these conditions. The radius of gyration calculated for this structure is $R_g= 20.2$Å, compared to the initial value of 22Å, indicating the inception of the compaction. The nature of this incipient compaction is evident in the comparison shown in Figure 4 between the starting (crystallographic) structure and the snapshot at 129ps.

4.2 CA²⁺ BINDING AND THE INTERDOMAIN ELECTROSTATIC INTERACTIONS IN CALMODULIN

The compaction of CAM resulting from the bend (or kink) developing in the central helix brings the two domains much closer together, as is seen in the reduction of the center of mass separation by 13Å, and increases their interaction. This appears surprising because at physiological pH both domains carry a large net negative charge which might be expected to contribute unfavorably to the total interaction energy, preventing the two domains from coming closer together. A number of factors may be operative in compensating the apparent interdomain electrostatic repulsion. These include the screening effect of the polar solvent and the Debye screening from ions. But the net interaction will depend crucially on the nature of the charge distribution in the two domains.

To gain further insight into this question we have applied a previously proposed model [44, 45] to calculate the electrostatic interactions between the domains. In this model the potential at a point r_i is calculated from the formula

$$\Phi(r_i) = \sum_{j \neq i} q_j / \varepsilon(r_{ij}) r_{ij}$$

and the electrostatic contribution to the free energy is given by

$$\Delta w = \sum_i q_i \bullet \Phi(r_i)$$

where q_i is the net charge at point r_i, ϕ is the potential at that point, and ε is a distance dependent dielectric permittivity. The form of ε and its parametrization have been given elsewhere [44].

The analysis is carried out by selecting a reference point in each domain, calculating the potential due to the charge distribution of the other domain at that point, and computing an effective electrostatic energy by assigning the net charge of the domain to the reference point. In the calculations reported here the center of mass (CM) of each domain has been used as the reference point. Although this selection is somewhat arbitrary, results in Table 1 show that for both the x-ray structure and the final compacted structure obtained from the simulation, the CM's and the centers of positive and negative charge in each domain are nearly coincident. This result indicates that both the positive and the negative charges are fairly evenly distributed throughout the domains, making the CM a reasonable choice for the reference point. For each CM the proximity of the residues of the other domain is different in the two structures so that some care has to be taken in selecting the portion of the protein defining the domain external to the domain of the CM. Here we simply exclude that part of the other protein domain which lies within a 15Å sphere around the CM.

The electrostatic potential and interaction energy have been calculated at each CM for the x-ray and simulation structures with and without the Ca²⁺ present. The calculations were done at an ionic strength of zero (the condition at which the molecular dynamics simulation was carried out), and at physiological ionic strength (I = 0.15). The results are given in Table 2.

At zero ionic strength removal of the Ca²⁺ leads to a reduction in the potential at the CM by 0.3-0.6 Kcal/mole. Although apparently small, these changes determine the differences in free energy contributions so that even small shifts can have a

TABLE 1: Centers of mass (CM) and charge (QM) in calmodulin

X-ray structure (optimized)		x	y	z
N-Domain	CM	-9.278	13.481	6.937
$(1\text{-}76 +2Ca^{2+})$	Q-M	-9.502	13.866	6.765
	Q+M	-9.318	14.004	6.685
C-Domain	CM	10.686	-14.715	-8.012
$(84\text{-}198 +2Ca^{2+})$	Q-M	11.180	-14.677	-8.347
	Q+M	11.426	-14.758	-7.760

Simulated structure (compacted)				
N-Domain	CM	-2.062	11.707	0.801
	Q-M	-2.406	11.431	1.202
	Q+M	-2.479	11.349	1.132
C-Domain	CM	1.805	-12.513	-1.455
	Q-M	1.785	-12.739	-2.003
	Q+M	2.195	-13.042	-1.817

TABLE 2: Interdomain electrostatic interaction in calmodulin

Interacting system	I	Q_{Ca}	Q_{CM}	X-ray (optimized)		Dynamics (compacted)	
				Φ	Δw	Φ	Δw
CM_N - C-Domain	0.0	1.626	-6.748	-2.516	16.98	-2.378	16.04
	0.15	1.626	-6.748	-0.159	1.07	-0.192	1.30
	0.0	0	-10.0	-2.883	28.80	-2.870	28.70
	0.15	0	-10.0	-0.162	1.62	-0.206	2.06
CM_C - N-Domain	0.0	1.626	-6.748	-1.895	12.79	-1.194	8.06
	0.15	1.626	-6.748	-0.141	0.95	-0.058	0.39
	0.	0	-10.0	-2.206	22.06	-1.691	16.91
	0.15	0	-10.0	-0.142	1.42	-0.073	0.73

substantial influence on the overall equilibrium. Removal of the Ca^{2+} has such a small effect because the ions are well over 20Å from the CM, and at such separations the full screening of the solvent is in effect. For example, at a distance of 20Å, the contribution to the potential of an ion with charge 1.6 is only 0.35 Kcal/mole. In the Ca-free system the net charge decreases to -10 from -6.75 in the Ca-loaded molecule. Thus, in spite of the small change in potential the Ca^{2+} contributes substantially to the stabilization of the systems. It is also apparent that at an ionic strength of 0.15 the effects are reduced by at least one order of magnitude. At ionic strengths between 0 and 0.15 the damping will be smaller but nonlinear, since the functional form of the ionic screening is exponential (see [46]).

The change in electrostatic potential at the CM between the crystallographic and the compacted structure generally favors the latter. In assessing these changes it is noted that the simulations were carried out at zero ionic strength with the fully Ca^{2+} loaded protein. Therefore the changes in potential and in energy obtained from the dynamics structure for the Ca-free system probably has only limited relevance since it is unlikely that the Ca-free system would compact. We therefore confine our remarks to the results obtained from the structure with conditions which correspond to those actually used in the molecular dynamics simulations.

Although both the CM_N- C-domain and CM_C-N-domain interactions change to stabilize the compacted structure, the effect is substantially more pronounced in the latter where the shift in Δw is -4.73 Kcal/mole as compared with only -0.94 Kcal/mole for the former. In this connection it is interesting to note that the RMS difference between the x-ray and simulation structure was smaller in the C-domain than in the N-domain [34]. Particularly the EF hands of the C-domain were very little changed. Thus it seems that for the particular simulation being considered here, the electrostatic interactions between the two domains is more effective in stabilizing the C-terminus domain than the N-terminus domain.

In spite of the fact that the two domains move closer together, the potentials at the CM increase, thereby decreasing the repulsive interaction at the CM. Examination of the structural changes occurring on compaction shows that the relative orientation of the two domains changes (see analysis in [43]). This reorientation results in differential changes in the distances of the charged groups to the CM. For example, some of the acidic residues in segment 1-10 move further away from CM_C in the compacted structure than in the x-ray structure, thereby decreasing their contribution to the potential. Thus it becomes clear that changes in the detailed structural arrangements of the charged groups is the ultimate origin of the calculated changes in the electrostatic interactions.

4.3 THE MECHANISM OF STRUCTURAL REARRANGEMENT IN CA^{2+}-BOUND CALMODULIN

Key elements in the Ca^{2+}–dependent mechanism of interaction between CAM and the proteins it activates are the selectivity for Ca^{2+} ions, discussed above, and the requirement for Ca^{2+}–dependent conformational changes (for a recent review see [47]). We reported recently on results from a detailed study based on computational simulations that identified discrete steps in the mechanism of structural rearrangement of Ca^{2+}-bound CAM [43, 48] that produces the compacted structure associated with the modulatory function of the protein. Using molecular dynamics simulations, we had explored the mechanism that can drive the molecule from the extended structure found in the crystal to the more compacted one described above. A series of steps identified in the process of structural rearrangement of CAM implicate the side chain of Arg74, and partially also those of Arg86 and Arg90, in the bending of the central *alpha* helix that tethers the two Ca^{2+}-binding lobes. This series of steps involves the formation of successive combinations of hydrogen bonds between the Arg74 side chain and proton acceptor sites which include the protein backbone, neighboring side chains, and water molecules. The structural and energetic considerations based on the results of the simulation [43] point to this dynamic hydrogen bonding pattern around the arginine residue as a ratcheting-type mechanism causing the kinking of the central helix. The structural rearrangements and the cognate energetics in the mechanism of compaction were analyzed to identify the role of the Ca^{2+} ions in this process, and the manner in which the target protein could modulate this process upon binding to CAM; the functional significance of the compaction was also explored from the results of simulations of the binding of trifluoperazine, a known "CAM-inhibitor" [43]. As evidenced by the details of this study presented briefly below, the practical importance of the resulting mechanistic insights is that they lead directly to propositions for experimental engineering of the molecular structure of CAM that can serve to probe the hypotheses and their consequences for the function of this important protein.

The computational simulations were carried out as described for the studies presented above, with the standard CHARMM parametrization and the Hori et al. parameters for Ca^{2+}. A standard factor (E14FACT) of 0.4 was used to reduce the interaction between atoms with relative position 1-4, and the standard shift function was used with a cutoff of 7.5 for both electrostatic and van der Waals interactions. The non-bonded list was built with a cutoff of 8.0, and was updated every 0.01ps. The extended atom approximation was used for aliphatic and aromatic carbons. Bonds involving hydrogens were restrained to the distance at the starting point by using the SHAKE algorithm. The integration made use of the Verlet algorithm with a step of 0.001ps. The coordinates were saved every 0.025 ps. The procedure for the dynamics started with a heating period of 6 ps, incrementing the temperature by 5° every 0.1ps up to 300°K. The temperature was increased with the velocities assigned as a gaussian distribution. During the subsequent equilibration process, the velocities were scaled as needed from determinations at 0.1 ps intervals, to maintain the temperature at 300°±10°K. The equilibration was examined by calculating the average structure every 20 ps, calculating the RMS fluctuation of the backbone as well as the RMS deviation of the *alpha*-carbons with respect to the starting point, and by monitoring the values of the radius of gyration and maximal distance in the molecule, as described before [34, 43]. The final structures used in the various analyses described below are the minimized average structures from the last 40 ps of each trajectory.

Several different starting points were used as described in detail [43], in an attempt to overcome some of the shortcomings of insufficiently long runs for a molecule as large as CAM. In the resulting structures, the variation of the radius of gyration (Rg) indicates that only two of the thirteen trajectories from different starting points failed to undergo compaction during the simulation. The compaction that was observed for all the other trajectories occurred at different times, but the compacted structures exhibited close structural relationships (cf. Fig. 2 in [43]).

The analysis of the process of compaction focused on the changes in local interactions and in the hydrogen bonding patterns that accompanied the reduction in the radius of gyration. Similar changes appeared in all the runs leading to a compacted structure. The first changes in tertiary structure related to the process of compaction appeared to be the changes in the relative orientation of the two Ca^{2+} binding domains before they start to move closer together. The calcium-binding lobes were found to change their orientation from a *trans*-like conformation to a *cis*-like arrangement, with compaction starting after this rearrangement was completed [43].

The most striking and consistent result of the analysis of the hydrogen bonding pattern in the beginning stages of the compaction process was the formation of a new hydrogen bond between Arg74 and the carbonyl oxygen of Val55. As a consequence, Arg74 becomes anchored by hydrogen bonds to a residue at a distal site, while maintaining the hydrogen bonding to another residue, in its immediate vicinity (Thr70, or Glu67) during the entire process. This consistent binding structure was observed [43] to produce a significant change in the tertiary structure of CAM during the simulation trajectory. It is noteworthy that even in the simulations performed in the absence of bulk solvent, with only the crystallographic waters present, discrete water molecules were observed around the Arg side chains consistently forming hydrogen-bonds throughout the simulation. Such a consistent pattern is recognizable around Arg74 at different times during the trajectory of a specific run identified as CAM10 (see [43]). Comparative analyses of the hydrogen bonding patterns in all the trajectories started from the various initial conditions indicate that for Arg74 to trigger the compaction of the protein, it must form *both* a proximal anchoring hydrogen bond, and a latching one with a distal site. The mechanism of initiation of the compaction was found to be the same for all the runs, with a complete correlation between the final compaction and the hydrogen bonding pattern corresponding to the proposed ratcheting mechanism for Arg74. The hydrogen bond (Arg - carbonyl O) that most frequently forms part of the anchor, may be essential for the distortion of the backbone of the tether that produces a kink in the area of the bond.

The detailed analysis of the mechanism by which hydrogen bonding of the Arg residues can produce a consistent bending of the tether has thus identified a sequential hydrogen bonding pattern that may constitute a reversible series of intermediate steps that determine the structural rearrangement. Since the hydrogen bonding pattern is easily reversible in aqueous solution, the process of compaction is likely to exhibit the dynamic equilibrium with the opposite process of erection of the helix to the conformation observed in the crystal, as discussed earlier [34]. This dynamic equilibrium is consonant with observations from experiments on CAM and TNC, and from the computational simulations.

5. Concluding Remarks

General considerations for molecular design and bioengineering of calcium-binding proteins emerged from these studies based on molecular simulation. For calmodulin, the analysis of the local structural changes related to the kinking of the tether helix have identified specific residues involved in the process of compaction that is intimately associated with the biological functions of this versatile protein. The results identified the special role of the arginine residues, and in particular Arg74, in the proposed ratcheting mechanism that determines the compaction of CAM. This mechanism is clearly attributable to the multiple hydrogen bonding potential of the Arg side chains. Their role as key elements in the compaction of calmodulin can be probed directly by specific mutations of the molecular sequence. In collaborative studies with the laboratory of Jurgen Brosius we are currently using the cDNA of rat brain calmodulin [49] in mutation experiments designed on the basis of these insights obtained from the simulations. The aim is to test the inferences, and to design new modulatory modalities and putative therapeutic agents. The use of the special attributes of arginine side chains to perform a specialized function is evidently more widespread, and has also been proposed recently in a case of protein-RNA interaction in which an "arginine fork", reminiscent of the network observed in the present simulations, was implicated in RNA recognition [50]. The approaches illustrated here for calmodulin should undoubtedly be directly applicable to this system, and to a large variety of bioengineering problems of great practical and theoretical significance.

6. Acknowledgements

The work was supported in part by NIH grant GM-41373, a Research Scientist Award DA-00060 from the National Institute on Drug Abuse (to HW), and a Swiss National Science Foundation Grant 31-8840.86 (to ELM). Computations were performed on the supercomputer systems at the Pittsburgh Supercomputer Center (sponsored by the National Science Foundation, and the Cornell National Supercomputer Facility (sponsored by the National Science Foundation and IBM) - as well as at the Advanced Scientific Computing Laboratory at the Frederick Cancer Research Facility of the National Cancer Institute (Laboratory for Mathematical Biology) and at the University Computer Center of the City University of New York.

7. References

1. McCammon, J.A. and Harvey, S.C. (1987) 'Dynamics of Proteins and Nucleic Acids', Cambridge University Press, New York.

2. McCammon, J.A. (1987) 'Computer-aided molecular design', Science 238, 486-491.

3. Brooks, C.L., Karplus, M. and Pettitt, B.M. (1988) 'Proteins: A Theoretical Perspective of Dynamics, Structure, and their Thermodynamics', John Wiley & Sons, New York.

4. Weinstein, H., (ed.) (1986) 'Computational Approaches to Enzyme Structure and Function' Enzyme, Vol. 36, Karger, Basel.

5. Jonsson, B., (ed.) (1989) 'Structure and dynamics in biological systems' Nobel Symposia, Cambridge University Press, Cambridge, New York, pp. 221.

6. Karplus, M. and Petsko, G.A. (1990) 'Molecular dynamics simulations in biology', Nature 347, 631-639.

7. Evered, D. and Whelan, J., (ed.) (1986) 'Calcium and the cell' John Wiley & Sons, New York.

8. Gerday, C., Bolis, L. and Gilles, R., (ed.) (1988) 'Calcium and calcium binding proteins' Springer-Verlag, Berlin, pp. 259.

9. Heizmann, C.W., (ed.) (1991) 'Novel Calcium-Binding Proteins' Springer-Verlag, New York, pp. 624.

10. Pantoliano, M.W., Whitlow, M., Wood, J.F., Rollence, M.L., Finzel, B.C., Gilliland, G.L., Poulos, T.L. and Bryan, P.N. (1988) 'The engineering of binding affinity at metal ion binding sites for the stabilization of proteins: Subtilisin as a test case.', Biochemistry 27, 8311-8317.

11. Pantoliano, M.W., Whitlow, M., Wood, J.F., Dodd, S.W., Hardman, K.D., Rollence, M.L. and Bryan, P.N. (1989) 'Large increases in general stability for subtilisin BPN' trough incremental changes in the free energy of unfolding', Biochemistry 28, 7205-7213.

12. Betzel, C., Pal, G.P. and Saenger, W. (1988) 'Three-dimensional structure of proteinase K at 0.15-nm resolution', Eur. J. Biochem. 178, 155-171.

13. Bajorath, J., Raghunathan, S., Hinrichs, W. and Saenger, W. (1989) 'Long-range structural changes in proteinase K triggered by calcium ion removal', Nature 337, 481-484.

14. Kuroki, R., Taniyama, Y., Seko, C., Nakamura, H., Kikuchi, M. and Ikehara, M. (1989) 'Design and creation of a Ca2+ binding site in human lysozyme to enhance structural stability', Proc. Natl. Acad. Sci. USA 86, 6903-6907.

15. Heizmann, C.W. and Hunziker, W. (1991) 'Intracellular calcium-binding proteins: more sites than insights', Trends in Biochem. Sci. 16, 98-103.

16. Toma, S., Campagnoli, S., Margarit, I., Gianna, R., Grandi, G., Bolognesi, M., DeFilippis, V. and Fontana, A. (1991) 'Grafting of a calcium-binding loop of thermolysin to Baccilus subtilis neutral protease', Biochemistry 30, 97-106.

17. Herzberg, O. and James, M.N.G. (1988) 'Refined crystal structure of Troponin C from turkey skeletal muscle at 2.0A resolution', J.Mol.Biol. 203, 761-779.

18. Babu, Y.S., Bugg, C.E. and Cook, W.J. (1988) 'Structure of calmodulin refined at 2.2 A resolution', J. Mol. Biol. 204, 191-204.

19. Szebenyi, D.M.E. and Moffat, K. (1986) 'The refined structure of vitamin D-dependent calcium-binding protein from bovine intestine', J.Biol.Chem. 261, 8761-8777.

20. Strynadka, N.C.J. and James, M.N.G. (1989) 'Crystal structures of the helix–loop–helix calcium–binding proteins', Ann. Rev. Biochem. 58, 951-998.

21. Kretsinger, R.H. (1982) 'Structure and evolution of calcium–modulated proteins', CRC. Crit. Rev. Biochem. 8, 118-174.

22. Herzberg, O. and James, M.N.G. (1985) 'Common structural framework of the two Ca2+/Mg2+ binding loops of troponin C and other Ca2+ Binding proteins', Biochemistry 24, 5298-5302.

23. Baum, P., Furlong, C. and Byers, B. (1986) 'Yeast gene required for spindle pole body duplication: Homology of its product with Ca2+-binding protein',

Proc.Natl.Acad.Sci.USA 83, 5512-5516.

24. Hori, K., Kushick, J.N., Factor, A. and Weinstein, H. (1987) 'Parameters and mechanisms of calcium binding to peptides and proteins', Int. J. Quantum Chem. QBS14, 341-345.

25. Weinstein, H., Hori, K., Kushick, J.N., Sussman, F. and Factor, A. (1989) 'Computer simulation studies of structure-function relations in calcium-binding proteins', in Kartashev, L.P. and Kartashev, S.I. (eds.), Proc. 4th Intl. Conf. Supercomputing Intl., Supercomputing Inst. Inc., 106-108.

26. Falke, J.J., Snyder, E.E., Thatcher, K.C. and Voertler, C.S. (1991) 'Quantitating and engineering the ion specificity of an EF-hand-like Ca2+ binding site', Biochemistry 30, 8690-8697.

27. Sussman, F. and Weinstein, H. (1989) 'On the ion selectivity in Ca-binding proteins: The cyclo (-L-Pro-Gly)3 peptide as a model', Proc.Natl.Acad. Sci.USA 86, 7880-7884.

28. Weinstein, H. and Osman, R. (1990) 'Molecular biophysics of specificity and function in enzymes, receptors and calcium binding proteins', in Beveridge, D.L. and Lavery, R. (eds.), Theoretical Biochemistry and Molecular Biophysics: A Comprehensive Survey, Adenine Press, NY 275-289.

29. Kartha, G., Varughese, K.I. and Aimoto, S. (1982) 'Conformation of cyclo (-L-Pro-Gly)3 and its Ca2+ and Mg2+ complexes', Proc.Natl.Acad. Sci. USA 79, 4519-4522.

30. Skelton, N.J., Kordel, J., Forsen, S., Chazin, W.J. (1990) 'Comparative structural analysis of the calcium free and bound states of the calcium regulatory protein calbindin D9k', J. Mol. Biol. 213, 593-598.

31. Heidorn, D.B. and Trewhella, J. (1988) 'Comparison of the crystal and solution structures of calmodulin and troponin C', Biochemistry 27, 909-915.

32. Ikura, M., Kay, L.E. and Bax, A. (1990) 'A novel approach for sequential assignment of 1H, 13C, and 15N spectra of larger proteins: Heteronuclear triple-resonance three-dimensional NMR spectroscopy. Application to calmodulin', Biochemistry 29, 4659-4667.

33. Brooks, B.R., Bruccoleri, R.E., Olafson, B.D., States, D.J., Swaminathan, S. and Karplus, M. (1983) 'CHARMM: A program for macromolecular energy, minimization and dynamics calculations', J. Comput. Chem. 4, 187-217.

34. Mehler, E.L., Pascual-Ahuir, J.L. and Weinstein, H. (1991) 'Structural dynamics of calmodulin and troponin C', Protein Eng. 4, 625-637.

35. Hori, K., Kushick, J.N. and Weinstein, H. (1988) 'Structural and energetic parameters of Ca2+ binding to peptides and proteins', Biopolymers 27, 1865-1886.

36. Kataoka, M., Head, J.F., Seaton, B.A. and Engelman, D.M. (1989) 'Melittin binding causes a large calcium-dependent conformational change in

calmodulin', Proc. Natl. Acad. Sci. USA 86, 6944-6948.

37. Matsushima, N., Izumi, Y., Matsuo, T., Yoshino, H., Ueki, T. and Miyake, Y. (1989) 'Binding of both Ca^{2+} and mastoparan to calmodulin induces a large change in tertiary structure', J. Biochem. 105, 883-887.

38. Heidorn, P.B., Seeger, P.A., Rokop, S.E., Blumenthal, D.K., Means, A.R., Crespi, H. and Trewhella, J. (1989) 'Changes in the structure of calmodulin induced by a peptide based on the calmodulin–binding domain of myosin light chain kinase', Biochemistry 28, 6757–6764.

39. Kataoka, M., Head, J.F., Vorherr, T., Krebs, J. and Carafoli, E. (1991) 'Small-angle X-ray scattering study of calmodulin bound to two peptides corresponding to parts of the calmodulin-binding domain of the plasma membrane Ca2+ pump', Biochemistry 30, 6247-6251.

40. LaPorte, D.C., Wierman, B.M. and Storm, D.R. (1980) 'Calcium-induced exposure of a hydrophobic surface on calmodulin', Biochemistry 19, 3814-3819.

41. Tanaka, T. and Hidaka, H. (1980) 'Hydrophobic regions function in calmoduline-enzyme(s) interaction', J. Biol. Chem. 255, 11078-11080.

42. O'Neil K.T., D., W.F. (1990) 'How calmodulin binds its targets: sequence independent recognition of amphiphilic alpha-helices', Trends Biochem. Sci 15, 59-64.

43. Pascual-Ahuir, J.-L., Mehler, E.L. and Weinstein, H. (1991) 'Calmodulin structure and function: Implication of arginine residues in the compaction related to ligand binding', Molec. Eng. 1, 231-247.

44. Mehler, E.L. and Eichele, G. (1984) 'Electrostatic effects in water-accessible regions of proteins', Biochemistry 23, 3887-3891.

45. Mehler, E.L. and Solmajer, T. (1991) 'Electrostatic effects in proteins: Comparison of dielectric and charge models', Protein Eng. 4, 905-910.

46. Mehler, E.L. (1990) 'Comparison of dielectric response models for simulating electrostatic effects in proteins', Protein Eng. 3, 415-417.

47. Strynadka, N.C.J. and James, M.N.G. (1991) 'Towards an understanding of the effect of calcium on protein structure and function', Current Opin. Struct. Biol. 1, 905-914.

48. Pascual-Ahuir, J.L., Weinstein, H. (1991) 'The bending of calmodulin may be modulated by its target protein.', Biophys. J. 59, 118a.

49. Sherbany, A.A., Parent, A.S. and Brosius, J. (1987) 'Rat calmodulin cDNA', DNA 6, 267-272.

50. Calnan, B.J., Tidor, B., Biancalana, S., Hudson, D. and Frankel, A.D. (1991) 'Arginine-mediated RNA recognition: The arginine fork', Science 252, 1167-1171.

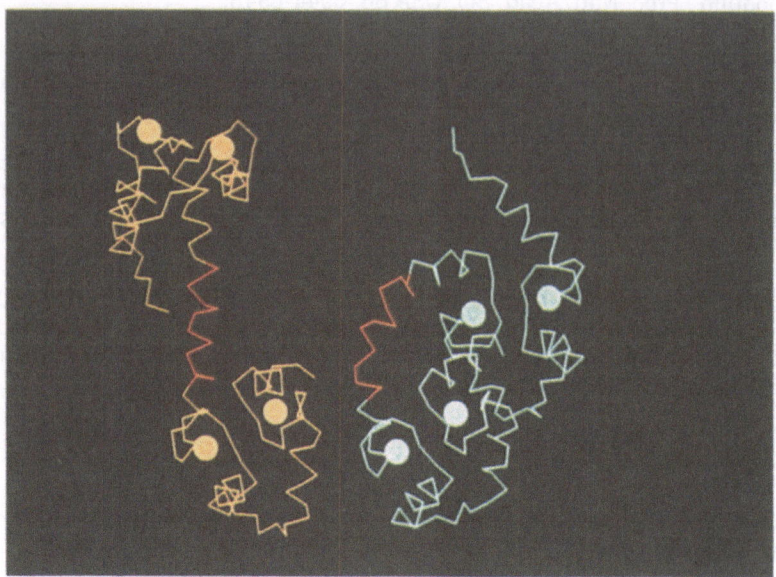

Figure 1. Comparison of calmodulin structure in the crystal (in yellow, at left) and as a result of molecular dynamics simulations (in blue, at right). The *alpha*-carbon ribbons are shown with the Ca^{2+} positions indicated in the binding loops. The tether portion of the central helix is indicated in red. The structures are oriented for optimal overlap of their C-terminal domains; note the near-identity of the structures in the calcium-binding lobes.

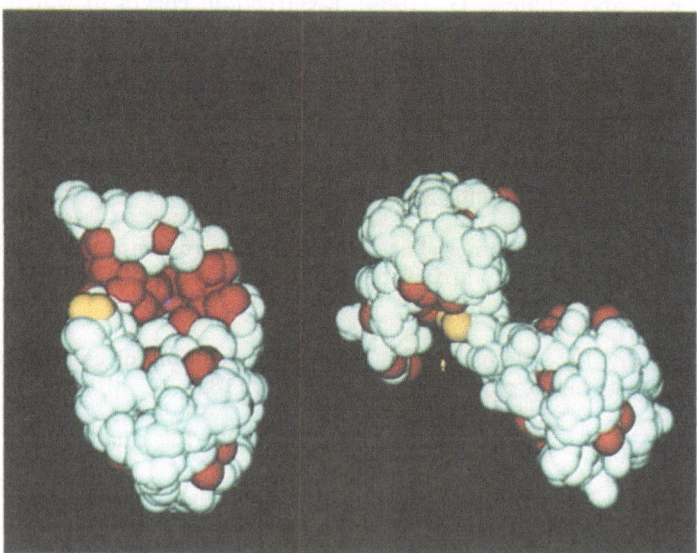

Figure 2. Comparison of the solvent-exposed surfaces in calmodulin from the crystal structure (right) and the compacted structure from molecular dynamics simulation (left). The red patches indicate the hydrophobic areas, and the yellow cursor is pointing at methionine 71 in the crystal structure. Note that Met71 is exposed for reactivity in the compacted structure, but much less so in the crystal structure.

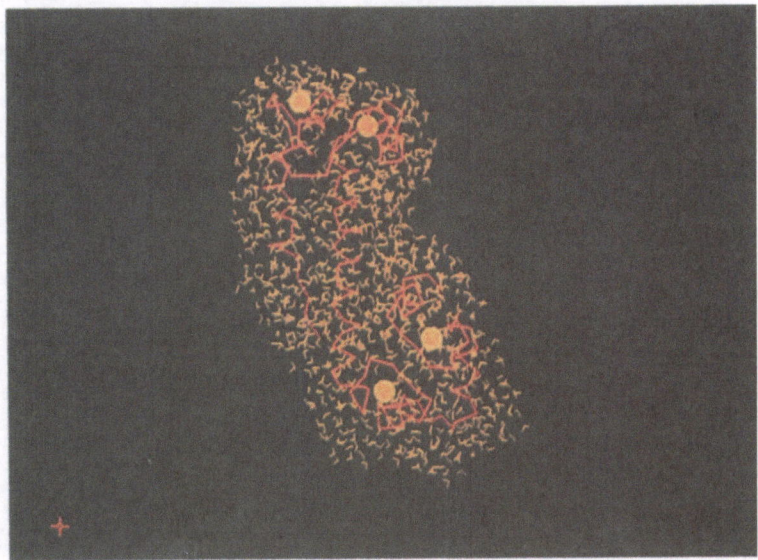

Figure 3. Snapshot at 129 ps of the simulation of calmodulin surrounded by a water shell of 4Å (see text for details of the simulation protocol). The *alpha*-carbon structure of CAM is shown in red, with the Ca^{2+} ions indicated as yellow spheres. Note that the waters in the hydration shell follow the protein.

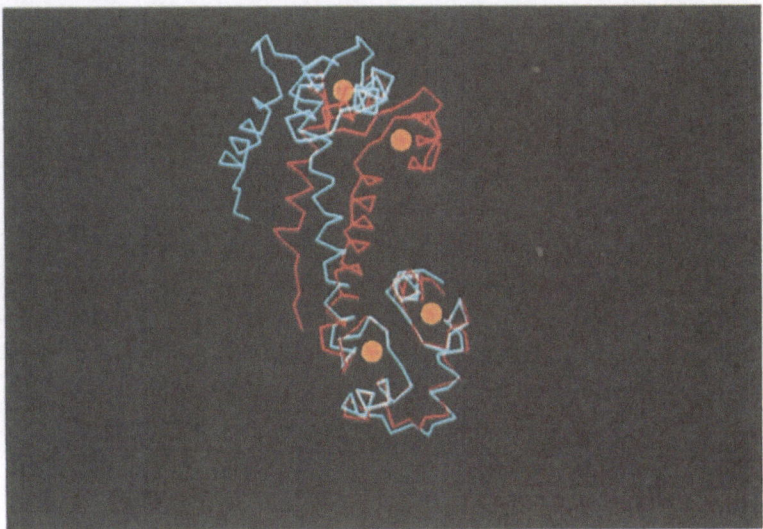

Figure 4. Comparison of the structures of calmodulin in the crystal (blue) and from the molecular dynamics simulation (in red) performed in a hydration shell of 4Å (not shown; see Fig. 3). The *alpha*-carbon ribbons of the structure are shown superimposed for best fit of the C-terminal domains.

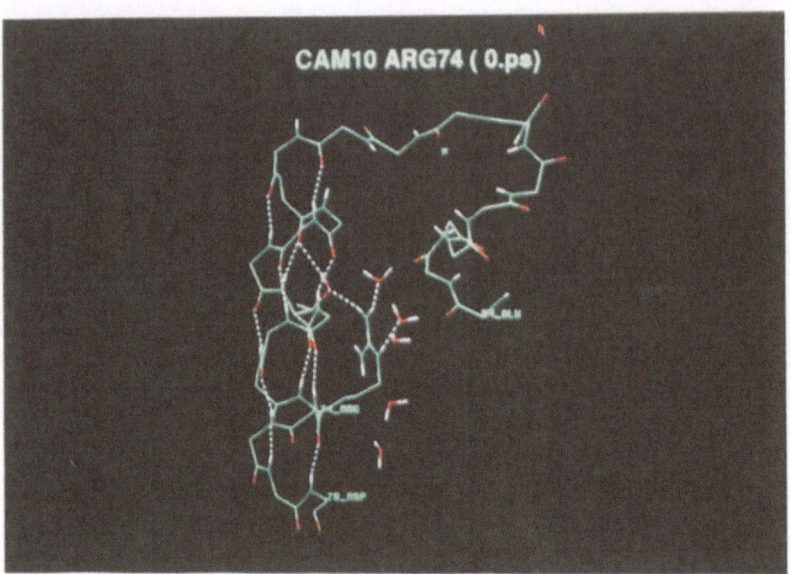

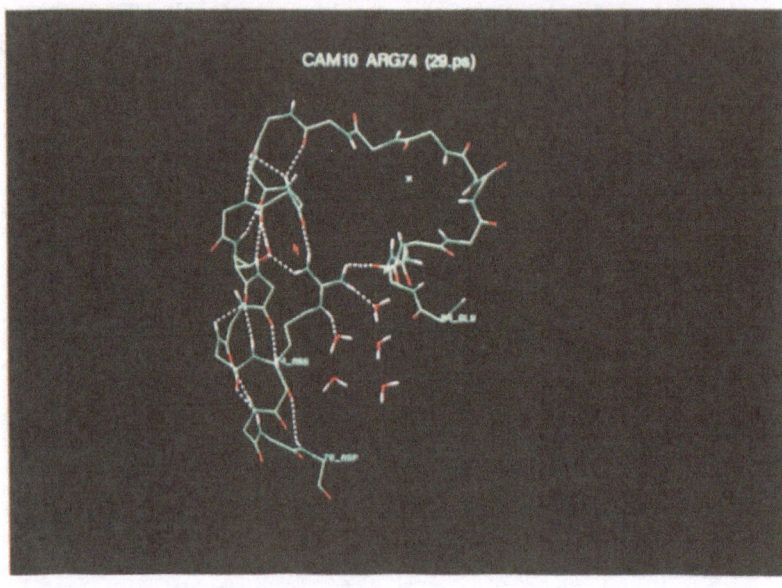

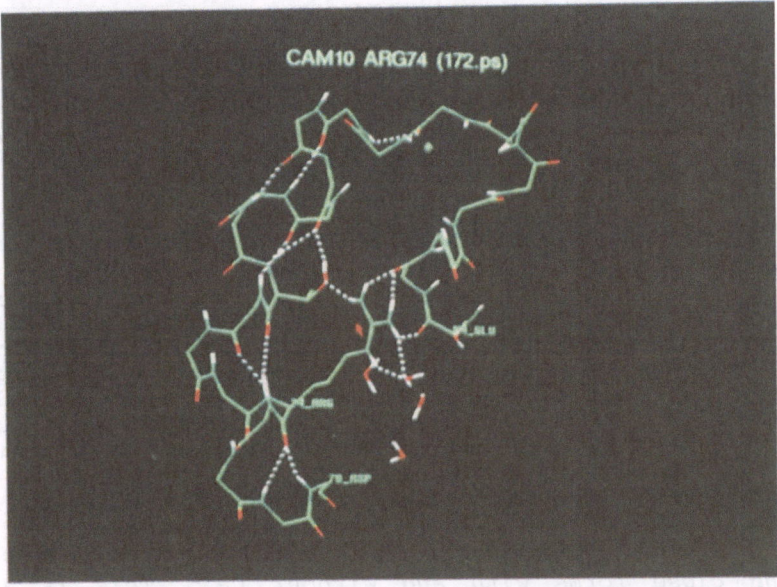

Figure 5. The hydrogen-bonding network of Arg74 in three snapshots from a molecular dynamics simulation of calmodulin during which the central helix bends and the structure compacts. Only resides 54-78 are shown, with Arg74 in the center of the picture. Note the lack of a distal bond (to Glu54) at 0 ps, compared to the complex H-bonding pattern to waters and to the distal part of the protein at 29 ps and 172 ps.

DISCUSSION

Prof. Lesyng: *Calmodulin and other calcium binding proteins exhibit an interesting cooperativity phenomenon. If the first calcium cation is bound the second binds easier. Could you comment this experimental observation?*

Prof. Weinstein: The cooperativity in Ca^{2+} binding has been studied experimentally in great detail by Forsen's group (e.g., see S. Linse et **al.**, Biochemitry 30:154-162, 1991). From a computational viewpoint this is a very complex problem that requires a degree of quantitative accuracy not yet achieved in our calculations. The methodological problem is somewhat complicated by the long range effects of mutations on the binding of Ca^{2+} observed for calbindin (S. Linse et al., Nature 335:651-652, 1988), and for troponin C (K. Fujimori et al., Nature 345:182-184, 1990) which indicate the solvent environment is likely to play an important role in the quantitative aspects of the observed cooperativity.

Prof. A. Warshel: *The main question is the relative energetics of the open and close forms with and without Ca^{+2}. This is a major electrostatic challenge which has nothing to do with the specific force field.*

Prof. Weinstein: I agree with the comment. As described in the text, we are currently examining the effects of solvation on the details of the molecular dynamics of CAM by simulation with a discrete description of the solvent environment. We are putting special emphasis on the analysis of the role of electrostatics in determining the properties of the CAM structures in the compacted and extended forms (see text).

Prof. Smith: *The possibility of obtaining complete crystallographic data sets on very short timescales using the Laue method is indeed mouth-watering. Do you have any evidence that Ca^{2+} renoval from your protein might constitute a workwhile process to study using this method?*

Prof. Weinstein: As you know, the answer to your questions depends on the timeframe of the association/dissociation of Ca^{2+} at the specific sites in the various proteins, and the attendant structural reorganization. Unfortunately, very little is known about these dynamic details. However, some information may be forthcoming for CAM from the NMR work of Ad Bax's group (see ref. 32 and *Biophys. J. 61:A403, 1992*)

Prof. Maggiora: *Has anyone studied the effect on the "open"/"close" calmodulin equilibrium as a function of the calmodulin inhibitor TFP?*

Prof. Weinstein: Yes, we have studied the effect of calmodulin compaction on the binding of trifluoperazine (TFP) to evaluate the functional implications of the bending. The first results have been published recently (J.L. Pascual-Ahuir, E.L Mehler and H. Weinstein, Molec. Engineering 1:231-247, 1991). We show that the compaction changes the shape of the interaction site for TFP with the CAM molecule. The most obvious change involves residues **Thr** 146 and **Ala** 147 which are part of the C-terminal domain of CAM. In the extended (crystallographic) structure these residues are too far from the binding site of TFP in the N-terminal domain of CAM to have any effect on the binding. However, in the compacted structure they have moved closer to the interaction site and form part of the interaction "pocket" for the polar head group of TFP. The work cited above provides a detailed analysis of the consequences of compaction for TFP binding and offers specific suggestions for site directed mutations that would test the effect of compaction on TFP binding as well as on the number and nature of the binding sites.

Prof. Maggiora: *Do you of other examples of Zn^{2+} metalloproteinases which also bind Ca^{2+}?*

Prof. Weinstein: As summarized in the chapter, both the enzymatic function and the thermal stability of a number of enzymes have been shown to be susceptible to modulation through binding of Ca^{2+}. Among the novel calcium-binding proteins described in ref. 9 (see text) some also incorporate or bind Zn^{2+}. In this respect, it is especially interesting that a Ca^{2+} binding site can be grafted onto a protein structure to produce a more stable molecule that preserves its enzymatic activity (see text). Thus, whether Zn^{2+} metalloproteinases, such as thermolysin, contain a native Ca^{2+}-binding site or not, the use or introduction of such a site to engineer new desired properties constitutes a very exciting proposition.

SIMULATIONS OF PROTON TRANSFER AND HYDRIDE TRANSFER REACTIONS IN PROTEINS

A. WARSHEL
Department of Chemistry
University of Southern California
Los Angeles, California 90089-1062
U.S.A.

ABSTRACT. Simulations of proton transfer and hydride transfer reactions by the Empirical Valence Bond method are described. The use of this method in examination of Linear Free Energy Relationships (LFER) and in obtaining quantum mechanical rate constants is illustrated. It is pointed out that LFERs are probably valid even in enzymatic reactions since computer simulations indicate that the electrostatic response of protein active sites can be described by the linear response approximation. The validity of LFERs in proteins should provide a powerful tool in protein design, although it is important to realize that such relationships are much more valid for transitions between different resonance structures than for transitions between reactants and product states.

1. Introduction

Proton Transfer (PT) and Hydride Transfer (HT) reactions play an important role in many chemical and biological processes[1,2]. Most theoretical studies in this challenging field have been based on macroscopic phenomological approaches (e.g. refs. 2-5). However, the studies of enzymatic reactions requires one to either obtain the macroscopic parameters from microscopic simulations, or to simulate the entire reaction on a microscopic level. In principle one would like to use ab-initio quantum mechanical approaches for such studies. Unfortunately, while ab-initio studies of PT and HT reactions can be accomplished at a reasonable level of accuracy for small donors and acceptors in the gas phase (e.g. ref. 6), such approaches cannot be implemented in studies of large macromolecules. Here one must resort to some approximations and represent a large part of the system classically. In fact, encouraging progress has been made in quantum mechanical/classical studies of PT reactions in solutions and proteins[2,7−10]. Perhaps the most extensive microscopic exploration of different aspects of enzymatic reactions have been conducted with the Empirical Valence Bond (EVB) method[7b,12] that has been applied to such problems as Linear Free Energy Relationships (LFER), and dynamical effects in PT reactions in proteins.

This work considers simulations of different aspects of PT and HT reactions in proteins as an illustration of the power of the EVB method. The hydride transfer step in the catalytic reaction of LDH is taken as a specific example and used to examine the validity of LFER in proteins and to examine nuclear tunnelling effects.

J. Bertrán (ed.), Molecular Aspects of Biotechnology: Computational Models and Theories, 175–191.
© *1992 Kluwer Academic Publishers.*

2. EVB Calculations of the Catalytic Reaction of LDH

Chemical processes in enzymes can be described, in principle, by SCF-MO approaches[2,7a,8-11], or by Valence Bond (VB) approaches[2,7b-7e,12], provided that the enzyme environment and the surrounding solvent are modelled consistently and that all the main contributions (e.g. the surrounding solvent is considered). However, most enzymatic reactions involve bond breaking and bond making processes as well as formation of ionized fragments. Thus it is easier to obtain reliable potential surfaces by VB approaches than by MO approaches. In particular it is much simpler to calibrate VB potential surfaces using experimental information about the reacting fragments in solution than to accomplish this by MO approaches. Similarly, it is simpler to incorporate VB approaches consistently in FEP approaches[12]. Thus we use here the previously developed EVB method (e.g. ref. 12). This method has been applied successfully to simulations of various enzymatic reactions (see for example refs. (7,12) and was found to reproduce reliably the effect of mutations on catalytic activity of enzymes and to provide reasonable approximations for the corresponding absolute free energy profiles.

The basic idea (see ref. 12 for more details) is to represent the overall reaction in terms of various valence bond structures with clear physical meaning and then to obtain the actual ground state by mixing these structures. Using VB type approach provides a convenient connection to useful linear free energy formulations and increases the reliability of calculations of environmental effects. In short the EVB ground state surface is obtained by solving the equation

$$HC_g = E_g C_g \qquad (1)$$

where the EVB Hamiltonian H is constructed from diagonal elements which describe the energies of the different resonance structures (these energies are described by force field type potential functions that include the interaction between the solute and the solvent) and off diagonal elements that are kept the same in different environments. The EVB potential surface is calibrated using accurate gas phase data (if available) or by using experimental information for the reaction in solution (for example, equilibrium constant for the reaction in solution, experimental solvation free energies for isolated ions etc.). It can also be calibrated conveniently using ab initio potential surfaces[12,13]. The main aspects of the present EVB treatment will be considered below.

2.1 CONSTRUCTING THE POTENTIAL SURFACES

LDH catalyzes the redox interconversion of pyruvate and lactate in the presence of reduced coenzyme nicotinamide adenine dinucleotide (NADH). The x-ray structure of LDH has been resolved to high resolution in its apo, binary and ternary forms[14-17]. The active site is located between two subunits. The substrate interacts with the active site residues and positioned such that it can accept a hydride ion from the nicotinamide ring of NADH. LDH from B. stearothermophilius has k_{cat}/K_M value of 4.2 x $10^6 M^{-1} s^{-1}$ for pyruvate[18].

The mechanism of this reaction can be described in terms of a transfer between the resonance structures of Fig. 1. The first resonance structure ψ_1 represents the reactant state having protonated His 195, substrate and the reduced NAD. The second resonance structure ψ_2 represents the intermediate where the hydride has been transferred to the

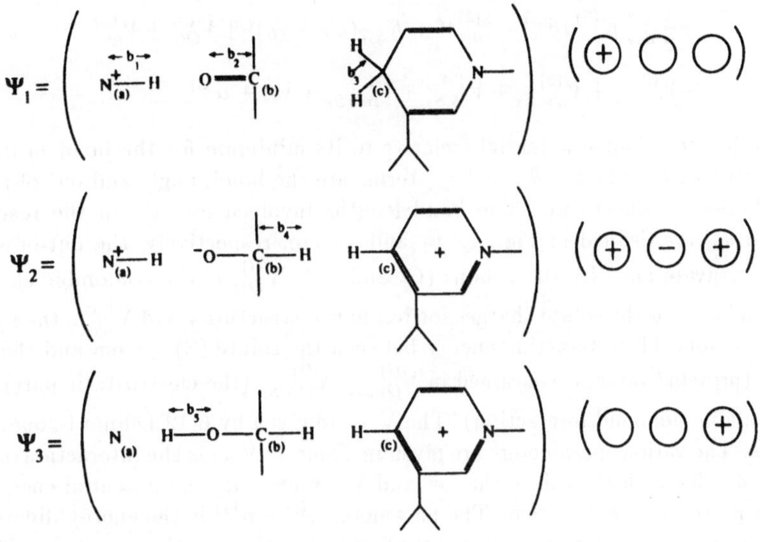

Figure 1: The valence bond resonance structures used to describe the catalytic reaction of LDH. $N_{(a)}$ represents the nitrogen of the Histidine residue.

pyruvate substrate and the third resonance structure ψ_3 represents the product that is deprotonated His, lactate and oxidized NAD. A step-wise mechanism involves a transfer from ψ_1 to ψ_2 and then to ψ_3. A concerted mechanism involves the transfer from ψ_1 to ψ_3. The first step of the step-wise mechanism is considered to be the rate limiting step although a concerted mechanism cannot be excluded.

Dividing the system to the quantum mechanical region (which is referred to as the 'solute' region) that includes all the atoms involved in the three resonance structures and the remaining protein/water environment (which is referred to as the 'solvent' system), we may now construct the EVB hamiltonian. The diagonal elements of this hamiltonian are simply given by the following force fields

$$
\begin{aligned}
\varepsilon_1 \quad = \quad H_{11} = {} & \Delta M(b_1) + \Delta M(b_2) + \Delta M(b_3) + \frac{1}{2}\sum_m K_{b,m}^{(1)}(b_m^{(1)} - b_0^{(1)})^2 \\
& + \frac{1}{2}\sum_m K_{\theta,m}^{(1)}(\theta_m^{(1)} - \theta_0^{(1)})^2 + K_{\chi,1}(\chi_1 - \chi_0)^2 + V_{QQ}^{(1)} + V_{nb}^{(1)} \\
& + V_{Q\mu,Ss}^{(1)} + V_{Q\alpha,Ss}^{(1)} + V_{nb,Ss}^{(1)} + V_{ind,Ss}^{(1)} + V_{ss} + \alpha^{(1)}
\end{aligned}
\tag{2}
$$

$$
\begin{aligned}
\varepsilon_2 \quad = \quad H_{22} = {} & \Delta M(b_1) + \Delta M(b_4) + \frac{1}{2}\sum_m K_{b,m}^{(2)}(b_m^{(2)} - b_0^{(2)})^2 \\
& + \frac{1}{2}\sum_m K_{\theta,m}^{(2)}(\theta_m^{(2)} - \theta_0^{(2)})^2 + K_{\chi,2}(\chi_2 - \chi_0)^2 + V_{QQ}^{(2)} + V_{nb}^{(2)} \\
& + V_{Q\mu,Ss}^{(2)} + V_{Q\alpha,Ss}^{(2)} + V_{nb,Ss}^{(2)} + V_{ind,Ss}^{(2)} + V_{ss} + \alpha^{(2)}
\end{aligned}
$$

$$
\varepsilon_3 \quad = \quad H_{33} = \Delta M(b_4) + \Delta M(b_5) + \frac{1}{2}\sum_m K_{b,m}^{(3)}(b_m^{(3)} - b_0^{(3)})^2
$$

$$+\frac{1}{2}\sum_m K_{\theta,m}^{(3)}(\theta_m^{(3)} - \theta_0^{(3)})^2 + K_{\chi,2}(\chi_2 - \chi_0)^2 + V_{QQ}^{(3)} + V_{nb}^{(3)}$$

$$+V_{Q\mu,Ss}^{(3)} + V_{Q\alpha,Ss}^{(3)} + V_{nb,Ss}^{(3)} + V_{ind,Ss}^{(3)} + V_{ss} + \alpha^{(3)}$$

where $\Delta M(b_m)$ denote a Morse potential (relative to its minimum for the bond m in the with resonance structure). The b_m, θ_m and χ_m terms are the bond, angle and out-of-plane bending contributions, respectively. The bond lengths involved directly in the reaction (b_1, b_2, b_3, b_4 and b_5) are defined in Fig. 1. χ_1 and χ_2 are, respectively, the out-of-plane angles for the pryruvate and NADH carbons (C_b and C_c). $V_{QQ}^{(i)}$ is the coulombic electrostatic interaction between the solute charges for resonance structure i and $V_{nb}^{(i)}$ is the solute nonbonded interaction. The interaction energy between the solute (S) system and the solvent (s) system (protein/water) is contained in $V_{Q\mu,Ss}^{(i)}$, $V_{Q\alpha,Ss}^{(i)}$ (the electrostatic part) and $V_{nb,Ss}^{(i)}$ (the remaining nonbond interaction). The V_{nb} are given by 6-12 Lennard-Jones potential functions. The various parameters are given in Table 1. V_{ind} is the interaction of the solvent induced dipoles with the solute charges and V_{ss} represents the potential energy of the surrounding protein/solvent system. The parameter $\alpha^{(i)} - \alpha^{(j)}$ is the energy difference between ψ_i and ψ_j with the solute fragments at infinite separation in the gas phase. These $\Delta\alpha$'s will be referred to as the 'gas phase shifts'.

The off diagonal elements of the EVB hamiltonian are given by a simple single exponential approximation

$$H_{ij} = f(r) = A_{ij}exp(-\mu r) \tag{3}$$

The parameter A_{ij} can be adjusted to reproduce the observed barrier for the reaction in solution. Once calibrated, the value of A_{ij} is kept fixed for the reaction in protein. With the H_{ii} of eq (2) and the H_{ij} of eq (3) we can solve eq (1) and obtain the ground state energy E_g.

2.2 THE CALIBRATION PROCEDURE

The $\alpha^{(i)}$ parameters of eq (2) (which are referred to as the 'gas phase shifts') are evaluated based on experimental solvation free energies for isolated ions and experimental equilibrium constants for reaction in solution. The overall equilibrium constant for

$$Pyruvate + NADH \longrightarrow Lactate + NAD^+ \tag{4}$$

in solution is 3.7 x 10^5 at pH 7[19]. Now let us consider the thermodynamic cycle shown in Fig. 2. From the above equilibrium constant in solution we obtain $\Delta G_{1\rightarrow3}^{\infty,w}$ = -6 kcal/mol. Using our preliminary knowledge about pK_a'[20] we obtain $\Delta G_{2\rightarrow3}^{\infty,w}$ = 2.3RT $[pK_a(ImH^+)$ - $pK_a(lactate)]$ = 1.38[7.5-13] = -7.6 kcal/mol. Using these ΔG's and our thermodynamic cycle gives

$$\Delta G_{1\rightarrow2}^{\infty,w} = \Delta G_{1\rightarrow3}^{\infty,w} - \Delta G_{2\rightarrow3}^{\infty,w} = 1.6 \text{ kcal/mol} \tag{5}$$

Now we can use the relationship

$$\alpha^{(j)} - \alpha^{(i)} \simeq \Delta G_{sol,w}^{(i),\infty} - \Delta G_{sol,w}^{(j),\infty} + (\Delta G_{i\rightarrow j}^{\infty})_{obs,w} \tag{6}$$

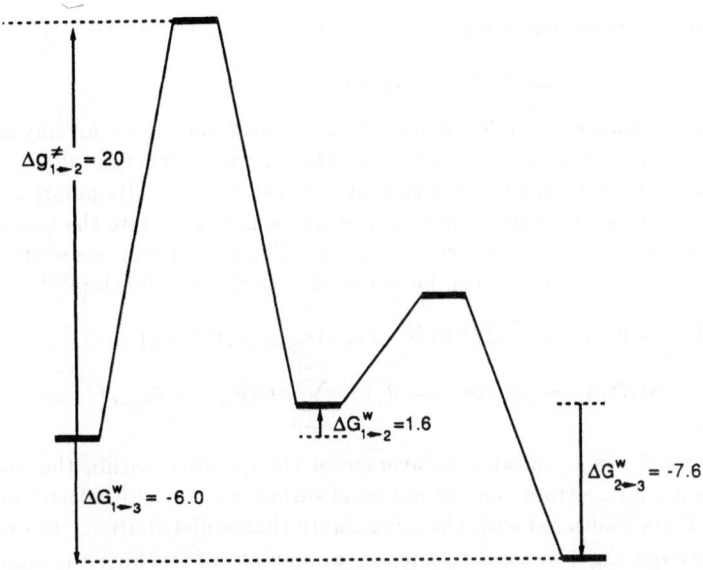

Figure 2: The thermodynamic cycle (in kcal/mol) for the reaction in solution. This cycle is used to determine the gas phase shifts (see text for more details).

where $\Delta G_{sol,w}^{(i),\infty}$ is the solvation free energy of the ψ_i configuration and $\Delta G_{i \to j}^{\infty}$ is the reaction free energy in solution for the conversion ψ_i to ψ_j when the relevant fragments of each resonance form are at infinite separation. This relationship is based on the fact that the energy difference between ψ_j and ψ_i in the gas phase (with the fragments at infinite separation) plus the difference between the corresponding solvation energy, gives the energy difference between ψ_i and ψ_j in solution. Once the gas phase shifts (the α's) are calibrated with the observed $\Delta G_{obs,w}^{\infty}$ for the reactions in solution (reference reaction) they are kept unchanged for the corresponding reaction in protein.

In constructing the solute off-diagonal elements we note that the second step of the reaction involves standard proton transfer from histidine to an alcohol. Thus for H_{23} we use the previously determined parameters of ref. 12b. The H_{12} for the hydride transfer step can be determined by noting that gas phase experiments[21] and ab initio calculations[6b,6c] for gas phase hydride transfer give no barrier or a very low barrier. Taking $\Delta g^{\neq} \simeq 0$ in the gas phase and using

$$\Delta g^{\neq} \simeq \Delta \epsilon^{\neq} - H_{12}(r^{\neq}) \qquad (7)$$

where $\Delta \epsilon^{\neq}$ is the value of our ϵ_1 at its intersection with ϵ_2, we obtained $H_{12}(r^{\neq}) \simeq 25$ kcal/mol. We also assumed that $H_{13} \simeq 0$ since we deal with shift of more than one electron in considering the coupling between ψ_1 and ψ_3. This assumption should, however, be examined more carefully in studies that concentrate on the concerted mechanism. The parameter set obtained with our calibration procedure is summarized in ref. 22.

2.3 CALCULATIONS OF FREE ENERGY FUNCTIONS

After constructing the analytical potential surface the major task is to obtain the activation free energy $\Delta g^{\ddagger}$. This is done by combining the EVB method with a free energy technique using a methodology that has been discussed in detail elsewhere[12,23]. The main features of the method are outlined below

First we perform a series of simulations using a mapping potential of the form

$$\varepsilon_m = (1 - \theta_m)\varepsilon_1 + \theta_m \varepsilon_2 \tag{8}$$

where ε_1 and ε_2 are the diabatic energies of state 1 and state 2 calculated for any solvent configuration generated by the simulation. Changing the parameter θ_m from 0 to 1 drives the system from state 1 to state 2, while the solvent is forced to adjust its polarization to the changing charge distribution of the solute. This enables us to calculate the free energy the solvent needs to adjust to the solute configuration. The free energy associated with changing ε_1 to ε_2 in n equal increments can be obtained with the relationship[24].

$$\delta G(\theta_m \longrightarrow \theta_{m'}) = -(1/\beta)ln[< exp(-(\varepsilon'_m - \varepsilon_m))\beta >_m] \tag{9}$$

$$\Delta G(\theta_n) = \Delta G(\theta_0 \longrightarrow \theta_n) = \sum_{m=0}^{n-1} \delta G(\theta_m \longrightarrow \theta_{m+1})$$

where $\beta = (1/k_B T)$ and $<>_m$ indicates an average of the quantity within the brackets calculated by propagating trajectories on the potential surface ε_m. $\Delta G(\theta)$ reflects the electrostatic solvation effects associated with changing solute charge distribution. To evaluate the activation free energy, $\Delta g^{\ddagger}$, we need to find the probability of reaching the transition state for trajectories that move on the actual ground state potential surface E_g. The relevant formulation has been described in detail elsewhere[23,29]. To summarize, we define the reaction coordinate X^n in terms of the energy gap $\varepsilon_2 - \varepsilon_1$. This is done by dividing the configuration space of the system into subspaces, S^n, that satisfy the relationship $\Delta\varepsilon(S^n) = X^n$, where the energy differences X^n are constants and $\Delta\varepsilon = \varepsilon_2 - \varepsilon_1$. The X^n are used as the reaction coordinate, giving

$$exp[-\Delta g(X^n)\beta] = exp[-\Delta G(\theta_m)\beta] < exp[-(E_g(X^n) - \varepsilon_m(X^n))\beta] >_m \tag{10}$$

where θ_m is the θ that keeps the system near the given X^n. Thus we calculate the energy difference $(E_g - \epsilon_m)$ between the mapping potential and the ground state potential at all the configurations evaluated during our mapping procedure and use these differences via eq. (10) to evaluate the ground state free energy function, $\Delta g(X)$, and the corresponding $\Delta g(X^{\ddagger})$.

In addition to evaluating $\Delta g(X^n)$ one can obtain the probability of being at X^n on ε_i. This gives us the free energy function Δg_i

$$\Delta g_i(X^n) = \int exp[-\varepsilon_i(X^n)\beta]dS^n \tag{11}$$

which can be evaluated by[23]

$$exp[-\Delta g_i(X^n)\beta] = exp[-\Delta G(\theta_m)\beta] < exp[-(\varepsilon_i(X^n) - \varepsilon_m(X^n))\beta] >_m \tag{12}$$

The Δg_i curves give the reorganization energy through the relationship[23b,29]

$$\Delta g_j(X_i^0) - \Delta g_i(X_i^0) - \Delta G_{i \to j} = \lambda \tag{13}$$

where X_i^0 is the minimum of the Δg_i curve. This energy is called reorganization energy since it represents the reduction in the free energy of the product state upon relaxation

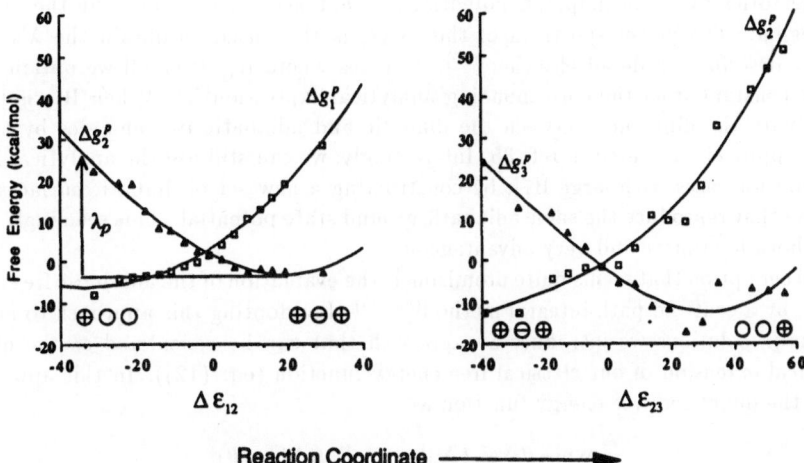

Figure 3: Showing the calculated solvent contribution to the free energy functions Δg_i for the hydride transfer and proton transfer steps in the stepwise mechanism of LDH. The Δg's are given as functions of the corresponding energy gaps (the Δg's) which are taken as the generalized reaction coordinates. The relevant configurations are designated by the corresponding charge distribution.

of the system from the minimum of the reactant surface to the minimum of the product surface.

As a demonstration of our approach we may consider the protein contributions to the free energy functions for LDH. The results of such calculations that are described in detail in ref. 22 are given in Fig. 3. These curves will be considered below in studies of Linear Free energy relationships.

2.4 CALCULATIONS OF NUCLEAR TUNNELLING AND ZERO-POINT ENERGY EFFECTS

The EVB framework appears to provide a convenient way for obtaining quantum mechanical corrections to the rate constants of PT and HT reactions in solutions and proteins. In using the EVB for calculations of quantized rate constants we adopt the hypothesis[33-35] that the quantum mechanical corrections can be attributed primarily to the corresponding activation barriers so that

$$k_{qu} \simeq k_{cl} \exp\{-\beta(\Delta g_{qu}^{\neq} - \Delta g_{cl}^{\neq})\} \tag{14}$$

where qu and cl designate quantum mechanical and classical quantities, respectively. In order to evaluate $\Delta g^{\neq}$ we developed and examined several strategies. The first one is based on the Dispersed Polaron (DP) model[30] which approximates the diagonal EVB energies by

$$H_{ii} = \epsilon_i \simeq \frac{1}{2} \sum_k \hbar w_k (q_k - \lambda_k^i)^2 + \alpha_i \tag{15}$$

The q's and the λ's for the entire solute-solvent or solute-protein system are obtained from the same MD simulations used to evaluate $\Delta g^{\neq}$. This is done by collecting the time dependence energy gap between the relevant diabatic states (e.g. $\Delta\epsilon(t)_{12} = \epsilon_2(t) - \epsilon_1(t)$)

for trajectories over the mapping potential ϵ_m that keeps the system at the transition state region. The power spectrum of this $\Delta\epsilon(t)$ is then used to obtain the λ's and ω's in a way described in detail elsewhere[30]. In cases where H_{12} is small we obtain directly the rate constant from the corresponding analytical expression[35]. When H_{12} is large we can evaluate the difference between the diabatic and adiabatic free energies by the path integral approach presented in ref. 35. Interestingly, we can still use the analytical DP rate expression for cases with large H_{12}, by constructing a new set of diabatic surfaces with a small H_{12} that reproduce the same adiabatic ground state potential. This seemingly strange trick is both legitimate and very advantageous.

Another option that seems quite promising is the evaluation of the adiabatic free function function by a centroid path integral method[33-35]. In adopting this approach to adiabatic reactions in solutions we introduce an approach that can be considered as the quantum mechanical extension of our classical free energy function (eq. (12)). In this approach we express the quantized free energy function as

$$\exp[-\beta\Delta g_{qu}(X)] \simeq \exp[-\beta\Delta G_{qu}(\theta)] \tag{16}$$

$$\times < \delta(X - \Delta\epsilon^c)\exp\{-(\beta/p)\sum_k[E_g(x_k) - \epsilon_m(\bar{x})]\} >_{\epsilon_m^{qu}}$$

where the potentials E_q and ϵ_m^{qu} of eq. (12) are replaced here by effective potentials which are given in the one dimensional case by

$$\epsilon_m^{qu} = \sum_k^p \{\frac{1}{2p}M\Omega^2(x_{k+1} - x_k)^2 + \frac{1}{p}\epsilon_m(x_k)\} \tag{17}$$

here we replace each classical particle by a ring of a quasiparticle, where M is the mass of our particle and $\Omega = p/\hbar\beta$. The δ function which is used to collect the contributions to the different values of the reaction coordinate involves $\Delta\epsilon^c = \epsilon_2(\bar{x}) - \epsilon_1(\bar{x}) =$ constant, where $\bar{x}$ is the "centroid" of the system defined by $\bar{x} = \sum_k x_k/p$. The evaluation of Eq. (17) is largely simplified by reexpressing it as an average over the original classical trajectories used to evaluate the $\Delta g_{cl}^{\neq}$ of eq. (12) and an extra average over a free particle distribution function (see ref. 36 for derivation). This type of expression and the approximate DP potential of eq. (15) were used to evaluate the quantum corrections for the hydride transfer step in the reaction of LDH. The DP "spectrum" for this system is given in Fig. 4 while the quantized and classical $\Delta g(X)$ functions are illustrated in Fig. 5. This treatment gave an isotope effect of $\sim$4.5 which should not yet be compared to the observed isotope effect in LDH since the calculations might not correspond to the rate limiting step and should only be considered as a preliminary feasibility study.

3. Considerations of Linear Free Energy Relationships

Linear Free Energy Relationships (LFER) are among the most fundamental concepts in physical organic chemistry. The validity of such relationships for chemical processes in solutions has been the subject of many experimental and theoretical studies (e.g. refs. 3, 12a, 23, 25, 26) and appears to be reasonably established. The concept of LFER can be

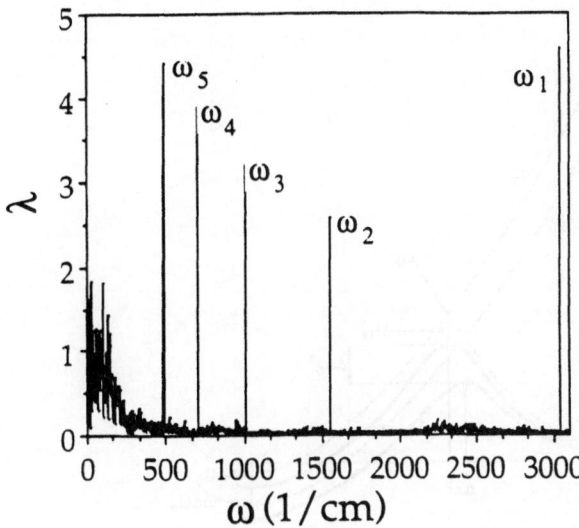

Figure 4: The $\lambda(\omega)$ of the DP model for the hydride transfer step of the catalytic reaction of LDH. The five shared lines correspond to the solute effective modes.

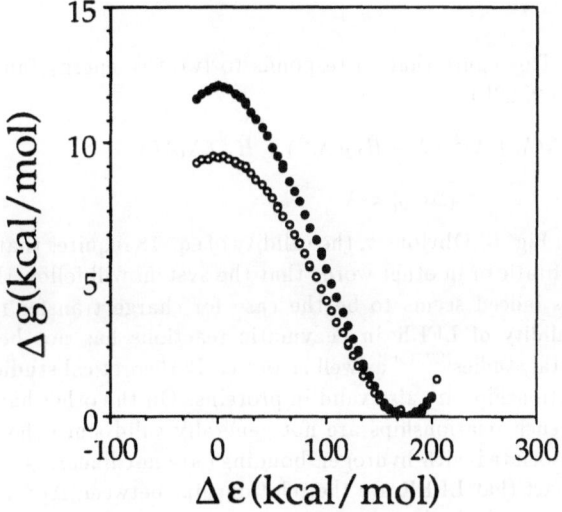

Figure 5: The classical free energy function, $\Delta g_{cl}(X)$ (filled circle), and the corresponding quantum mechanical function, $\Delta g_{qu}(X)$ (open circle), for the hydride transfer step in the catalytic reaction of LDH.

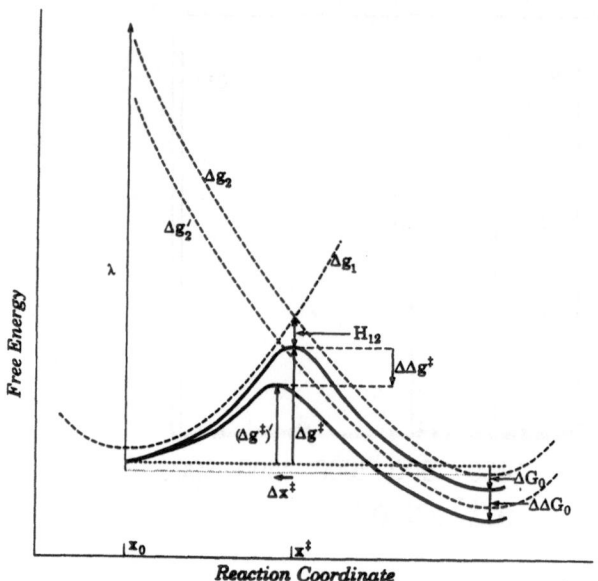

Figure 6: A schematic description of the relationship between the free-energy difference ΔG_0 and the activation free energy $\Delta g^{\neq}$. The figure illustrates how a shift of Δg_2 by $\Delta\Delta G_0$ (that changes Δg_2 to $\Delta g_2'$ and ΔG_0 to $\Delta G_0 + \Delta\Delta G_0$) changes $\Delta g^{\neq}$ by a similar amount.

best understood from Fig. 6. This figure that corresponds to two free energy functions of equal curvature gives (see refs. 2, 23b)

$$\Delta g^{\neq} \approx (\Delta G_0 + \lambda)^2/4\lambda - \bar{H}_{12}(X^{\neq}) + \bar{H}_{12}^2(X_0)/\lambda \tag{18}$$

$$|\Delta G_0| < \lambda$$

where X_0 and $X^{\neq}$ are defined in Fig. 6. Obviously, the validity of eq. 18 requires that the free energy functions would be quadratic or in other words that the system will follow the linear response approximation. This indeed seems to be the case for charge transfer reactions in solutions[23b,29,32]. The validity of LFER in enzymatic reactions has not been fully established. Recent experimental studies[37,38] as well as our early theoretical studies[23a,39] seem to indicate that such relationships are also valid in proteins. On the other hand it has been argued recently[40] that such relationships are not generally valid since the relevant interactions (e.g. the forces associated with hydrogen bonding) are not linear. Apparently, this argument overlooked the fact that LFER are the relationships between $\Delta g^{\neq}$ and ΔG_0 and they are not related to the issue whether the forces that determine ΔG_0 are linear or not (see also ref. 41). Nevertheless, it is important to examine the validity of LFER in proteins in a systematic way.

In principle it would be tempting to examine LFER in LDH by correlating the effect of mutations on ΔG_0 with their effect on $\Delta g^{\neq}$. However, a mutation can change both λ and ΔG_0, thus complicating the analysis. Here one may use simulations to estimate the effect of the mutations on λ, but even this option is not straightforward since it is not clear whether the mechanism is step-wise or concerted. Furthermore, $\Delta G_{1\to 2}^0$ is not known experimentally.

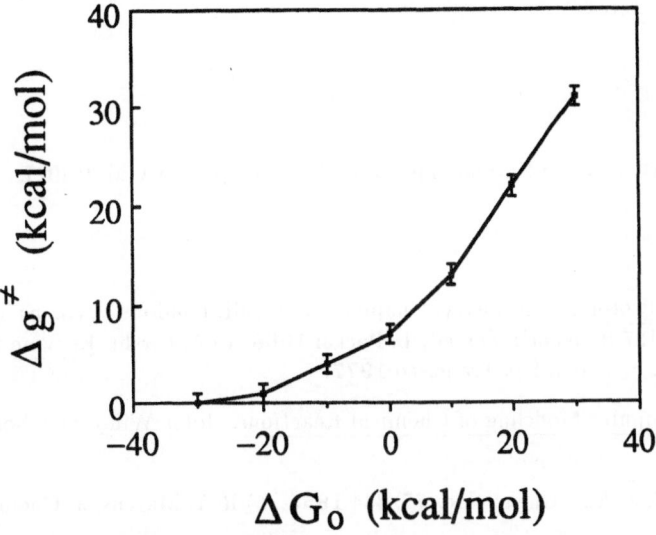

Figure 7: The relationship between $\Delta g^{\neq}$ and ΔG_0 for the protein contribution to the free energy function for the hydride transfer step in the stepwise mechanism of LDH. The calculations are done with $H_{12} = 0$, since the solute contribution is not included.

Fortunately , as was demonstrated in our studies of electron transfer (ET) and $S_N 2$ reactions in solutions[12,29] it is possible to explore the validity of LFER by computer simulation approaches. This can be accomplished by calculating the free energy functions and the corresponding $\Delta g^{\neq}$ for different assumed values of ΔG_0. Thus, we may take the protein contribution to the free energy functions of the hydride transfer step in LDH as a generic test case and examine the relationship between $\Delta g^{\neq}$ and ΔG_0 by repeating the calculations of Fig. 3 for different values of ΔG_0 (as was done in ref. 29 in an examination of the Marcus' relationship for ET reactions). The results of such a study are summarized in Fig. 7. As seen from the figure we obtain quite regular LFER for a model hydride transfer reaction in a protein active site. The reason for this is associated with the fact that the free energy functions of Fig. 3 are almost quadratic. This is so because the electrostatic response of the protein active site follows basically the linear response approximation. If this is a general phenomenon then LFERs should be a general rule in enzymatic reactions.

One may yet wonder whether the contribution of the solute to the free energy functions can change the LFER obtained in Fig. 7. Previous studies in of PT reactions solutions (12a) and proteins (39) gave LFER of the type predicted by eq. (18) when the solute contribution was considered. It is still possible that the quantum mechanical corrections will lead to some deviations from eq. (18) but preliminary studies using our path integral approach have indicated that this is not the case. In fact, it seems that the quantum mechanical corrections lead to a similar trend in the enzyme and the reference reaction in solution and that one can use the classical LFER without quantum corrections to examine the catalytic effect of the enzyme.

If the use of LFER in enzymes will be established as a reliable approximation then it would be possible to focus on the factors that change ΔG_0 rather than the more complicated issue of the change of $\Delta g^{\neq}$. This may provide a useful tool in protein engineering. Here, however, it is important to realise that the true LFER exists when one considers the energetics of the different resonance structures and not the energy difference between the reactant and product states[2].

Acknowledgment

This work was supported by the National institute of Health (Grant GM 24492).

References

1. a) R.P. Bell, The Proton in Chemistry, Chapman and Hall, London, **1973**. b) Methods in enzymology, 127 Biomembranes, ed. L. Packer **1986**. c) A. Fersht, Enzyme Structure and Mechanism, Freeman, San Francisco **1977**.

2. A. Warshel, Computer Modeling of Chemical Reactions, John Wiley and Sons, Inc. **1991**.

3. a) G.S. Hammond, J. Am. Chem. Soc. 77, 334 **1955**. b) R.A. Marcus, J. Chem. Phys. 72, 891 **1958**.

4. a) E.D. German and A.M. Kuznetsov, J. Chem. Soc., Faraday Trans I **1980** 76, 1128. b) E.D. German and A.M. Kuznetsov, J. Chem. Soc. Faraday Trans I **1981** 77, 397. c) J. Suhnel and K. Gustav, Chem. Phys. **1984** 87, 179.

5. J. Ulstrup, Charge Transfer Processes in Condensed Media, Springer, Berlin, **1979** .

6. a) Hillenbrand, E.A. and Scheiner, S. J. Am. Chem. Soc. **1984** 106 , 6206. b) Wu, Y. and Houk, K.N. J. Am. Chem. Soc.**1987a** 109 , 906. c) Wu, Y. and Houk, K.N. J. Am. Chem. Soc.**1987b** 109 , 2226. d) Tapia, O., Cardenas, R., Andres, J. and Colonna-Cesari, F. J. Am. Chem. Soc. **1988** 110, 4046.

7. a) Warshel, A. and Levitt, M. J. Mol. Biol. **1976** 103 , 227. b) Warshel, A. and Weiss, R.M. J. Am. Chem. Soc. **1980** 102 , 6218. c) Warshel, A. and Russell, S.T. J. Am. Chem. Soc. **1986** 108, 6569. d) Warshel, A. J. Phys. Chem....1982. e) Åqvist, J. and Warshel, A. (1989)Biochemistry **1989** 28 , 4680.

8. Kollman, P.A. and Hayes, D.M. J. Am. Chem. Soc. **1981** 103, 2955.

9. Van Duijnen, P. Th., Thole B. Th., Broer, R. and Nieuwpoort, W.C. Int. J. Quant. Chem. **1980** 17 , 651.

10. Singh, U.C., Proc. Natl. Acad. Sci. USA **1988** 85, 4280.

11. Tapia, O., Andres, J., Aullo, J.M. and Branden, C.I. J. Chem. Phys. **1985** 83 , 4673.

12. a) Hwang, J.K., King, G., Creighton, S. and Warshel, A. J. Am. Chem. Soc. **1988** 110 , 5297. b) Warshel, A., Sussman, F. and Hwang, J.K. J. Mol. Biol.**1988** 201, 139.

13. Chang, Y.T. and Miller, W.H. J. Phys. Chem. **1990** 94, 5884.

14. Holbrook, J.J., Liljas, A., Steindel, S.J. and Rossmann, M.G. in *The Enzymes (Ed. Boyer, P.D.)* **1975** 3rd Ed. Vol XI, 191, Academic Press, New York.

15. Adams, M.J., Buehner, M., Chandrasekhar, K., Ford, G.C., Hackert, M.L., Liljas, A., Rossmann, M.G., Smiley, I.E., Allison, W.S., Everse, J., Kaplan, N.O. and Taylor, S.S. *Proc. Natl. Acad. Sci. USA* **1973** ,*70* , 1968.

16. Barstow, D.A., Clarke, A.R., Chia, W.N., Wigley, D., Sharman, A.F., Holbrook, J.J., Atkinson, T., Minton, N.P. *Gene* **1986** *46* , 47.

17. Piontek, K., Chakrabarti, P., Schar, H.P., Rossmann, M.G. and Zuber, H. *Proteins* **1990** 7 , 74.

18. Wilks, H.M., Hart, K.W., Feeney, R., Dunn, C.R., Muirhead, H., Chia, W.N., Barstow, D.A., Atkinson, T., Clarke, A.R. and Holbrook, J.J. *Science* **1988** *242* , 1541.

19. Gutfreund, H. *Biophys. Molec. Biol.* **1975** *29* , 161.

20. Perrin, D.D., Dempsey, B. and Serjeant, E.P. in *pK$_a$ prediction for organic acids and bases* , Chapman and Hall, New York, **1981**

21. Meot-Ner (Mautner), M. *J. Am. Chem. Soc.* **1987** *109* , 7947.

22. Yadav, A., Jackson, R.M., Holbrook, J.J., and Warshel, A. *J. Am. Chem. Soc.* **1991** *113* , 4800.

23. a) Warshel, A. *Pontif. Acad. Sci. Script. Var.* **1984** *55* , 59. b) Hwang, J.K. and Warshel, A. *J. Am. Chem. Soc.* **1987** *109*, 715.

24. Valleau, J.P. and Torrie, G.M. *In Modern Theoretical Chemistry (Berne, B.J. ed.)* **1977** *5* , 169, Plenum, New York.

25. Marcus, R.A. *Annu. Rev. Phys. Chem.* **1964** *15* , 155.

26. Albery, W.J. *Annu. Rev. Phys. Chem.* **1980** *31* , 227.

27. Kreevoy, M.M. and Kotchevar, A.T. *J. Am. Chem. Soc.* **1990** *112* , 3579.

28. Kreevoy, M.M., Ostovic, D., Lee, I.H., Binder, D.A. and King, G.W. *J. Am. Chem. Soc.* **1988** *110* , 524.

29. King, G. and A. Warshel, *J. Chem. Phys.*, **1990**, *93*, 8682.

30. (a) Warshel, A. and Chu, Z.T. *J. Chem. Phys.* **1990** *93*, 4003. (b) Warshel, A., Chu, Z.T. and Parson, W.W. *Science* **1989** *246* 112.

31. Clarke, A.R., Wilks, H.M., Barstow, D.A., Atkinson, T., Chia, W.N. and Holbrook, J.J. *Biochemistry* **1988** *27* , 1617.

32. Kuharski, R.A., Bader, J.S., Chandler, D., Sprik, M., Klein, M.L. and Impey, R.W. *J. Chem. Phys.* **1988** *89* , 3248.

33. Gilan, M.J. *Phys. Rev. Lett.* **1987** *58* , 563.

34. Voth, G.A., Chandler, D. and Miller, W.W., *J. Chem. Phys.* **1989** *91* , 7749.

35. Warshel, A. and Chu, Z.T., *J. Chem. Phys.* **1990** *93* , 4003.

36. Hwang, J.-K., Chu, Z.T. and Warshel, A., *J. Phys. Chem.* (submitted).

37. Fersht, A.R., Leatherbarrow, R.J. and Wells, T.N.C., *Nature* **1986** *322* , 284.

38. Toney, M.D. and Kirsch, J.F., *Science* **1989** *243* , 1485.

39. Warshel, A., Russell, S.T., and Sussman, F., *Israel, J. Chem.,* **1987** *27* , 217.

40. Straub, J.E. and Karplus, M. *Protein Engineering* **1990** *3* , 673.

41. Fersht, A.R. and Wells, T.N.C., *Protein Engineering* **1991** *4* , 229.

DISCUSSION

Prof. Kollman: *Why can't enzymes increase tunneling by reducing the A-H...B distance in a proton transfer system, which would decrease the barrier and increase tunneling rate?*

Prof. Warshel: There are two ways your suggestion can be presented and I believe both can be shown to be ineffective. The first suggestion would be that the enzyme pulls the donor and acceptor away to say 5A and then leads to enormous barrier and large tunneling. This, however, will be a <u>bad</u> enzyme. Good enzymes will have the donor and acceptor at 2.6A exactly as in solution. The second suggestion is that the enzyme compresses the donor and acceptor. However, now the enzyme has to invest strain energy in "compression" which is usually not done by flexible enzymes.

Prof. Lesyng: *I would like to refer to the Dr. Kollman's question. The energy fluctuations in an enzyme of the serine proteinase size are of the order of 40 kcal/mol. This value can be estimated from the heat capacity, the mass of the enzyme, and the temperature. A part of this energy can be accomodated in squeezing vibrational modes of the protein. Such effects have already been observed in classical MD simulations. This can lead to a decrease of the barrier for the proton transfer, and in consequence can result in an increase of the proton transfer probability (quantum tunneling is very sensitive to the barrier height). Could your please comment this?*

Prof. Warshel: Sorry, the fluctuations in energy are the wrong quantity to examine. One really has to look at the free energy profile or potential of mean force. We have done this frequently and demostrated, going back to the 1976 lysozyme work, that strain is not effective in catalysis. Our studies of real enzymatic reactions show that energy flow form the protein and dynamical effects are not likely to be important in catalysis (see our J. Mol. Biol. 1988). As much as tunneling is concerned, I again argue (see response to Kollman) that while tunneling is a real effect it should be similar in solutions and in proteins. In fact, our approach for calculating nuclear tunneling in Proton Transfer in solutions (J.C.P. 1990) was recently used in enzymes and gave the same trend.

Prof. Kollman: *Role of polarization energy - since the force field is calibrated to include it, taking it out will appear more important than if the force field were calculated- what about it?*

Prof. Warshel: This is an interesting comment. We always take into account the polarization energy by a self-consistent approach but one may argue that this is not needed. However, we find that the induced dipoles are quite important for excited states and for VB excited states. These effects cannot be parametrized by ground state properties. In fact the activation barrier for Electron Transfer process is drastically reduced by polarizability effects.

Prof. Duran: Which are the VB structures you choose for hydride transfer reactions?

Prof. Warshel: The one chosen are those with the hydrogen on the NADH and the one where the hydrogen is on the pyrovate (C-H C=O) and (C^+H-C-O⁻)please see the paper, Yadav et, al., J.Am. Chem. Soc. 113, 1991

Prof. Lesyng: During the parametrization procedure of the free energy surfaces you have to select "slow" degrees of freedom. This is always a kind of arbitrary procedure which you have to make, and which in some cases can be difficult.
What do you do to make the optimal choice?

Prof. Warshel: Sorry. Please read Hwang et. al. J. Am. Chem. Soc. 1988. We do not select slow and fast coordinate but ask what is the probability to have different values of the generalized reaction coordinate - the energy gap between the diabatic surfaces. Finding the corresponding free energy functions does not involve an assumption about adiabatic coordinate. After finding the generalized transition state we use linear response approximation to see if there are slow and fast coordinates (e.g. solute and solvent).

Prof. Lluch: What is your opinion about the validity of Marcus relationship for outer-sphere electron transfer reactions. Do you think that this equation hold for inner-sphere electron transfer reactions or for proton transfer?

Prof. Warshel: Our studies in the last ten years seem to repeatedly indicate that enzymes can be treated with the linear response theory in both proton transfer and electron transfer reactions. Thus we feel we demostrated that linear free energy relationship and marcus relationship are reasonable approximations in proteins.

Prof. Miller: *It is not clear how your VB treatment gives a correct stabilization of the transition state charge distribution since it seems to interact with the full charge of each state.*

Prof. Warshel: Not really. Let consider the SN2 case (Cl⁻ C–Cl – Cl–C Cl⁻) and see how the EVB interact correctly with the (-1/2 -1/2) charge of the transition state. The calculations are done with a mapping potential $E_m = 0.5(E_1 + E_2)$. So that the solvent is polarized with charges which correspond to -1/2 and -1/2 on each Cl. Now this solvent polarization is allowed to interact with the full charge in each state, but the final ground state energy is obtained by solving the 2X2 VB matrix and giving $E_g = 1/2 (E_1 + E_2) - H_{12}$ which is exactly what we need.

REACTION DYNAMICS IN POLYATOMIC MOLECULAR SYSTEMS: SOME APPROACHES FOR CONSTRUCTING POTENTIAL ENERGY SURFACES AND INCORPORATING QUANTUM EFFECTS IN CLASSICAL TRAJECTORY SIMULATIONS

WILLIAM H. MILLER
Department of Chemistry, University of California, and
Chemical Sciences Division of the Lawrence Berkeley Laboratory,
Berkeley, California 94720

ABSTRACT. This paper deals with the two essential tasks necessary to model chemical reactions theoretically: obtaining the potential energy surface (i.e., the electronic energy of the molecular system as a function of nuclear positions) and then determining the dynamical motion of the nuclei/atoms governed by it.

It is sometimes possible to model the potential energy surface for a chemical reaction *locally*, e.g., a harmonic valley along the *reaction path* which passes through the transition state (saddle point on the potential energy surface) connecting reactants and products. More generally, though, it is necessary to have a *global* potential energy function that is not restricted to the vicinity of reaction path. Many completely empirical potential functions have been developed for non-reactive molecular motions, and it is shown here how the *empirical valence bond* (EVB) idea can be used to combine these non-reactive potential functions, which describe reactant and product regions individually, with *ab initio* calculations for the transition state region and thus obtain a global potential energy surface for a chemical reaction.

Although the dynamics of nuclear motion should in principle be treated quantum mechanically, the difficulty of doing so is prohibitive for polyatomic systems. Most dynamical treatments employ classical mechanics which, though an excellent approximation in many regards, has some serious deficiencies. One such problem discussed here has to do with zero point vibrational energy and the other with tunneling of light (e.g., hydrogen) atoms. Though quite rigorous semiclassical theories exist that would in principle solve these problems, we describe in this paper more approximate approaches that are intended to be suitable for direct incorporation into classical simulation algorithms.

J. Bertrán (ed.), Molecular Aspects of Biotechnology: Computational Models and Theories, 193–235.
© 1992 *Kluwer Academic Publishers.*

1. Introduction

This lecture will not deal explicitly with biological applications, but rather will describe some recent developments in theoretical/computational methodology that could have practical implications for dynamical simulations of such processes. All of the theoretical approaches discussed here are "trajectory based", i.e., geared to be utilized within the framework of a classical trajectory simulation of the atomic/molecular motion. Though the applications carried out in my research group are for medium size polyatomic molecules — e.g., the intramolecular isomerization in malonaldehyde,

$$(1.1)$$

—- I believe that the issues involved, and the methodologies being developed, also have relevance for the more complex molecular systems of biological interest.

The following three sections deal with three separate issues involving classical trajectory simulations in polyatomic systems: (1) convenient (and accurate!) representations of a global potential energy surface for *reactive* systems; (2) how to deal with the problem of zero point (vibrational) energy in polyatomic molecules; and (3) how to incorporate tunneling (primarily in H atom transfer reactions) into trajectory simulations in an accurate, yet practically manageable way.

2. An Empirical Valence Bond Model for Reactive Potential Energy Surfaces

One of the most difficult steps in theoretical treatments of chemical reactions in polyatomic molecular systems is representing the potential energy surface.[1] Ideally, of course, one would like to be able to compute the Born-Oppenheimer electronic energy $V(q_1,...,q_{3N-6})$ from first principles for any values of the 3N-6 coordinates that are necessary to specify the configuration of the N atom system. Though *ab initio* quantum chemistry calculations[2] are becoming increasingly possible for polyatomic molecular systems, the number of such calculations needed for more than 3 or 4 atom systems tends to make this direct approach unfeasible.

One of the ways used for dealing with the situation has been to exploit the idea of a reaction path.[3-5] Here one computes the potential energy surface only along a one-dimensional curve (the reaction path) in the 3N-6 dimensional space that connects reactant and product configurations. This is often the steepest descent path (in mass-weighted cartesian coordinates) that passes through the transition state for the reaction under study — the "intrinsic" reaction path[5] — but other paths are possible[6] and sometimes more useful.[7] One typically also determines the force constant matrix along this path, thus providing a local harmonic approximation to the potential energy surface along the reaction (or reference) path.

Though reaction path approaches have been very useful, particularly for qualitative and approximate dynamical treatments, and will certainly continue to be so, there are times when a global potential energy surface is needed. This is true, for example, for highly vibrationally excited molecules, where the dynamics tends not to be localized about any one reaction path, and also for cases with a number of low frequencies modes orthogonal to the reaction path, which allows for large amplitude motion far away from any reference path.

For vibrational motions about stable molecular geometries a standard normal mode expansion — harmonic plus perhaps anharmonic corrections — provides an adequate global potential function. There also exist a number of completely empirical potential functions[8-12] that describe a variety of non-reactive motions and interactions. Unless special alterations are made, however, these potential functions are not capable of modeling the potential energy surface for a chemical reaction.

In this paper we wish to pursue and develop an approach used by Warshel[13] that is especially designed to model reactive potential functions, namely the empirical valence bond (EVB) model. To illustrate the basic idea, consider an isomerization reaction such as Eq. (1.1), which is characterized by a multi-dimensional double well potential function. One *imagines* that this Born-Oppenheimer potential energy surface results from a quantum chemistry calculation with a 2-state valence bond electronic wavefunction

$$|\psi> = c_1|\phi_1> + c_2|\phi_2>, \tag{2.1}$$

where $|\phi_1>$ is a valence bond wavefunction that describes the electronic structure of the reactants (1) in Eq. (1.1), and $|\phi_2\rangle$ the corresponding wavefunction that describes the electronic structure of the products (2). The lowest electronic eigenvalue, i.e., the Born-Oppenheimer potential energy surface, is then given by the lower root of the 2x2 secular equation, specifically

$$V = 1/2 \, (V_{11}+V_{22}) - \sqrt{\left(\frac{V_{11}-V_{22}}{2}\right)^2 + V_{12}^2}, \tag{2.2}$$

where

$$V_{11} = \langle\phi_1|H_{el}|\phi_1\rangle$$

$$V_{22} = \langle\phi_2|H_{el}|\phi_2\rangle$$

$$V_{12} = \langle\phi_1|H_{el}|\phi_2\rangle,$$

and H_{el} is the electronic Hamiltonian. V is a function of the nuclear coordinates $q \equiv (q_1, ..., q_{3N-6})$ because the electronic Hamiltonian depends on q, and thus V_{11}, V_{22}, and V_{12} also do.

In the *empirical* valence bond approach, however, no electronic matrix elements are actually calculated. $V_{11} \equiv V_{11} \, (q_1, ..., q_{3N-6})$ is identified as the potential energy surface for the reactants and thus taken as a non-reactive (i.e., single minimum) potential energy surface that describes non-reactive motion about the reactant geometry. The simplest imaginable model for $V_{11}(q)$ would be a harmonic normal mode approximation about the reactant equilibrium geometry. At a more sophisticated level, one could use one of the non-reactive empirical potential models[8-12] that has the bonding designated as in (1) of Eq. (1.1). $V_{22}(q)$ is similarly a non-reactive (i.e., single minimum) potential energy surface that describes motion about the product geometry. V_{11} and V_{22} are often referred to as *diabatic* potential surfaces, in contrast to V itself which is the Born-Oppenheimer, or *adiabatic* potential surface.

The most crucial part of the EVB model is the exchange matrix element (or resonance integral) $V_{12} = V_{12}(q)$, for it is less obvious how it should be chosen. Warshel[13] has used some very simple approximations in his (very complex) applications, while here we describe a way of choosing it so that the EVB potential $V(q)$ of Eq. (2.2) exactly reproduces a given harmonic force field about a given transition state geometry. We envision that the transition state quantities (geometry, energy, and force constant matrix) will be obtained by *ab initio* quantum chemistry calculations. I.e., the logic of the approach is that *ab initio* quantum calculations of useful accuracy can be carried out for a few selected features of the reactive potential surface, and the most important of these are the transition state parameters since this is the least well known region of the potential and also the most important region for describing the reaction. The reactant and product regions are described reasonably well by simple (non-reactive) empirical potential functions[8-12] for stable molecules. The EVB model that we present is thus a way of incorporating *ab initio* calculations for the transition state parameters with simple diabatic potential functions that describe reactants and products separately.

The potential energy surface $V(q)$ is thus taken to be in the form of Eq. (2.2), where the diabatic potentials $V_{11}(q)$ and $V_{22}(q)$ are non-reactive (i.e., single minimum) potential functions that correctly describe the regions near the equilibrium geometries q_1 and q_2, respectively. V_{11} and V_{22} are assumed to be known, and the goal is to find a useful way of determining the exchange matrix element $V_{12}(q)$. It is clear that in the reactant or product regions themselves, i.e., for q near q_1 or q_2, one will have

$$V_{12}^2 \ll \left(\frac{V_{11}-V_{22}}{2}\right)^2 , \tag{2.3a}$$

and in this limit it is easy to see that Eq. (2.2) gives

$$V(q) \cong \text{Min}[V_{11}(q),V_{22}(q)], \tag{2.3b}$$

which is clearly correct in these regions. It is thus only necessary to know $V_{12}(q)$ in the intermediate region between reactants and products, and to determine it in this region we appeal to *ab initio* quantum chemistry.

Eq. (2.2) can be used to express V_{12} in terms of V_{11}, V_{22}, and V as follows,

$$V_{12}(q)^2 = [V_{11}(q)-V(q)][V_{22}(q)-V(q)]. \tag{2.4}$$

Near the transition state geometry one has

$$V(q) \cong V_0 + 1/2 \ (q\text{-}q_0)\cdot K_0\cdot(q\text{-}q_0), \tag{2.5}$$

where the transition state geometry q_0, energy V_0, and force constant matrix K_0 are obtained from an independent *ab initio* calculation. Since the non-reactive potential functions $V_{11}(q)$ and $V_{22}(q)$ are known, they can also be expanded in a Taylor's series about the transition state geometry

$$V_{nn}(q) = V_n + D_n\cdot\Delta q + 1/2 \ \Delta q\cdot K_n\cdot\Delta q , \tag{2.6}$$

where $\Delta q = q\text{-}q_0$,

$$V_n = V_{nn}(q_0) ,$$

$$D_n = \left(\frac{\partial V_{nn}(q)}{\partial q}\right)_{q=q_0}$$

$$K_n = \left(\frac{\partial^2 V_{nn}(q)}{\partial q \partial q}\right)_{q=q_0}$$

for n=1,2. With Eqs. (2.5) and (2.6), Eq. (2.4) thus gives the following power series expansion for V_{12}^2, correct through quadratic order in $\Delta q \equiv q\text{-}q_0$,

$$V_{12}^2 = (V_1\text{-}V_0)(V_2\text{-}V_0) + (V_2\text{-}V_0) \ D_1\cdot\Delta q + (V_1\text{-}V_0) \ D_2\cdot\Delta q$$

$$+ 1/2 \ (V_1\text{-}V_0)\Delta q\cdot(K_2\text{-}K_0)\cdot\Delta q + 1/2 \ (V_2\text{-}V_0)\Delta q\cdot(K_1\text{-}K_0)\cdot\Delta q$$

$$+ (D_1\cdot\Delta q)(D_2\cdot\Delta q). \tag{2.7}$$

A cumulant resummation,[14] though, gives better extrapolation properties; therefore $V_{12}^2(q)$ is taken to be a generalized Gaussian

$$V_{12}^2(q) = A \ \exp\ [B\cdot\Delta q\text{-}1/2 \ \Delta q\cdot C\cdot\Delta q] , \tag{2.8}$$

and this function expanded through quadratic order in Δq and equated to the corresponding terms on the RHS of Eq. (2.7) to determine the parameters $\mathbf{A}$, $\mathbf{B}$ (a vector) and $\mathbf{C}$ (a matrix). The arithmetic is straight-forward and one obtains

$$A = (V_1 - V_0)(V_2 - V_0) \tag{2.9a}$$

$$\mathbf{B} = \frac{\mathbf{D}_1}{(V_1 - V_0)} + \frac{\mathbf{D}_2}{(V_2 - V_0)} \tag{2.9b}$$

$$\mathbf{C} = \frac{\cdot \mathbf{D}_2 \mathbf{D}_2 \cdot}{(V_2 - V_0)^2} + \frac{\cdot \mathbf{D}_1 \mathbf{D}_1 \cdot}{(V_1 - V_0)^2} + \frac{(\mathbf{K}_0 - \mathbf{K}_1)}{V_1 - V_0} + \frac{(\mathbf{K}_0 - \mathbf{K}_2)}{V_2 - V_0} . \tag{2.9c}$$

For completeness, we note that if the intermediate position q_0 is actually not the transition state geometry, so that Eq. (2.5) has a linear term $\mathbf{D}_0 \cdot \Delta q$, then Eqs. (2.7)-(2.9) still apply if the following change is made in (2.9b) and (2.9c),

$$\mathbf{D}_n \rightarrow \mathbf{D}_n - \mathbf{D}_0 , \tag{2.9d}$$

for n=1,2.

Eqs. (2.8)-(2.9) are the basic theoretical content of the model. They give a very simple prescription for the exchange matrix element that will cause the EVB potential Eq. (2.2) to reproduce a given harmonic force field about a given transition state (or any other intermediate) geometry. Because of its Gaussian form, V_{12} is damped out away from this region so that the EVB expression [Eq. (2.2)] reduces to V_{11} or V_{22} in the reactant and product regions. It thus provides a useful way to incorporate *ab initio* quantum chemistry calculations for the transition state with simple empirical potential functions which model the non-reactive motions of the reactants and products.

We now apply this version of the EVB model to some simple, but non-trivial, test problems to illustrate its capabilities (and limitations) in a variety of situations. The first example is a two-dimensional double well potential function that has been used previously[15] as a test of various dynamical theories and also as a model for isomerization reactions such as Eq. (1.1). The specific form of the potential function is

$$V(s,Q) = V_0(s) + 1/2 \, m\omega^2 \left(Q - \frac{cs^n}{m\omega^2} \right)^2 , \tag{2.10}$$

where $V_0(s)$ is a one-dimensional symmetric double well potential, and c is a coupling constant which characterizes the strength of the coupling between the "reaction coordinate" s and the "bath mode" Q. Written in this re-normalized form, the barrier height is independent of the coupling constant. n=1 or 2 in Eq. (2.10) determines the symmetry of the coupling. In all cases the mass m is that of a hydrogen atom and the one-dimensional double well potential is

$$V_0(s) = 1/2 \ (v_{11}(s) + v_{22}(s)) - \sqrt{\left(\frac{v_{11}(s)-v_{22}(s)}{2}\right)^2 + v_{12}(s)^2} \qquad (2.11a)$$

where

$$v_{11}(s) = 1/2 \ m\omega_0^2(s+s_0)^2 \qquad (2.11b)$$

$$v_{22}(s) = 1/2 \ m\omega_0^2(s-s_0)^2 \qquad (2.11c)$$

$$v_{12}(s) = a \ exp(-bs^2) , \qquad (2.11d)$$

with parameters $\omega_0 = 1600$ cm-1, $s_0 = 1$, a = 0.036065963, b = 1.781678095 (all distances in atomic units). These parameters yield a barrier height of ~8.2 kcal/mole, which is typical of H atom transfer.

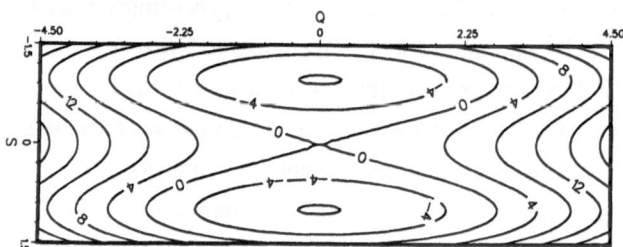

Figure 1. The double well potential energy surface of Eq. (2.10), for the uncoupled case (c=0) and a low frequency (ω=300 cm⁻¹) bath mode Q. The coordinates are in atomic units and the contour values in kcal/mole.

The examples below consider the case of a low frequency ($\omega = 300$ cm⁻¹) or a high frequency ($\omega = 3000$ cm⁻¹) bath mode, and in all cases here the diabatic potentials V_{11}

and V_{22} are taken as the harmonic normal mode potentials for reactants and products; i.e.,

$$V_{11}(s,Q) = 1/2 \, m\omega_1^2 \, s'^2 + 1/2 \, m\omega_2^2 \, Q'^2 \tag{2.12a}$$

$$V_{22}(s,Q) = 1/2 \, m\omega_1^2 \, s''^2 + 1/2 \, m\omega_2^2 \, Q''^2 \tag{2.12b}$$

where s' and Q' are the normal mode coordinates (linear combinations of s and Q) about the reactant minimum on the potential surface, and s" and Q" are the product normal mode coordinates. (The normal mode frequencies ω_1 and ω_2 are the same for reactants and products in this example because of symmetry.) As discussed in the Introduction, this is the simplest possible choice for the diabatic potentials.

Figure 1 shows a contour plot of the uncoupled (c=0) potential surface, Eq. (2.10)-(2.11), for the case of a low frequency ($\omega \cong 300 \text{ cm}^{-1}$) bath mode. Since the one-dimensional double well function $V_0(s)$ of Eq. (2.11) is of EVB form, it is clear the general EVB model, Eq. (2.2) and (2.8)-(2.9), will exactly reproduce the potential in the uncoupled limit. It is thus of interest to see how the EVB model performs as the coupling c is increased.

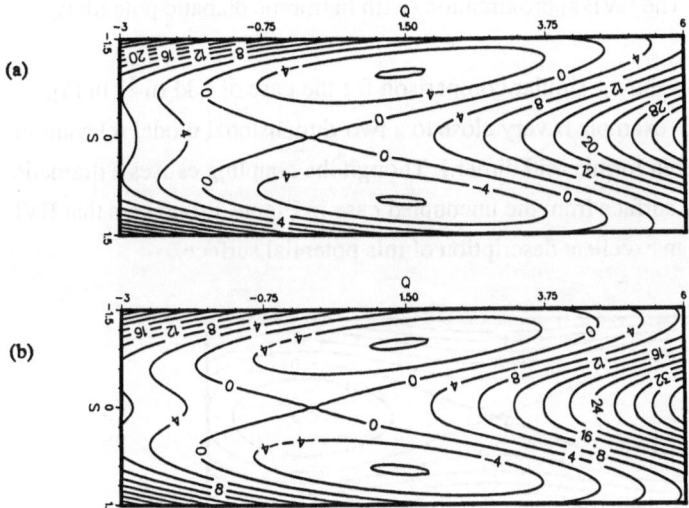

Figure 2. Same as Fig. 1, but for even (n=2) coupling with the constant c=0.005. (a) The original potential of Eq. (2.10). (b) The EVB approximation given by Eqs. (2.2) and (2.8)-(2.9), with the harmonic diabatic potentials of Eq. (2.12).

202

Figures 2a and 2b show contour plots of the original potential and the EVB approximation to it, respectively, for a modest size even (n=2 in Eq. (2.10)) coupling constant. Though some quantitative differences are apparent, on the whole the EVB model does an excellent job in representing the important regions of the potential energy surface.

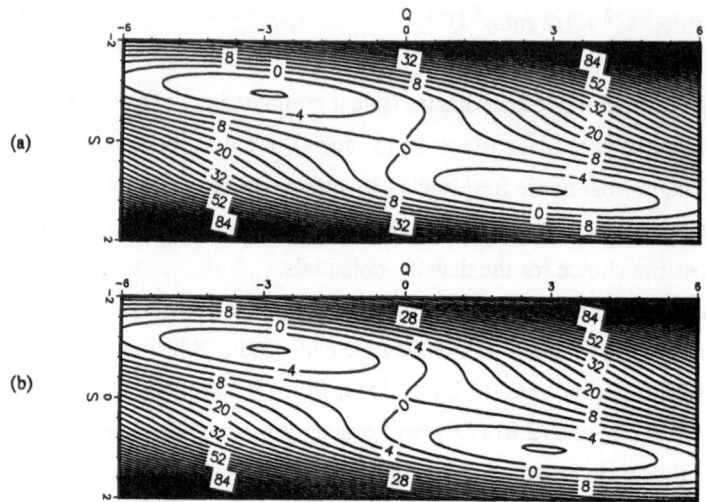

Figure 3. Same as Fig. 1, but for odd (n=1) coupling with the constant c=0.01. (a) The original potential. (b) The EVB approximation (with harmonic diabatic potentials).

Figures 3a and 3b show a similar comparison for the case of odd (n=1 in Eq. (2.10)) coupling. (This example is very close to a two-dimensional model relevant to double H atom transfer in formic acid dimer.) Though the coupling causes a dramatic change in the potential surface from the uncoupled case in Figure 1, one sees that EVB model again provides an excellent description of this potential surface.

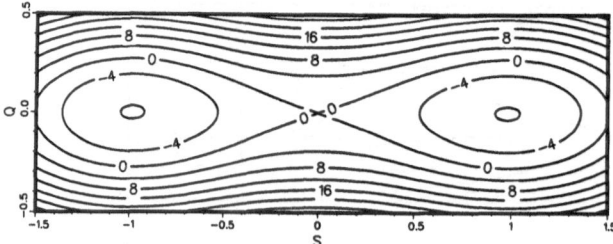

Figure 4. The double well potential function of Eq. (2.10), for the uncoupled case (c=0) and a high frequency (ω=3000 cm^{-1}) bath mode Q.

High frequency bath modes are usually easier to describe correctly than low frequency ones because the steeper harmonic potential does not allow for as large excursions in such degrees of freedom. Figure 4 shows the uncoupled (c=0) double well potential function of Eq. (2.10) for the case of a high frequency ($\omega = 3000$ cm^{-1}) bath mode. Again, the EVB model exactly reproduces the potential in the uncoupled limit, so we consider its behavior for non-zero coupling.

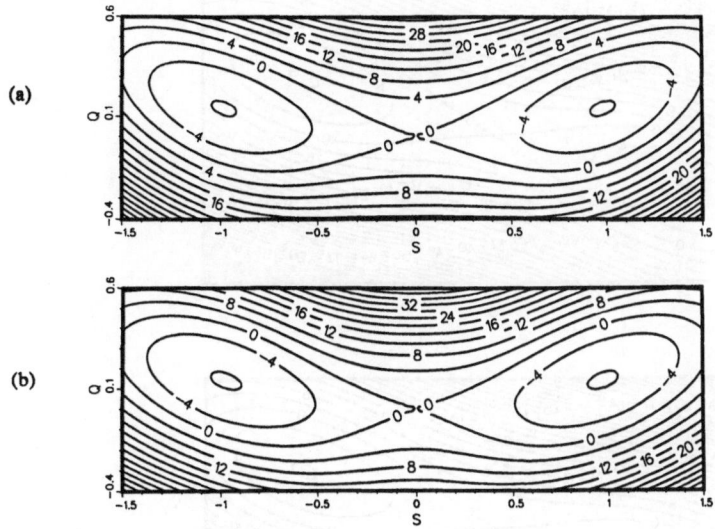

Figure 5. Same as Fig. 4, but for even coupling with the constant c=0.05. (a) The original potential. (b) The EVB approximation (with harmonic diabatic potentials).

Figures 5a and 5b show the original potential and its EVB approximation, respectively, for the case of even coupling, and Figures 6a and 6b show a similar comparison for odd coupling, both for fairly large coupling constants. (The potential wells are displaced less drastically from their uncoupled position than for the low frequency case because the high frequency of the bath mode makes the potential "stiffer" with regards to perturbation in the Q-direction.) In both cases one sees that the EVB model provides an excellent description of the true potential.

These examples show that the EVB model, with the exchange potential V_{12} chosen to reproduce the transition state region of the potential energy surface, is able to provide an excellent global description of reactive potential surfaces for a wide variety of situations. These examples have all used the simplest possible choice for the diabatic potential V_{11} and V_{22}, namely a harmonic normal model approximation about their respective minima. One can improve the model by using more accurate diabatic potentials, ones that represent the true potential more accurately over wider regions, for in this case the exchange potential is required to describe matters in a smaller region about the transition state.

(a)

(b)

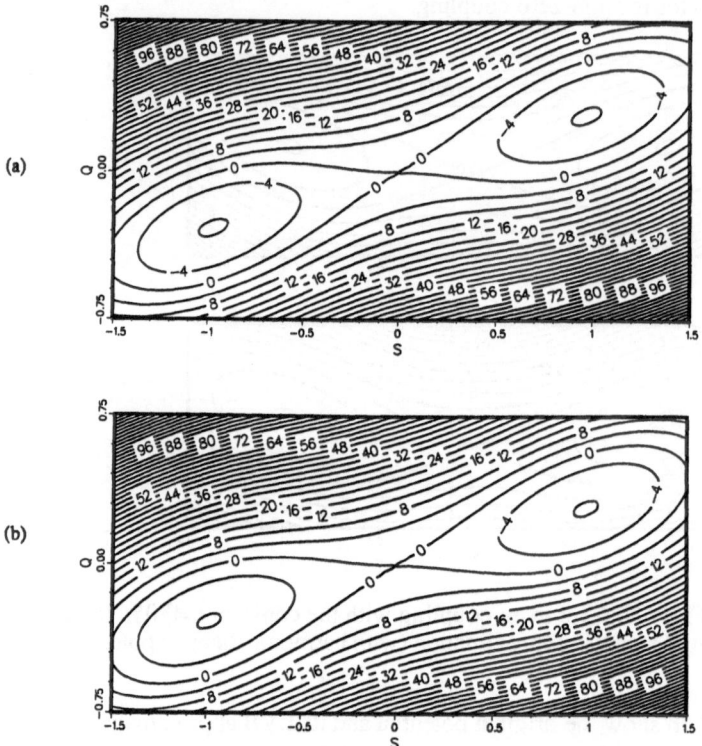

Figure 6. Same as Fig. 4, but for odd coupling with the constant c=0.1. (a) The original potential. (b) The EVB approximation (with harmonic diabatic potentials).

3. How to Deal with the Zero Point Energy Problem in Classical Mechanics

One of the frustrating shortcomings in using classical mechanics to simulate dynamical processes in polyatomic molecules has to do with a problem involving the zero point energy of vibrational degrees of freedom.[16-18] To describe the problem, recall first the simpler situation of an atom-diatom (gas phase) bimolecular reaction,

$$A + BC(v_l=0) \rightarrow AB(v_f) + C, \tag{3.1}$$

where it is indicated that the reactant diatom BC is initially in its ground vibrational state. It is well-known that a classical trajectory simulation[19] of this process works best if initial conditions for the trajectories are chosen to have the correct zero point vibrational energy in the diatom, with the initial phase of the vibrational motion selected at random (i.e., averaged over), so-called "quasiclassical" initial conditions. Agreement with (the correct) quantum reaction probabilities, or cross sections, would be much worse if the trajectory were begun with *no* vibrational energy. A problem can arise even for this simple process if the reaction is endoergic and most of the product is produced in $v_f=0$. It is possible classically to obtain reactive trajectories with less than the zero point vibrational energy in the product molecule AB, clearly an unphysical result since this permits reaction below the quantum threshold for the reaction! This problem is usually dealt with[20] by performing the classical simulation always in the *exoergic* direction and then using microscopic reversibility to obtain probabilities or cross sections in the reverse direction.

One thus believes that a classical simulation of a *polyatomic* molecular system will mimic nature (i.e., quantum mechanics) more closely if trajectories are begun with (at least) zero point vibrational energy in all vibrational degrees of freedom, with the phases of the vibrational motion selected at random (i.e., averaged over). To simulate vibrational relaxation of CH local mode overtones in benzene,[21,22] for example, it would seem most reasonable to begin trajectories with the appropriate vibrational energy in the CH stretch and zero point vibrational energy in all the other normal modes.

Since the potential energy function for the polyatomic system is in general anharmonic, energy can flow between various degrees of freedom; often, in fact, it is this intramolecular vibrational energy redistribution (IVR) that one is wishing to simulate. The "zero point energy problem"[16-18] mentioned in the first paragraph is that the energy in some vibrational modes may fall below the (quantum) zero point energy ($1/2\hbar\omega_k$, where ω_k is the harmonic frequency for mode k). This may not at first seem like a serious problem, but in even a medium size polyatomic molecule (e.g., benzene) the zero point energy is a sizeable amount of energy (52.2 kcal/mole in benzene). It is a particularly serious problem if the zero point energy flows out of several modes and "pools" into a specific weak bond. For large molecules it may even happen that the classical mechanics is chaotic at its zero point energy. These are all clearly unphysical effects that arise because classical mechanics cannot prevent the energy in each vibrational mode from dipping below its zero point value.

Here we describe a model for modifying the classical equations of motion in order to remedy this situation, i.e., to prevent the vibrational energy in each mode from at any time dipping below its zero point value. The algorithm we present affects the classical trajectory only when the vibrational energy of a mode attempts to decrease below its zero point value; otherwise the trajectory is the ordinary classical one. The algorithm conserves the total energy of the polyatomic system, and since it prevents the energy in each mode from decreasing below its zero point value, there can be no unphysical "energy pooling" of the zero point energy from many modes into one bond.

The model is first described in its simplest version, and then a more general version. The resulting algorithm is actually quite simple: if the energy in any mode k, say, decreases below its zero point value at time t, then at this time the (Cartesian) momentum p_k has its sign changed, and the trajectory continues; this is essentially a *time reversal* for mode k (only!). One can think of the model as supplying impulsive "quantum kicks" to a mode whose energy is trying to fall below its zero point value, i.e., a kind of "Planck demon" analogous to a Brownian-like random force.

The Hamiltonian is assumed to be of standard Cartesian form

$$H(\mathbf{p},\mathbf{x}) = 1/2\ \mathbf{p}^2 + V(\mathbf{x}),\tag{3.2}$$

where $\mathbf{x} \equiv \{x_k\}$ are the mass-weighted Cartesian coordinates of the system. The potential energy function consists of a harmonic part plus an anharmonic coupling,

$$V(\mathbf{x}) = V_0(\mathbf{x}) + V_1(\mathbf{x})\tag{3.3a}$$

where

$$V_0(\mathbf{x}) = \sum_k 1/2\ \omega_k^2 x_k^2\tag{3.3b}$$

It is useful to introduce the usual harmonic action-angle variables (n_k, q_k), in terms of which the Cartesian variables are

$$\chi_k = \sqrt{(2n_k+1)\hbar/\omega_k} \, \cos q_k \tag{3.4a}$$

$$p_k = -\sqrt{(2n_k+1)\hbar\omega_k} \, \sin q_k \tag{3.4b}$$

In terms of the action-angle variables, the Hamiltonian is

$$H(n,q) = H_0(n) + V_1[(x(n,q)], \tag{3.5a}$$

where

$$H_0(n) = \sum_k \hbar\omega_k \, (n_k+1/2). \tag{3.5b}$$

The actions $\{n_k\}$ are the classical counterparts to the harmonic quantum numbers, and the angles $\{q_k\}$ are the phases of the vibrational motion. [The actions $\{n_k\}$ are actually the classical action in units of $\hbar$, with the added constant "1/2" so that integer values of $\{n_k\}$ correspond to the quantum numbers; i.e., $n_k = $ (classical action)$/\hbar$ - 1/2.]

We seek a modified classical mechanics that maintains

$$n_k(t) \geq 0 \tag{3.6}$$

at all times and for all modes k. This is accomplished by adding to the Hamiltonian hard wall terms $W(n_k)$, such that

$$W(n_k) = \begin{cases} 0, & n_k > 0 \\ +\infty, & n_k < 0 \end{cases}. \tag{3.7}$$

This is analogous to a hard sphere repulsive potential $V(r)$ that keeps an interparticle distance r(t) greater than a hard sphere radius r(t), i.e.,

$$V(r) = \begin{cases} 0, & r > r_0 \\ +\infty, & r < r_0 \end{cases} \tag{3.8}$$

In the hard sphere case it is well-known how to deal with the situation: the hard sphere potential (3.8) has no effect unless r(t) decreases to the value r_0, at which time one makes the instantaneous change

$$\begin{pmatrix} r(t) \\ p_r(t) \end{pmatrix} \rightarrow \begin{pmatrix} r(t) \\ -p_r(t) \end{pmatrix} \tag{3.9}$$

i.e., the hard spheres experience an impulsive collision and are reflected. We follow this same course for the hard wall term in the action variable, Eq. (3.7).

Thus the hard wall potentials of Eq. (3.7) have no effect so long as $n_k(t) \geq 0$, but if $n_k(t)$ dips below zero at some time along the trajectory, one makes the replacement

$$\begin{pmatrix} n_k(t) \\ q_k(t) \end{pmatrix} \rightarrow \begin{pmatrix} n_k(t) \\ -q_k(t) \end{pmatrix} \tag{3.10}$$

Eq. (3.10) results from integrating the classical equations of motion with the hard wall potential $W(n_k)$ over the infinitesimal time increment of the impulsive interaction. By using Eq. (3.4), this can be expressed in terms of the original Cartesian variables as

$$\begin{pmatrix} x_k(t) \\ p_k(t) \end{pmatrix} \rightarrow \begin{pmatrix} x_k(t) \\ -p_k(t) \end{pmatrix} \tag{3.11}$$

Eq. (3.11) shows that the modification made to mode k at time t is essentially a *time reversal* for that mode. Only the mode for which $n_k(t)$ dips below 0 is modified as in (3.11).

One can verify more directly that (3.11) will indeed keep $n_k(t) > 0$. Hamilton's equations show that

$$\frac{d}{dt} \hbar n_k(t) = - \frac{\partial H(n,q)}{\partial q_k} = - \frac{\partial V_1}{\partial q_k}$$

$$= - \frac{\partial V_1}{\partial x_k} \frac{\partial x_k}{\partial q_k}$$

$$= - \frac{\partial V_1}{\partial x_k} p_k/\omega_k.$$

$$(3.12)$$

Thus if $n_k(t)$ is decreasing through 0 at time t_0 — i.e., $n_k(t_0) = 0$ and $\dot{n}_k(t_0) < 0$ — making the replacement $p_k \rightarrow -p_k$ will change the sign of $\dot{n}_k$ [cf. Eq. (3.12)] so that $\dot{n}_k(t_0) > 0$, and $n_k(t)$ will then increase and remain above 0. Since the Hamiltonian (3.2) is Cartesian, it is also clear that the instantaneous modification (3.11) conserves the value of H, i.e., the total energy.

The algorithm may be summarized as follows, all in terms of the original Cartesian variables:

1) Start the trajectory in the appropriate manner (e.g., quasiclassical initial conditions).

2) At the end of each time step in the trajectory, insert the Fortran statement

$$p_k = p_k * \text{Sign} (1/2 \, p_k^2 + 1/2\omega_k^2 x_k^2 - 1/2\hbar\omega_k) \qquad (3.13)$$

for all k.

3) Keep on computing!

Eq. (3.13) clearly accomplishes the modifications discussed above.

The simple version of the model described above prevents the zeroth order actions $\{n_k(t)\}$ of the fixed-frequency reference Hamiltonian H_0, Eq. (3.5), from becoming negative. If the dynamics of interest involves motion about a relatively well-defined equilibrium geometry, then this treatment may be adequate. In the more extreme case of a fragmentation process, e.g., unimolecular decomposition, however, it will not be a reasonable description because the physically relevant modes of the system change radically (the frequencies of some vibrations even going to zero as they evolve into rotations). In this section we show how the simple model above can be generalized by applying it to the zero point energy of the *instantaneous* normal modes.

Thus at every time t along the classical trajectory $x(t)$ we determine the energy in the various instantaneous normal modes, and then require that this not be below their respective zero point values. At the arbitrary time t, therefore, the potential is expanded through quadratic terms about the instantaneous position $x(t) \equiv x_t$,

$$V(x) \cong V(x_t) + f(x_t)^T \cdot (x-x_t) + 1/2(x-x_t)^T \cdot K(x_t) \cdot (x-x_t), \qquad (3.14a)$$

where

$$f(x) = \frac{\partial V(x)}{\partial x} \qquad (3.14b)$$

$$K(x) = \frac{\partial^2 V(x)}{\partial x \partial x} \qquad (3.14c)$$

If $\{L_k\}$ are the eigenvectors of the force constant matrix $K(x_t)$ and Ω_k^2 the eigenvalues, then the local normal coordinates Q and momenta P are related to the original Cartesian variables (x,p) by

$$x - x_t \equiv \sum_k L_k Q_k = L \cdot Q, \qquad (3.15a)$$

$$p = L \cdot P \qquad (3.15b)$$

Within this local quadratic approximation about x_t, the Hamiltonian is given in terms of the local normal mode variables (Q,P) by

$$H(P,Q) \cong V(x_t) + \sum_k (1/2 P_k^2 + D_k Q_k + 1/2 \Omega_k^2 Q_k^2), \qquad (3.16a)$$

where $\{D_k\} \equiv D$ is given by

$$D = L^T \cdot f(x_t). \qquad (3.16b)$$

Eq. (3.16a) is a sum of one-dimensional harmonic oscillator Hamiltonians, each with a shifted equilibrium position,

$$1/2P_k^2 + D_k Q_k + 1/2\Omega_k^2 Q_k^2 = 1/2P_k^2 + 1/2\Omega_k^2 [Q_k + (D_k/\Omega_k^2)]^2 - (D_k^2/2\Omega_k^2), \tag{3.17a}$$

so that the zero point energy constraint is

$$1/2P_k^2 + D_k Q_k + 1/2\Omega_k^2 Q_k^2 \geq 1/2\hbar\Omega_k - (D_k^2/2\Omega_k^2), \tag{3.17b}$$

for each mode k. We are actually interested in imposing this constraint at the instantaneous position $x = x_t$, i.e., $Q = 0$, so that the constraints we wish to impose are

$$1/2P_k^2 + (D_k^2/2\Omega_k^2) \geq 1/2\hbar\Omega_k \tag{3.18a}$$

for all k. In terms of the original Cartesian coordinates and momenta, this reads

$$1/2(L_k^T \cdot p)^2 + 1/2(L_k^T \cdot \frac{\partial V}{\partial x})^2 / \Omega_k^2 \geq 1/2\hbar\Omega_k. \tag{3.18b}$$

The left hand side of Eq. (3.18) may be thought of as the instantaneous (i.e., at time t) energy in the local, harmonic mode k: the first term of (3.18b) is clearly the kinetic energy in mode k, and the second term is the potential energy which is due to the fact that the position $x(t)$ is not at the minimum of the instantaneous harmonic potential. If some of the frequencies Ω_k are zero, or imaginary (i.e., mode k corresponds to moving over a barrier rather than in a well), then one does not apply a zero point energy constraint because there is no quantum zero point energy for such modes. We note also that more rigorously, one should diagonalize the *projected force constant matrix*,[4a] so that the six degrees of freedom corresponding to overall translations and rotations of the polyatomic system will yield zero frequencies and thus have no zero point energy constraint.

We thus summarize, in terms of the original Cartesian variables, the more general algorithm as follows:

1. Begin the trajectory in the appropriate manner to describe the process of interest.
2. After each time step diagonalize the force constant matrix $\partial^2 V/\partial x \partial x$ (more

rigorously, the *projected* force constant matrix) at the current position $x(t)$, yielding the eigenvalues Ω_k^2 and eigenvectors $\{L_k\}$.

3. For all modes k for which $\Omega_k^2 > 0$, make the replacement

$$p \rightarrow p - 2L_k(L_k^T \cdot p) \tag{3.19a}$$

if

$$1/2(L_k^T \cdot p)^2 + 1/2(L_k^T \cdot \frac{\partial V}{\partial x})^2 / \Omega_k^2 < 1/2\hbar\Omega_k. \tag{3.19b}$$

[It is not hard to show that Eq. (3.19a) corresponds to $P_k \rightarrow -P_k$ for mode k, and no change for all other modes.]

The model described above clearly does the job of not allowing the vibrational energy of any mode to dip below its zero point value. It is thus not possible for the zero point energy to "pool" in a single degree of freedom and cause unphysical behavior. Also, the imposition of the zero point energy constraint within a harmonic approximation should not be a serious limitation because the algorithm affects the classical mechanics only when the energy in a vibrational mode is near its zero point value, and most such degrees of freedom should be well described harmonically at this low level of excitation.

Various applications of this model are in progress to test its usefulness and reliability.

4. A Semiclassical Tunneling Model for use in Classical Trajectory Simulation

One of the most serious limitations of classical mechanics, which hinders its application to many interesting chemical problems, is its inability to describe tunneling effects. However the quantum mechanical phenomenon of tunneling is often quite prominent in chemical reactions that involve significant motion of light atoms. Typical examples include unimolecular dissociation or isomerization, e.g., the $H_2CO \rightarrow H_2 + CO$ decomposition, the isomerization reaction (1.1), as well as bimolecular reactions that involve H-atom transfer, e.g.,

$$H_2 + H \rightarrow H + H_2,$$

$$H_2 + F \rightarrow H + HF.$$

There do exist "rigorous" semiclassical theories that describe how classical trajectories tunnel e.g., classical S-matrix theory[23] and the "instanton" (periodic orbit in pure imaginary time) model,[24,25] but they are difficult to apply routinely to sizeable (e.g., more than three atom) molecular systems. There also exist a host of simple tunneling corrections to transition state theory[26] expressions for thermal rate constants; these often work well for this purpose, but they are not applicable to more general dynamical phenomena.

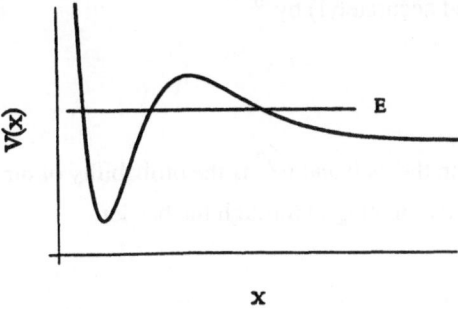

Figure 7. A typical one-dimensional potential for unimolecular decay. An energy level that corresponds to a quasi-bound state is indicated.

What we seek is a semiclassical model, as generally applicable as possible, for including tunneling in a classical trajectory simulation of the full molecular dynamics; the purpose of this paper is to present such a model. The model we have developed is similar in spirit to the Tully-Preston[27] surface hopping model for electronically non-adiabatic processes. In the Tully-Preston model a classical trajectory moving on one potential energy surface (i.e., Born-Oppenheimer electronic state) has a probability of making "hops", i.e., instantaneous transitions, to another potential energy surface at certain times. In the tunneling model presented herein the classical trajectory evolving in one classically allowed region of space will, at specific times, have a probability for making an instantaneous (in real time) transition to another classically allowed region of space. The model may also be viewed as the *classical* version of the semiclassical branching model of Waite and Miller,[28] but generalized to allow for a more general tunneling path. This more general tunneling path is very closely related to that used by Heller and Brown[29] in their semiclassical treatment of radiationless transitions.

We first give a qualitative discussion/motivation for the model, and then define it more precisely. The results of some test calculations on model Hamiltonians illustrate some of its quantitative features. In general the model is seen to provide an excellent description of tunneling phenomena over a wide range of conditions (e.g., coupling constants, different symmetries of coupling).

4.1 ONE-DIMENSIONAL CASE

It is well known that the rate of unimolecular decay from a one-dimensional well, as in Figure 7, is given semiclassically (and accurately!) by[30]

$$k = \frac{\omega}{2\pi} e^{-2\theta}, \tag{4.1a}$$

where ω is the vibrational frequency in the well and $e^{-2\theta}$ is the probability of tunneling through the barrier; θ is the classical action integral through the barrier,

$$\theta = \frac{1}{\hbar} Im \int_{barrier} p(x) dx. \tag{4.1b}$$

Since $\omega/2\pi$ is the frequency that the trajectory experiences a classical turning point at the barrier, one may interpret Eq. (4.1) as a classical trajectory that oscillates in the well, tunneling out with probability $e^{-2\theta}$ every time it "hits" the barrier. If the particle is considered to be in the well at time $t = 0$, the *net probability that the particle has tunneled*, $P_{net}(t)$ is[31]

$$P_{net}(t) = \sum_{n} h(t - t_n) P_n, \tag{4.2a}$$

where $h(\xi)$ is the usual step function ($= 1$ if $\xi > 0$, $= 0$ if $\xi < 0$), t_n are the various "tunneling times," the times that the classical trajectory $x(t)$ is at its outer turning point (i.e., "hits the barrier"), and

$$P_n = e^{-2\theta} \tag{4.2b}$$

is the tunneling probability for time t_n (here the same at each tunneling time). Figure 8 indicates the "staircase" character of $P_{net}(t)$. Averaging over the initial phase of the vibrational motion in the well will smooth out these steps, and it is not hard to see that the tunneling rate of Eq. (4.1a) is equivalently given by the slope of the averaged net tunneling probability $P_{net}(t)$, i.e.,

$$k = \frac{d}{dt} \langle P_{net}(t) \rangle. \tag{4.2c}$$

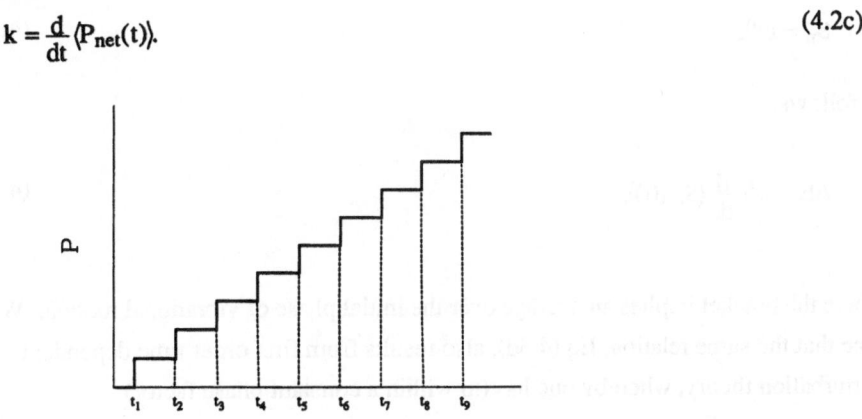

Figure 8. The net tunneling probability, $P_{net}(t)$, as a function of the time t, for a one dimensional potential of the type shown in Fig. 7. The "tunneling times" t_1, t_2,..., i.e., the times at which the trajectory experiences an outer turning point, are indicated.

The above discussion, which involves tunneling *probabilities*, applies to unimolecular dissociation, as pictured in Figure 7, and also to isomerization in an asymmetric double well potential that is irreversible on the time scale of physical interest. (Without the pre-exponential factor, it also gives the reaction probability for tunneling in bimolecular reactions, either symmetric or asymmetric.) The above description is *not* appropriate, however, to resonant tunneling in a *symmetric* double well potential, for the quantity of interest there, the splitting of the two degenerate energy levels, involves the tunneling amplitude.[30,32] For the present one-dimensional case this tunneling splitting is given semiclassically by

$$\Delta E = \frac{\hbar\omega}{\pi} e^{-\theta}, \tag{4.3a}$$

where ω and θ are the same quantities as above. Following the same analysis as in the preceding paragraph, however, one can express ΔE in terms of the *net tunneling amplitude*

$$S_{net}(t) = \sum_n h(t - t_n)S_n, \qquad (4.3b)$$

$$S_n = e^{-\theta}, \qquad (4.3c)$$

as follows:

$$\Delta E = 2\hbar \frac{d}{dt} \langle S_{net}(t) \rangle, \qquad (4.3d)$$

where the bracket implies an average over the initial phase of vibrational motion. We note that the same relation, Eq.(4.3d), also results from first order time dependent perturbation theory, whereby one has (to within a constant phase factor)

$$S(t) = H_{ab} \, t/\hbar,$$

$$\Delta E = 2H_{ab}$$

where H_{ab} is the exchange matrix element between states localized in wells "a" and "b".

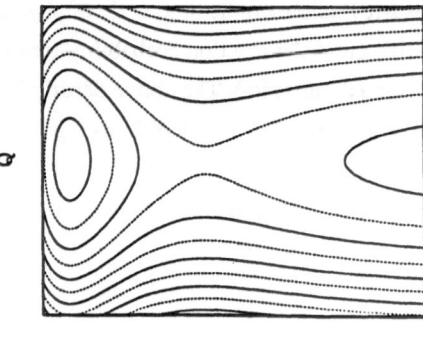

Figure 9. Contour plot of a separable two-dimensional potential. The potential in the s coordinate is that of Fig. 1, while the Q coordinate is a simple harmonic oscillator.

4.2 SEPARABLE MULTIDIMENSIONAL CASE

Now consider a separable N-dimensional potential

$$V(s,Q_1,...,Q_{N-1}) = V_0(s) + \sum_{i=1}^{N-1} V_i(Q_i), \tag{4.4}$$

where the potential for the s coordinate has a barrier, and the potentials V_i are simple oscillators. Figure 9 shows a contour plot of such a potential for N=2, and a classical trajectory for energy E below the top of the barrier is shown in Figure 10a. A separable potential is a special case of an integrable system, with the N constants of the motion E_i, i=1,...,N specified by energy conservation in each degree of freedom individually. All trajectories that correspond to the same constants of the motion are confined on an N-dimensional manifold embedded in 2N-dimensional phase space. If the motion is bounded, this manifold has the topology of a torus and its projection onto configuration space is an N-dimensional parallel piped (a "box"), with the sides (caustic surfaces) traced out by the trajectories as they go through turning points in each degree of freedom. A typical trajectory gives rise to a Lissajous-type figure (see Figure 10a). For N=2, the trajectory manifold touches the boundary of the energetically allowed region at four points, the "corners".

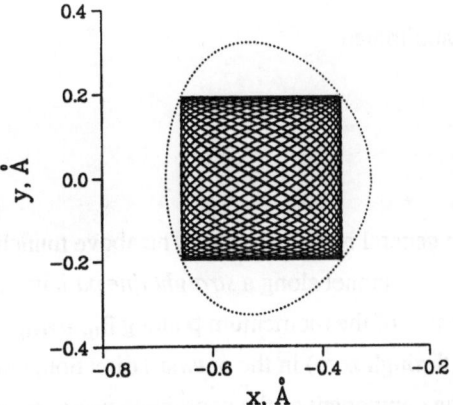

Figure 10a. A trajectory corresponding to the semiclassical ground state of a separable two-dimensional potential. The trajectory (or a set of trajectories that correspond to the same semiclassical state) defines a rectangular region bounded by the caustics. The dotted line shows the energy contour.

Since the potential is separable, tunneling involves only the s degree of freedom and is described as in Section IIa. Thus, the tunneling times $\{t_n\}$ occur whenever the s degree of freedom experiences an outer turning point, and the tunneling path is a straight line — the s axis — perpendicular to the trajectory at time t_n. The passage through the barrier takes place in pure imaginary time, during which the N-1 momenta $P_1,...,P_{N-1}$ remain constant. The decay rate is the same as that of the one-dimensional potential V_0, i.e., given by Eq. (4.2). Note that the action integral is

$$\theta = \frac{1}{\hbar}\text{Im}\int p_s ds = \frac{1}{\hbar}\text{Im}\int_{\text{tunneling path}} \mathbf{p} \cdot d\mathbf{q}, \tag{4.5}$$

where

$$\mathbf{q} \equiv (s,Q_1,...,Q_{N-1}) \text{ and } \mathbf{p} \equiv (p_s,P_1,...,P_{N-1}).$$

Finally we note that the trajectory reaches the products region with the same momentum $\mathbf{p}$ with which it began to tunnel and at the *same real time* (i.e., the tunneling involves only a pure imaginary time increment). The semiclassical picture is thus that the tunneling process is instantaneous (in real time) and conserves the momentum $\mathbf{p}$.

4.3 GENERAL (NON-SEPARABLE) CASE

We consider a generic Cartesian Hamiltonian

$$H(\mathbf{q},\mathbf{p}) = \sum_{i=1}^{N} \frac{p_i^2}{2m} + V(\mathbf{q}), \tag{4.6}$$

where the potential function V is in general non-separable. The above tunneling model is generalized by allowing the trajectory to tunnel along a *straight line path* in a specified direction $\hat{n}_0$, every time the component of the momentum $\mathbf{p}$ along $\hat{n}_0$, $\mathbf{p}\cdot\hat{n}_0$, experiences a classical turning point (i.e., goes through zero) in the outward direction; equivalently, this corresponds to the times that the component of the coordinate vector $\mathbf{q}$ along the direction $\hat{n}_0$, $\mathbf{q}\cdot\hat{n}_0$, goes through a relative maximum.

The choice of the tunneling direction $\hat{n}_0$ will be discussed more fully below. Requiring the tunneling path to be a straight line (in the full dimensional coordinate space) is, of course, an approximation in the general non-separable case, but a reasonable one. Calculations based on the more rigorous classical S-matrix[23] and "instanton" [24,25] theories show that the optimum tunneling path is relatively straight in the tunneling region.

To describe the tunneling model more specifically, it is useful to make a change of coordinates (a point transformation) $\{q_i\} \rightarrow \{x, y_1,...,y_{N-1}\}$, where x is the component of q along the tunneling path,

$$x = q \cdot \hat{n}_0 \qquad (4.7)$$

and $\{y_1,... y_{N-1}\}$ define N-1 orthogonal directions perpendicular to $\hat{n}_0$. The Hamiltonian is then expressed in the new coordinate system:

$$\widetilde{H}(x,y,p_x,p_y) = \frac{p_x^2}{2m} + \sum_{i=1}^{N-1} \frac{p_{y_i}^2}{2m} + \widetilde{V}(x,y), \qquad (4.8)$$

where $y \equiv (y_1,...,y_{N-1})$ and p_y is the vector of conjugate momenta. Since by our choice of tunneling path all components of y and p_y remain constant during the straight line tunneling process, the tunneling integral is given by

$$\theta = \frac{1}{\hbar} \mathrm{Im} \int_{\text{tunneling path}} p \cdot dq = \frac{1}{\hbar} \mathrm{Im} \int p_x dx. \qquad (4.9)$$

Furthermore, due to energy conservation along the tunneling path, we have

$$\frac{p_x(t)^2}{2m} + \frac{p_y(t)^2}{2m} + \widetilde{V}[x(t),y(t)] = \frac{p_x(t_0)^2}{2m} + \frac{p_y(t_0)^2}{2m} + \widetilde{V}(x_0,y_0) \qquad (4.10)$$

where (x_0,y_0) are the coordinates of the trajectory at the tunneling time t_0, and $t-t_0$ is pure imaginary. But $p_x(t_0) = 0$, $p_y(t) = p_y(t_0)$, and $y(t) = y_0$, so we obtain

$$p_x = i\sqrt{2m[\widetilde{V}(x,y) - \widetilde{V}(x_0,y_0)]} = i\sqrt{2m\{V[q_0 + (x-x_0)\hat{n}_0] - V(q_0)\}}. \qquad (4.11)$$

The tunneling integral is thus given

$$\theta_0 = \frac{1}{\hbar} \int_0^{\xi_{max}} \sqrt{2m[V(q_0 + \xi \hat{n}_0) - V(q_0)]} \, d\xi \qquad (4.12)$$

and the probability for tunneling at time t_0 is

$$P_0 = e^{-2\theta_0}, \qquad (4.13)$$

where ξ_{max} is the value of ξ at which the integrand of Eq. (4.12) equals zero, i.e., the value for which the tunneling trajectory reaches another classically allowed region of space.

It is useful to emphasize that the above algorithm is easily implemented in the original coordinates and momenta (q,p) of Eq. (4.6) without actually having to make the canonical transformation used above to describe it. Thus one monitors the quantity $x(t)$ of Eq. (4.7) while the trajectory $(q(t), p(t))$ is being computed; t_0 is a time at which $x(t)$ experiences a local maximum, and $q_0 \equiv q(t_0)$, $p_0 \equiv p(t_0)$. The tunneling integral is then given by Eq. (4.12), where ξ_{max} is the value of the integration variable at which the integrand vanishes, and the tunneling probability is given by Eq. (4.13). (If the integrand of Eq. (4.12) never vanishes for $\xi > 0$, then one has $\theta_0 \to +\infty$ and thus $P_0 \to 0$; i.e., the tunneling path never finds another classically allowed region.) If one wishes to follow the trajectory in the new classically allowed region — e.g., in order to determine the product state distribution — then the initial conditions for it are

$$q_{new}(t_0) = q_0 + \hat{n}_0 \xi_{max}, \qquad (4.14a)$$

$$p_{new}(t_0) = p_0. \qquad (4.14b)$$

4.4 CHOICE OF THE TUNNELING PATH

To complete the description of the model, we must specify the "tunneling direction" $\hat{n}_0$ introduced in the preceding Section. We have actually investigated a variety of choices and describe here the one which has proved most satisfactory in general and which seems most justifiable on theoretical grounds.

Consider first the initial conditions for a trajectory in the reactant potential well. For the semiclassical picture to be meaningful, we assume that the motion in the reactant potential well does not explore all energetically accessible regions of phase space, but is constrained to lie on an N-dimensional KAM torus,[33] as in the separable case, for the energy that corresponds to the desired initial conditions. This is a reasonable assumption, especially if one is interested in tunneling from low lying initial states (e.g., from the ground state). More specifically, we will be considering trajectories that start out with initial conditions (ϕ, J) in action-angle variables,[33(a)] where the actions

$$J_i = \frac{1}{2\pi} \int_{\Gamma_i} \mathbf{p} \cdot d\mathbf{q} = (n_i + \frac{1}{4}\mu_i)\hbar \qquad (4.15)$$

are defined along the N topologically distinct paths (basis contours) Γ_i on the torus and are quantized, and μ_i is the corresponding Maslov index.[34] Averaging over initial conditions then corresponds to averaging over the angles $\{\phi_i\}$, i=1,...,N. These are the polyatomic version of the standard quasiclassical initial conditions (EBK quantization).[35]

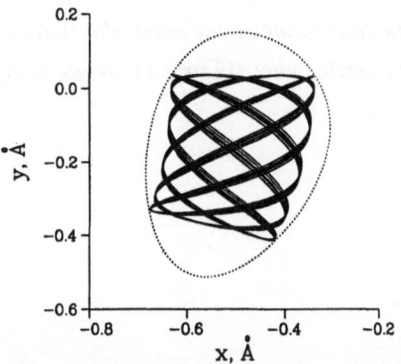

Figure 10b. A similar trajectory for a non-separable two-dimensional potential. The classical motion is regular, i.e., the trajectory lies on a KAM torus and will, over time, trace out two-dimensional region which is a subset of the energy shell. This region (the "trajectory manifold") is the projection of the KAM torus onto configuration space; it touches the energy contour at four points, the "corners", and is bounded by caustics. The dotted line shows the energy contour.

Since the motion is assumed to sweep out a torus, its projection onto configuration space still gives rise to "box"-like shapes (see Figure 10b). Unlike the separable case, the edges of such a "box" need no longer be straight lines, though one can show that they still cross at right angles near a corner;[36] i.e., the motion is locally separable about a corner. We now employ the concept of the semiclassical wavefunctions that correspond to these box-like trajectories and a simple argument from quantum mechanics to motivate our choice of the tunneling path. Quantum mechanically, the amplitude for a transition from the initial state $|\psi_i\rangle$ to the final state $|\psi_f\rangle$ is given by

$$\langle \psi_f | \hat{T} | \psi_i \rangle,$$

where T is the appropriate transition operator, in our case the Hamiltonian. The semiclassical wavefunction in the region near the caustic surfaces but *outside the trajectory manifold* can be defined (by generalizing the well known one-dimensional WKB results) as the analytic continuation of the WKB wavefunction on the manifold.[37] The wavefunction will be proportional to the exponential of a properly defined action integral, and is largest near the edges of the manifold. The tunneling amplitude accumulates its magnitude from the regions of space where the initial and final state wavefunctions overlap the most. The overlap of these wavefunctions is clearly maximized along the shortest straight line that joins these manifolds. We thus choose the tunneling direction $\hat{n}_0$ as *the straight line that connects the manifolds* that correspond to the initial and final state *in the shortest possible way*. Figure 11 shows three typical cases, and the tunneling path for each case.

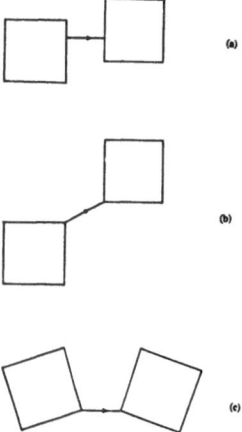

Figure 11. The boundary of the trajectory manifolds and the "tunneling path" according to the definition of Section 4.5 in three typical cases.

We note that the above choice for the tunneling direction is the same in spirit as that of Heller and Brown[29] in their treatment of semiclassical matrix elements for radiationless transitions. The primary difference is that our model does not restrict tunneling only to the "corner to corner" (or corner to edge) path, but allows for tunneling in the "corner to corner" *direction* every time the component of the motion along this direction experiences a classical turning point. We also note that this "corner to corner" tunneling direction is also very similar to that yielded in some applications of "rigorous" semiclassical theories, i.e., classical S-matrix theory[23] and the instanton model.[24,25]

4.5. APPLICATION: TUNNELING RATES IN MODEL POTENTIALS

We envision the present model to be most useful for describing tunneling processes in complex molecular systems involving unimolecular decomposition and isomerization. To provide a qualitative test, however, here we show how it performs on some simple two-dimensional Hamiltonians of the following form

$$H = \frac{p_s^2}{2m} + \frac{p_Q^2}{2m} + V(s,Q). \tag{4.16}$$

We performed calculations of the tunneling splitting in symmetric double wells and of uni-molecular decay rates from quasi-bound initial states. The specific form of the model potential that we used for this application is the same as that in our earlier papers,[15]

$$V(s,Q) = V_0(s) + \frac{1}{2} m\omega^2 \left[Q - \frac{f(s)}{m\omega^2} \right]^2. \tag{4.17a}$$

Here s is the reaction coordinate, $V_0(s)$ is a symmetric double well,

$$V_0(s) = -\frac{1}{2} a_0 s^2 + \frac{1}{4} c_0 s^4, \tag{4.17b}$$

f(s) is the coupling function, and Q is the orthogonal harmonic degree of freedom, which is linearly coupled to the reaction coordinate. Notice that the term $[f(s)]^2/2m\omega^2$ in Eq. (4.17a) renormalizes the height of the barrier so that it is independent of the strength of the coupling, and the dependence of the tunneling splitting on the coupling is merely due to the distortion of the potential surface away from the separable case. The constants were chosen such that the barrier height is 7.8 kcal/mol and the two potential minima are located at $s_\pm = \pm 0.53$Å. The mass was chosen to be that of a hydrogen atom, and the frequency of the harmonic oscillator was 298 cm^{-1}. These parameters are the same as those in Ref. 15 and are typical of H-atom transfer processes.

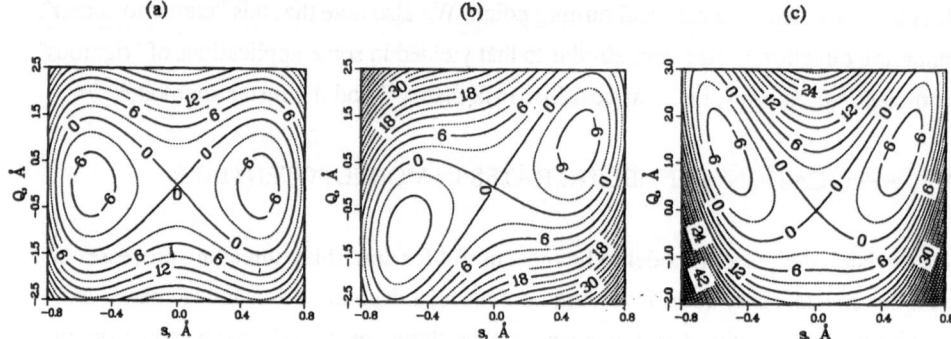

(a)   (b)   (c)

Figure 12. Contour plot of the two-dimensional potential used in the calculations [cf. Eq. (4.17)]. The potential in the s coordinate is a symmetric double well, while the Q coordinate is a harmonic oscillator of frequency 298 cm^{-1}. (a) Separable case (c=0); (b) Linear coupling, f(s)=cs, for c=0.004 hartree/bohr^2; (c) Quadratic coupling, f(s)=cs^2, for c=0.004 hartree/bohr^3.

We considered two different forms of the coupling function f(s) which give rise to different symmetries in the full (coupled) potential: (i) linear coupling, f(s)=cs, for which V possesses inversion symmetry, and (ii) quadratic coupling, f(s)=cs^2, in which case V has reflection symmetry with respect to the Q axis. Figure 12 shows contour plots of the potential in these two different cases for typical values of the coupling constant c.

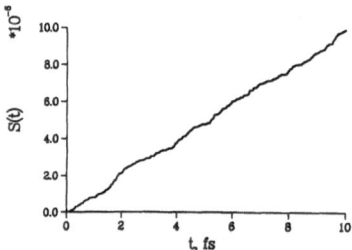

Figure 13. The net tunneling amplitude [cf. Eq. (2.3a)], averaged over 1000 trajectories, for a trajectory in the potential of Fig. 12b.

In the case of linear coupling, the two wells move apart in the Q direction as the coupling increases. For small values of c, our rule for choosing the tunneling path gives $\hat{n}_0 = \hat{s}$, i.e., the trajectories tunnel purely in the s direction (cf. Figure 11a). As the coupling constant gets larger, however, the shortest line that connects the "boxes" becomes the line that connects the nearest corners passing through the transition state (cf. Figure 11b). In the case of quadratic coupling, the tunneling path is always the line that connects the nearest corners, which is in this case the s direction (cf. Figure 11c).

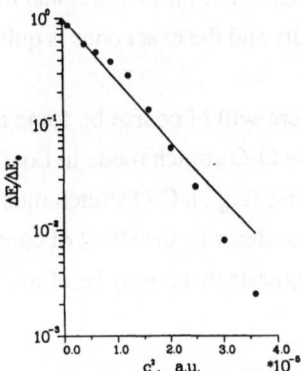

Figure 14. The tunneling splitting ΔE, normalized by the exact quantum value ΔE_0 of the splitting in the one-dimensional double well, as a function of the square of the coupling constant c, for the case of linear coupling, f(s)=cs. Solid line: exact quantum results, obtained by a basis set calculation. Circles: results obtained by using the semiclassical model presented in this paper.

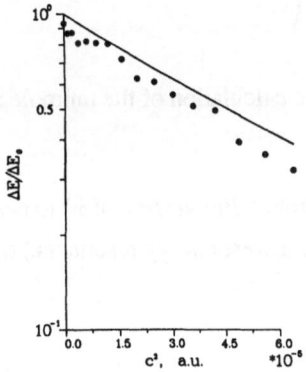

Figure 15. Same as Fig. 14, except for quadratic coupling $f(s)=cs^2$.

Since the double well is symmetric, we use the amplitude version of the model, as described at the end of Sec. 4.1. Figure 13 shows a typical graph of the net tunneling amplitude $S_{net}(t)$ defined by Eq. (4.3b), with $S_n = e^{-\theta_n}$, averaged over 1000 trajectories. The linear character of this function is obvious, and the slope gives the tunneling splitting according to Eq. (4.3c). Figures 14 and 15 show the tunneling splitting ΔE as given by this semiclassical model, normalized by the exact (quantum) value ΔE_0 of the tunneling splitting at zero coupling, as a function of the square of the coupling constant c, for the cases of linear and quadratic coupling. Also shown are the exact quantum mechanical values of $\Delta E/\Delta E_0$, obtained by numerical diagonalization of the Hamiltonian in a basis set. The agreement between the semiclassical results and the exact ones is quite good, even when the coupling is very strong.

In a real molecular system, e.g., Eq. (1.1), there will of course be some modes most akin to the even coupling case above (e.g., the O-O stretch mode in Eq. (1.1) and other modes that are typified by the odd coupling use [e.g., a C-O stretch mode in (1.1)]. The fact that the above semiclassical model is able to describe the effect of coupling on the tunneling dynamics for both kinds of modes suggests that it may be of useful generality for more complex molecular systems.

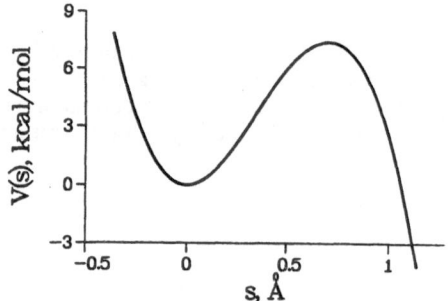

Figure 16. The potential V_0 which was used in the calculation of the unimolecular decay rate [cf. Eq. (4.18b)].

Finally, we apply the tunneling model (the probability version of it) to calculate the decay rate from the ground quasibound state (i.e., lowest energy resonance) of a two-dimensional potential which has the form

$$V(s,Q) = V_0(s) + 1/2m\omega^2 Q^2 - csQ, \qquad (4.18a)$$

$$V_0(s) = 1/2a_0s^2 - 1/3b_0s^3. \tag{4.18b}$$

The one-dimensional potential V_0 is shown in Figure 16; the local maximum occurs at $s=0.71$Å and the barrier height is 7.4 kcal/mol. The mass was chosen to be that of a hydrogen atom, and the frequency of the harmonic oscillator was 298 cm^{-1}. Figure 17 shows a contour plot of the two-dimensional potential for a typical value of the coupling parameter c. The semiclassical decay rate was calculated according to Eq. (4.2).

In order to generate accurate quantum mechanical results for comparison, we computed the width of the resonance using the method of complex scaling.[38] The s coordinate was rotated as $s \rightarrow se^{i\alpha}$, while the Q coordinate remained real. The complex scaled Hamiltonian was then diagonalized in a basis set of (real) particle in a box basis functions and the complex eigenvalues

$$E = E_R + iE_I \tag{4.19a}$$

which were stable under a change of the scaling angle α were identified as resonances, whose width is

$$\Gamma = -2E_I \tag{4.19b}$$

The decay rate k is obtained in terms of the resonance width according to

$$k = \frac{\Gamma}{\hbar}. \tag{4.19c}$$

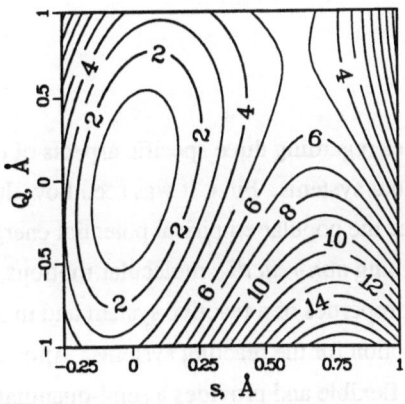

Figure 17. Contour of the two-dimensional potential Eq. 4.18, for c=0.004 hartree/bohr^2. The potential in the s coordinate is shown in Fig. 16, while the Q coordinates is a harmonic oscillator of frequency 298 cm^{-1}.

Figure 18 compares the results of the semiclassical model for the decay rate with the quantum mechanical ones, i.e., those of the complex scaling calculation. Plotted is the decay rate k (normalized by the exact quantum value of the rate for the uncoupled potential) as a function of the square of the coupling constant c. The agreement is of the same quality as in the previous applications — within a factor of 2 in the worst case, while the coupling is so strong that it has increased the value of k by two orders of magnitude.

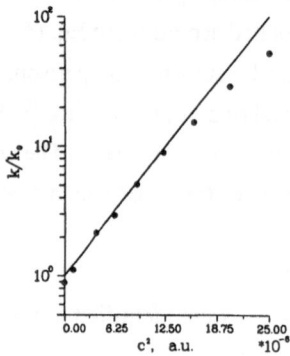

Figure 18. The decay rate k, normalized by the exact quantum value k_0 of the rate in the one-dimensional potential V_0, as a function of the square of the coupling constant c. Solid line: exact quantum results, obtained by the method of complex scaling. Circles: results obtained by using the semiclassical model described in this paper.

5. Concluding Remarks

This lecture has described recent advances regarding three specific aspects of chemical reaction dynamics in polyatomic molecular systems. First, it was seen how the empirical valence bond idea can be used to combine the popular empirical potential energy functions, which do a good job of describing *non-reactive* molecular motions, with *ab initio* calculations of the transition state properties of a reactive system and in a simple way obtain a *global* potential energy function for the reacting systems. Applications to test problems suggests that this model is flexible and provides a semi-quantitative description in a variety of situations. Recent applications to the formaldehyde dissociation

$$H_2CO \rightarrow H_2 + CO$$

has given an excellent description of this reaction.

Second, it was shown how a relatively simple procedure can be used to correct (approximately) classical trajectories for zero point energy effects. This model prevents the vibrational energy in each individual mode from dropping below its instantaneous zero point value and thus prevents unphysical effects that could result otherwise. Finally, a very general semiclassical model was described for including quantum tunneling effects (approximately) in a classical trajectory simulation. Applications of this model to a variety of examples showed the model to be quite accurate in a variety of topologically different situations.

All of these methodologies should be applicable (in some form) to carrying out theoretical simulations of reaction dynamics in biomolecular systems.

Acknowledgment

This work has been supported by the Director, Office of Energy Research, Office of Basic Energy Sciences, Chemical Sciences Division of the U.S. Department of Energy under Contract No. DE-AC03-76SF00098.

References

1. Good overviews of the problem of representing potential energy surfaces are the reviews (a) D. G. Truhlar, R. Steckler, and M. S. Gordon, Chem. Rev. 87, 217 (1987); and (b) G. C. Schatz, Rev. Mod. Phys. 61, 669 (1989).
2. See, for example, W. J. Hehre, L. Radom, P. v. R. Schleyer, and J. A. Pople, Ab Initio Molecular Orbital Theory, Wiley, N.Y., 1986.
3. R. A. Marcus, J. Chem. Phys. 49, 2610 (1968).
4. (a) W. H. Miller, N. C. Handy and J. E. Adams, J. Chem. Phys. 72, 788 (1980).
   (b) W. H. Miller, J. Phys. Chem. 87, 3811 (1983).
5. K. Fukui, Acct. Chem. Res. 14, 363 (1981).
6. J. T. Hougen, P. R. Bunker, and J. W. C. Johns, J. Mol. Spectrosc. 34, 136 (1970).

230

7. (a) B. A. Ruf and W. H. Miller, J. Chem. Soc. Faraday Trans. 2 $\underline{84}$, 1523 (1988).

    (b) W. H. Miller, B. A. Ruf, and Y.-T. Chang, J. Chem. Phys. $\underline{89}$, 6298 (1988).

8. (a) N. L. Allinger, J. Am. Chem. Soc. $\underline{99}$, 8127 (1977).

    (b) N. L. Allinger, Y. H. Yuh, and J.-H. Lii, J. Am. Chem. Soc. $\underline{111}$, 8551 (1989).

    (c) J.-H. Lii and N. L. Allinger, J. Am. Chem. Soc. $\underline{111}$, 8566 (1989).

    (d) J.-H. Lii and N. L. Allinger, J. Am. Chem. Soc. $\underline{111}$, 8576 (1989).

9. B. R. Brooks, R. E. Bruccoleri, B. D. Olafson, D. J. States, S. Swaminathan, and M. Karplus, J. Comput. Chem. $\underline{4}$, 187 (1983).

10. (a) P. K. Weiner and P. A. Kollman, J. Comput. Chem. $\underline{2}$, 287 (1981).

    (b) S. J. Weiner, P. A. Kollman, D. A. Case, U. C. Singh, C. Ghio, G. Alagona, S. Profeta, Jr., and P. Weiner, J. Am. Chem. Soc. $\underline{106}$, 765 (1984).

    (c) S. J. Weiner, P. A. Kollman, D. T. Nguyen, and D. A. Case, J. Comput. Chem. $\underline{7}$, 230 (1986).

11. W. L. Jorgensen and J. Tirado-Rives, J. Am. Chem. Soc. $\underline{110}$, 1657 (1988).

12. (a) A. Warshel and S. Lifson, J. Chem. Phys. $\underline{49}$, 5116 (1968).

    (b) A. Warshel, in Modern Theoretical Chemistry, Vol. 7, ed. G. A. Segal, Plenum, 1977, p. 133.

13. (a) A. Warshel and R. M. Weiss, J. Am. Chem. Soc. $\underline{102}$ 6218 (1980).

    (b) A. Warshel, Biochemistry $\underline{20}$, 3167 (1981).

    (c) A. Warshel, Acc. Chem. Res. $\underline{14}$, 284 (1981).

14. See, for example, R. Kubo, J. Phys. Soc. Japan $\underline{17}$, 1100 (1962).

15. N. Makri and W. H. Miller, J. Chem. Phys. $\underline{86}$, 1451 (1987); $\underline{87}$, 5781 (1987); $\underline{91}$, 4026 (1989).

16. R. A. Marcus, Ber. Bunsenges. Phys. Chem. $\underline{81}$, 190 (1977) .

17. W. L. Hase and D. G. Buckowski, J. Comput. Chem. $\underline{3}$, 335 (1982).

18. G. C. Schatz, J. Chem. Phys. $\underline{79}$, 5386 (1983) .

19. For review, see

    (a) R. N. Porter and L. M. Raff, in Dynamics of Molecular Collisions, B, ed. W. H. Miller, Plenum, NY, 1976, p. 1;

    (b) L. M. Raff and D. L. Thompson, in Theory of Chemical Reaction Dynamics, Vol. III, ed. M. Baer, CRC Press, Boca Raton, FL, 1985, p. 1.

20. See, for example, J. M. Bowman, G. C. Schatz, and A. Kuppermann, Chem. Phys. Lett. 24, 378 (1974) .

21. D.-H. Lu and W. L. Hase, J. Chem. Phys. 89, 6723 (1988).

22. D.-H. Lu and W. L. Hase, J. Chem. Phys., submitted.

23. W. H. Miller, (a) Adv. Chem. Phys. 25, 69 (1974); (b) ibid 30, 74 (1975); (c) Science 233, 171 (1986).

24. W. H. Miller, J. Chem. Phys. 62, 1899 (1975).

25. (a) S. Colemam, Uses of Instantons, in The Whys of Subnuclear Physics, edited by A. Zichichi, Plenum, N.Y., 1979, pp. 805-916;

    (b) A. 0. Caldiera and A. J. Leggett, Ann. Phys. (N.Y.) 149, 374 (1983).

26. Some examples are:

    (a) D. G. Truhlar and B. C. Garrett, Ann. Rev. Phys. Chem. 35, 159 (1984);

    (b) G. C. Lynch, D. G. Truhlar, and B. C. Garrett, J. Chem. Phys. 90, 3102 (1989);

    (c) R. A. Marcus and M. E. Coltrin, J. Chem. Phys. 67, 2609 (1977);

    (d) C. J. Cerjan, S. Shi, and W. H. Miller, J. Phys. Chem. 86, 2244 (1982).

27. J. C. Tully and R. K. Preston, J. Chem. Phys. 55, 562 (1971).

28. B. A. Waite and W. H. Miller, J. Chem. Phys. 76, 2412 (1982).

29. E. J. Heller and R. C. Brown, J. Chem. Phys. 79, 3336 (1983).

30. See, for example, K. W. Ford, D. L. Hill, M. Wakano, and J. A. Wheeler, Ann. Phys. (N.Y.) 7, 239 (1959).

31. Eq. (4.2a) is correct for small tunneling probabilities, $P_n \ll 1$. More generally, one must include a "survival probability" factor, so that Eq. (4.2a) becomes

$$P_{net}(t) = \sum_{n=1} h(t - t_n)P_n, \text{ where } p_1 = P_1 \text{ and } p_n = P_n x\left[1 - \sum_{n'=1}^{n-1} P_{n'} \right] \text{ for } n > 1.$$

32. Also see the discussion by W. H. Miller, J. Phys. Chem. 83, 960 (1979).

33. (a) A. J. Lichtenberg and M. A. Lieberman, Regular and Stochastic Motion (Springer, New York, 1983);

    (b) M. V. Berry, Regular and irregular motion, in Topics in Nonlinear Dynamics, S. Joma, ed. A.I.P. Conference Proceedings 46, 1976; V. I. Arnold, Mathematical Methods of Classical Mechanics (Springer, New York, 1978).

34. V. P. Maslov and M. V. Fedoriuk, Semiclassical Approximation in Quantum Mechanics, (Reidel, Boston, 1981).

232

35. (a) A. Einstein, Verh. Dtsch. Phys. Ges. $\underline{19}$, 82 (1917);
    (b) M. L. Brillouin, J. Phys. $\underline{7}$, 353 (1926);
    (c) J. B. Keller, Ann. Phys. $\underline{4}$, 180 (1958).
36. P. Pechukas, J. Chem. Phys. $\underline{57}$, 5577 (1972).
37. (a) M. Wilkinson, Physica $\underline{21D}$, 341(1986);
    (b) N. De Leon and E. J. Heller, Phys. Rev. $\underline{A\ 30}$, 5 (1984).
38. (a) B. R. Junker, Adv. Atom. Mol. Phys. $\underline{18}$, 207 (1982);
    (b) W. P. Reinhardt, Ann. Rev. Phys. Chem. $\underline{33}$, 223 (1982);
    (c) B. A. Waite and W. H. Miller, J. Chem. Phys. $\underline{74}$, 3910 (1981).

DISCUSSION

Prof. Lluch: *In order to calculate the ZPE in a non-stationary point, one can obtain the hessian matrix, project it in the direction orthogonal to the gradient (and to traslations and rotations) and then diagonalize the projected Hessian matrix. Is this procedure equivalent to one described here?*

Prof. Miller: Diagonalizing the projected Hessian matrix as you have described in your question will give the zero point energy in all directions **except** the gradient direction. (This is the type of calculation done in constructing the reaction path Hamiltonian, for example, when the gradient direction is along the reaction path; see ref. 4.) For present purposes, however, one needs to have the <u>total</u> zero point energy in each mode, including that associated with the gradient direction also, and this is given by Eq. (3.18b) of the text.

Prof. Lluch: *We have applied the semiclassical tunneling model proposed by Makri and yourself (JCP, 91, 4026 (1989) to the study of tunneling in the intramolecular proton transfer in malonaldehyde and hydrogenoxalate anion. An important problem is the choice of the tunneling direction in each classical turning point. If a straight path is chosen we obtain a tunneling splitting that is clearly lower than the one obtained using the basis set method on the same potential energy surfaces. Do you think that it would be necessary to optimize the tunneling path direction in each turning point in order to improve the tunneling splitting?*

Prof. Miller: You are absolutely correct that the choice of tunneling direction is the most critical part of this tunneling model. A straight line path is clearly the simplest, and easiest to generalize to a variety of applications, but it may not always be the best. The "best" choice of a tunneling path is the semiclassical "instanton" result (see ref. 24), but this is not simple to apply. The best compromise between accuracy and applicability is still, I believe, an open question.

Prof. Mezey: *My fundamental problem with an intuitive understanding of dynamics including zero point energy and tunneling considerations is with the subdivision of the model into primarily quantum mechanical and primarily classical mechanical components. I see the concept of trajectory as one motivated primarily on classical mechanics. In the spirit of the topological (non-geometrical) model of potential energy surfaces one may incorporate, in an approximate way, some components of zero point energy and tunneling contributions within a "trajectory-free" approach by defining an approximate, "intrinsic kinetic energy", which is inseparable from the actual potential energy surface (P.G. Mezey, Theor. Chim. Acta, 67, 115 (1985)). The intrinsic kinetic energy is defined locally, containing zero point energy and formal "tunneling reproducing" energy terms, with the motivation of generating an approximate, "topologically classical" surface that mimics the quantum mechanical behaviur on a classical surface. This model, however, misses the original concept of trajectories. In your opinion, if one retains trajectories is it possible to assign some parts of zero point energy and tunneling effects to some "intrinsic kinetic energy" term added to the potential?*

Prof. Miller: I do not see how one can mimic zero point and tunneling effects with reference only to the potential energy surface itself. This "static" picture misses the feature of kinetic energy. I.e., a particle can be at the minimum of a potential surface if it has suffient kinetic energy, but not if it has more. Thus I believe that the trajectory, i.e., dynamical picture is necessary. But of course you may find a way to do it!

Prof. Löwdin: *Your fine lecture shows that it is possible to refine "molecular dynamics" to include many quantum effects. Prof London pointed out more than 40 years ago that the minimum energy required to take care of the uncertainty by relations is the zero-point energy, and your scheme leads to a good estimate of the zero-point vibrations.*

As to the simple model of the transition state theory based on a generalization of the valence bond model associated with a second-order Hamiltonian matrix- as described by Arieh Warshel- it may very well happen that the perturbation treatment is divergent, but even in this case the matrix describes a "resonance state" with a finite life-time, as pointed out by Jack Smons in another connection.

Your semi classical treatment of the tunneling phenomen is very nice, and one knows -particularly thanks to the work by Hans Franenfelder- that molecular tunneling is of great importance also in molecular biology. In molecular physics, the tunneling through 3-dimensional barriers is very important, whereas in the standard quantum-mechanical text books only the tunneling through one-dimensional barriers is treated. Do you have any idea how to treat the more general problems?

Prof. Miller: The question is do I have any ideas of how to treat tunneling in more general poblems. All of the problems discussed in the lecture were **multidimensional** tunneling treatments, so in a sense this is dealing with these more general (than one dimensional) tunneling problems. The approximation we used was that the tunneling takes place along a straight line but **in a multimensional space**. It is thus not at all a one dimensional tunneling model. The straight line along which the system tunnels, for example, is different for different trajectories (even though the **direction** of the straight line in the multimensional space is the same).

Prof. Naray Szabo: *What might be the effects in biomecular systems that need to be treated by the correct reaction dynamics approach? Yesterday we heard about tunneling but are there others?*

Prof. Miller: A very good question, i,e., what practical manifestations do these more rigorous dynamical treatments have in biomolecular systems? I really do not know. As you noted, tunneling can certainly be important, particularly in H atom transfer reactions, and many of these are involved in biomolecular reactions. I believe that a proper treatment of zero point energy effects is also necessary if one is going to be able to stimulate chemical reactions reliably. Of course most current simulations of non-reactive processes (protein folding, docking, etc.) freeze out the high frequency motions (i.e., assume rigid bonds) so that zero point energy is not involved. But if these degrees of freedom are involved in a reaction that one tries to simulate, then it will be necessary to deal with this issue in a reliable manner.

A Strategy for Modelling of Chemical Reactivity using MC-SCF and MM-VB Methods

Massimo Olivucci, Ioannis N. Ragazos, and Michael A. Robb
Department of Chemistry
King's College London
Strand
London WC2R 2LS

F. Bernardi
Dipartimento Chimico 'G. Ciamician'
Universita di Bologna
Bologna

ABSTRACT. The problem of modelling chemical reactivity for larger molecular systems is discussed. A strategy involving VB theory implemented via a Heisenberg Hamiltonian is described. The central features of such an approach are a) it is possible to combine the quantum mechanics with force field methods for the study of reactivity and b) it is compatable with being parametrized using the best *state of the art* ab-initio methods that are available. The utility of the method is illustrated on the photochemistry of a steroid.

1. Introduction

While reactivity in larger molecular systems including systems of biological interest is often discussed on the basis of structure reactivity correlations this approach obviously has its limitations. The alternative, determining reaction paths together with studying transition structure and associated energetics using methods of quantum chemistry has until recently been prohibitively expensive. In the last few years, the availability of dedicated workstations with a peak performance close to 200 mflops (millions of floating point operations per second), together with the development of "direct" methods [1] in quantum chemistry (where integrals are used directly in the the SCF or CI method and not written to disk) which enable such workstations to be used efficiently, have changed this situation completely. Now ab-initio computations with several hundred atoms are feasible. The objective of the present article is to review the strategy for the study of reactivity in larger chemical systems in the light of recent advances in computing technology and software development.

In the modelling of larger molecular systems, semi-empirical (SE) methods (see for example reference [2]) and molecular mechanics (MM) [3-5] are the current interim standard. The commonly used SE methods are parametrized to reproduce experiment within the SCF approach. Similarly MM methods are parametrised relative to experiment. However in problems of chemical reactivity, there is simply no direct experimental data on the transition state region; thus such SE methods and MM approaches are severely

J. Bertrán (ed.), Molecular Aspects of Biotechnology: Computational Models and Theories, 237–250.
© 1992 *Kluwer Academic Publishers.*

limited. In particular, in SE methods, the SCF method is used in most cases; yet this method cannot properly describe homolytic bond fission to give bi-radicals so it sometines gets *the correct answer for the wrong reason*. Similarly, MM requires one to identify *atom types* (sp^2 sp^3 carbon) from the outset and has no mechanism for describing a transition state (which for a carbon atom might be $sp^{2.5}$). Further, even if one confined oneself to the transition state region, the necessary data for describing structure (reference bond lengths and stretching potentials) in order to reparametrize MM can only be obtained from theoretical computations[5]. Thus it seems as if a rather fundamentally approach is required in which the *de facto* standard for modelling structural problems, MM, is combined with quantum mechanics. The purpose of this report is to summarize some aspects of our current work in this area with some examples.

In the progression to larger molecules, while one may run *benchmark* ab-initio computations for prototypical examples, some sort of semi-empirical model must be used to investigate individual reactivity problems on a routine basis. Further, it seems obvious that a hybrid approach will be necessary where the *active* sites (ie those involved in bond breaking) are treated using a quantum mechanical model and the remainder of the system is treated by force field methods. There are then two issues: What type of quantum mechanical model should be used? How is this model to be parametrized by recourse to experiment or via ab-initio computations? It seems as if the most desirable quantum mechanical model to use should satisfy the following criteria: a) it must be possible to combine the quantum mechanics with force field methods and b) it must be compatible with being parametrized using the best *state of the art* ab-initio methods that are available. We will argue that the valence bond (VB) method implemented using the Heisenberg Hamiltonian methods pioneered in chemistry by the Toulouse group [6] provides a model that satisfies these two criteria and we shall demonstrate the efficiency of the model on a model reactivity problem.

2. The VB Model Implemented via a Heisenberg Hamiltonian

A VB formulation of chemical reactivity problems is attractive because one can focus ones attention directly on the bond making/breaking phenomena. This arises because one can attach a physical significance to the various covalent VB configurations. The description of a covalent bond is familiar to all readers and we briefly illustrate the basic approach with this simple example. For a 2 electron 2 orbital system, one has a covalent wavefunction formed from 2 *active* orbitals Φ_a and Φ_b (for the H_2 molecule these would be H 1s orbitals)

$$\psi_1^3 = |\Phi_a\overline{\Phi_b}| \pm |\overline{\Phi_a}\Phi_b|$$

[1]

and the energy simply

$$E_1^3 = Q \, {\textstyle \frac{-}{+}} \, K$$

[2]

Where the parameters Q and K are two centre coulomb and exchange integrals

$$\left(1 + \langle \Phi_a | \Phi_b \rangle^2\right) Q_{ab} = \langle \Phi_a | \hat{h} | \Phi_a \rangle + \langle \Phi_b | \hat{h} | \Phi_b \rangle + [\Phi_a \Phi_a \frac{1}{r_{12}} | \Phi_b \Phi_b] + V_{NN}$$

[3]

$$\left(1 + \langle \Phi_a | \Phi_b \rangle^2\right) K_{ab} = \langle \Phi_a | \Phi_b \rangle \left\{ \langle \Phi_a | 2\hat{h} + V_{NN} | \Phi_b \rangle \right\} + [\Phi_a \Phi_b \frac{1}{r_{12}} | \Phi_a \Phi_b]$$

[4]

where h is the one-electron kinetic + electron attraction operators, V_{NN} is the nuclear-nuclear repulsion and r_{12}^{-1} is electron-electron repulsion operator. In matrix notation, taking as basis states the two neutral VB determinants,

$$|\Phi_a \overline{\Phi_b}|$$

and

$$|\Phi_b \overline{\Phi_a}|$$

the VB wavefunction is an eigenfunction of a Heisenberg hamiltonian with the matrix representation shown below.

$$\hat{H}_s = \begin{bmatrix} Q & K \\ K & Q \end{bmatrix}$$

[5]

If we invert these relationships one has

$$\frac{Q}{K} \equiv \frac{\left(E^1 \pm E^3\right)}{2}$$

[6]

Thus it is clear that one could fit the singlet and triplet energies as a function of distance with two parameters Q and K. Thus H_s becomes a *model* hamiltonian with 2 parameters. Since the singlet and triplet energies could be computed to any level of sophistication in the quantum mechanical model, H_s could be determined to any level of accuracy required. But what has happened to the orbitals Φ_a and Φ_b? They seem to have vanished from the problem! Why are there no explicit ionic configurations of the form

$$|\Phi_a \overline{\Phi_a}| \text{ or } |\Phi_b \overline{\Phi_b}|$$

One seems to be getting the right answer for the wrong reason. Thus one needs to give some further justification as to why such a model has a firm foundation.

Let us discuss the orbitals Φ_a and Φ_b first. In fact they are undefined! Rather one is free to rationalize the behaviour of Q and K in terms of the undefined non-orthogonal orbitals in eq 3 and 4. This in turn will suggest a suitable form of function for representing Q and K. Modern VB computations of the type carried out by Geratt and his co-workers[7] have demonstrated in many examples that Φ_a and Φ_b are distorted atomic-like hybrid orbitals. In this case Q and K have a simple physical interpretation. The exchange integral K will be dominated by the one electron term

$$k_{ab} = 2 \langle \Phi_a | \Phi_b \rangle \langle \Phi_a | \hat{h} | \Phi_b \rangle$$

[7]

which simply represents the nuclear electron attraction of an "overlap density" between Φ_a and Φ_b and should behave much like the overlap itself. The coulomb term Q for H_2 is well understood and is the net result of electron-nuclear attraction and nuclear repulsion.

Now we can understand the reason why ionic configurations do not occur. In fact, orbital distortion (or delocalization) and the superposition of ionic configurations perform the same function. As Gerratt and co-workers [7] have shown in their spin-coupled VB model, as long as the orbitals are allowed to distort in the optimization process, the ionic configurations are redundant.

How are we to interpret the preceding formulae for an isolated covalent bond in a general molecular situation? Clearly K remains unchanged but Q must now incorporate the energy of the remaining fragment of the molecule as well as a contribution from the *active* orbitals Φ_a and Φ_b. If we are to describe the energy of a reaction in which a single bond is broken or made then the above model is adequate provided we develop a very general functional form for Q that includes the framework of the molecule as well as a contribution from the *active* orbitals Φ_a and Φ_b. For this purpose the most sensible strategy would appear to use the molecular mechanics approach for the *framework* part of Q to yield an MM-VB hybrid approach.

How does one extend this idea to a multi-bond reaction that involves more than a single bond making or breaking ? The only difference is clearly that the number of covalent determinants increases with the number of possible arrangements of covalent bonds (good discussion can be found in reference [8-9]). The general form of H_s can be written in terms of Q and a K_{ij} for each pair of active orbitals as

$$\langle K | \hat{H_S} | L \rangle = \delta_{KL} a^{KL} Q + \sum_{ij} b_{ij}^{KL} K_{ij}$$

[8]

where K and L index the covalent configurations and ij runs over the orbitals. The a^{KL} and b_{ij}^{KL} are simply numerical coefficients that can be computed easily [8,9] and in automatic way (eg. using Unitary Groups [10]). By analogy with the previous discussion

Q can be assumed to have the form

$$Q = Q_{\text{framework}} + \sum_{ij} Q_{ij}$$

[9]

where Q_{ij} is assumed to have the same physical significance as in eq 3 and we envision replacing $Q_{\text{framework}}$ by a molecular mechanics potential.

From a practical point of view the central question relates to the parametrization of the Q_{ij} and K_{ij}. Can the K_{ij} be transferred from computations involving a single covalent bond? This is unlikely because the orbitals must distort differently in a new molecular environment. However, how are we to study this phenomenon? Since the Q_{ij} and K_{ij} are matrix elements of a model hamiltonian H_S and we need some general method of extracting H_S from any ab-initio computation. The practical answer lies in effective hamiltonian theory [6,11] which provides us with a route to computing H_S rigorously and we now discuss this approach briefly.

Let us start with a basis of localized atomic-like orthogonal orbitals and assume that we can do a full CI. However the CI expansion in localized orbitals is an orthogonal VB wavefunction. The configuration space of covalent configurations can be partitioned into a space spanned by $\{\Lambda_R\}$, the set of covalent configurations and a set $\{\phi_S\}$ spanned by the remainder of the ionic configurations. The CI eigenvalue problem can then be written in partitioned form as

$$\begin{bmatrix} H^0 & Z \\ Z^\dagger & W \end{bmatrix} \begin{bmatrix} A & X \\ B & Y \end{bmatrix} = \lambda \begin{bmatrix} A & X \\ B & Y \end{bmatrix}$$

[10]

where H^0 is the projection onto $\{\Lambda_R\}$ the set of covalent configurations and W the projection onto the set $\{\phi_S\}$ spanned by the ionic configurations. We can then define an effective hamiltonian H_{eff} on the space $\{\Lambda_R\}$ spanned by the set of covalent configurations by block diagonalization with U

$$U^{-1} \begin{bmatrix} H^0 & Z \\ Z^\dagger & W \end{bmatrix} U = \begin{bmatrix} H_{\text{eff}} & 0 \\ 0 & \overline{W} \end{bmatrix}$$

[11]

From the diagonalization of H_{eff} we obtain a subset of the eigenvalues exactly (those with the largest projection on $\{\Lambda_R\}$ the set of *covalent* configurations). The set of equations 11 has a simple solution. If we define

$$C = B\,A^{-1}$$

and

$$S = (1 + C)^\dagger (1 + C)$$

then we have

$$H_{eff} = S^{-1}(1\text{-}X)\,H\,(1+X)$$

with

$$X = \begin{bmatrix} 0 & -C^\dagger \\ C & 0 \end{bmatrix}$$

We immediately recognise the role of the metric S in equation 12 (bi-orthogonality) and thus we can interpret

$$H_{eff}^{NO} = (1\text{-}X)\,H\,(1+X)$$

[12]

as the effective hamiltonian in a non-orthogonal basis. This can be symmetrically orthogonalized to give

$$H_{eff} = S^{-1/2}(1\text{-}X)\,H\,(1+X)\,S^{-1/2}$$

[13]

The effective hamiltonian H_{eff} defined in eq. 13 provides us with a rigorous definition of H_S. Thus, in principle, we can always parametrise H_S from an ab-initio computation and the accuracy is limited only by the accuracy of the ab-initio computation of H_{eff} and the accuracy of the fitting. Of course this would not be very useful per se because one would need to fit each potential surface individually. Thus a practical solution involves fitting data for Q_{ij} and K_{ij} for model covalent bonding situations for single covalent bonds and then using the ab-initio computation of H_{eff} to develop an algorithm for modelling the change of Q_{ij} and K_{ij} in general molecular environments.

Thus it can be seen that the use of a VB model in the form of a Heisenberg Hamiltonian is an ideal model for use in the study of reactivity problems. It can be parametrized using the best state of the art ab-initio computations and via Q as defined in eq. 9 it can be combined with force field methods. We now turn our attention to an example that is closer to the main theme of this volume.

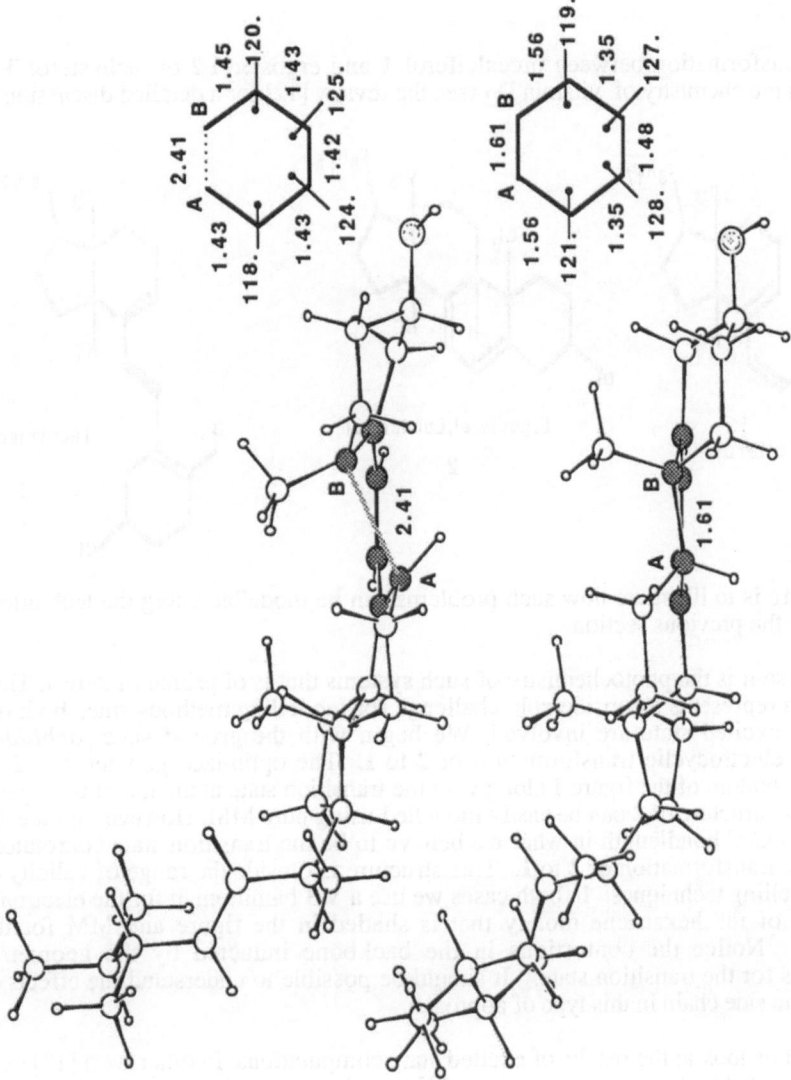

Figure 1 MM-VB Structures for Ergosterol and the Transition Structure for Thermal Conrotatory Ring Opening

3. Electrocyclic and Cis-trans Isomerization Reactions or Precalciferol

The transformation between precalciferol **1** and ergosterol **2** or tachysterol **3** is important in the chemistry of vitamin D_2 (see the review [12] for a detailed discussion).

Precalciferol
1

Ergosterol, Lumisterol
2

Tachysterol
3

Our aim here is to illustrate how such problems can be modelled using the techniques discussed of the previous section.

Of course it is the photochemistry of such systems that is of primary interest. Thus this problem represents a considerable challenge for modelling methods since both the ground and excited state are involved. We begin with the ground state *forbidden* conrotatory electrocyclic transformation of **2** to **1**. The optimized geometry of **2** is shown at the bottom of the figure 1 along with the transition state at the top of the figure. Of course the structure of **2** can be easily modelled using pure MM. However, notice the long (2.41Å) C-C bondlength in what we believe to be the transition state conrotatory electrocyclic transformation of **2** to **1**. This structure is outside the range of validity of normal modelling techniques. In both cases we use a VB hamiltonian for the electronic interactions of the hexatriene moiety that is shaded in the figure and MM for the framework. Notice the contortions in the backbone inducted by the geometric requirements for the transition state. It should be possible to understand the effects of modifying the side chain in this type of approach.

Now let us look at the results of excited state computations. In other work[13], we have demonstrated that conical intersections[14] can play a central role in photochemistry by providing a pathway for a fully efficient return from the excited state. An oversimplified view of this process is illustrated below in which the reactant system emerges on the ground state surface at touching of a double cone.

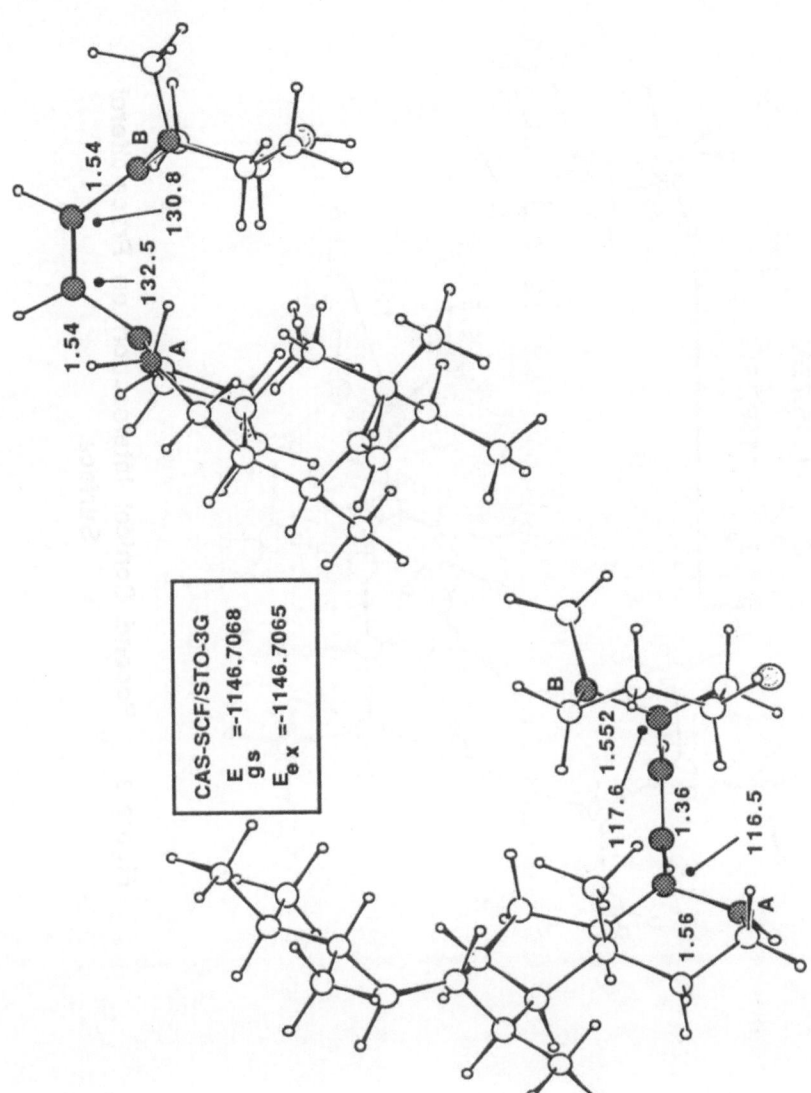

CAS-SCF/STO-3G

$E_{gs} = -1146.7068$

$E_{ex} = -1146.7065$

Figure 2   First Conical Intersection on Precalciferol Surface

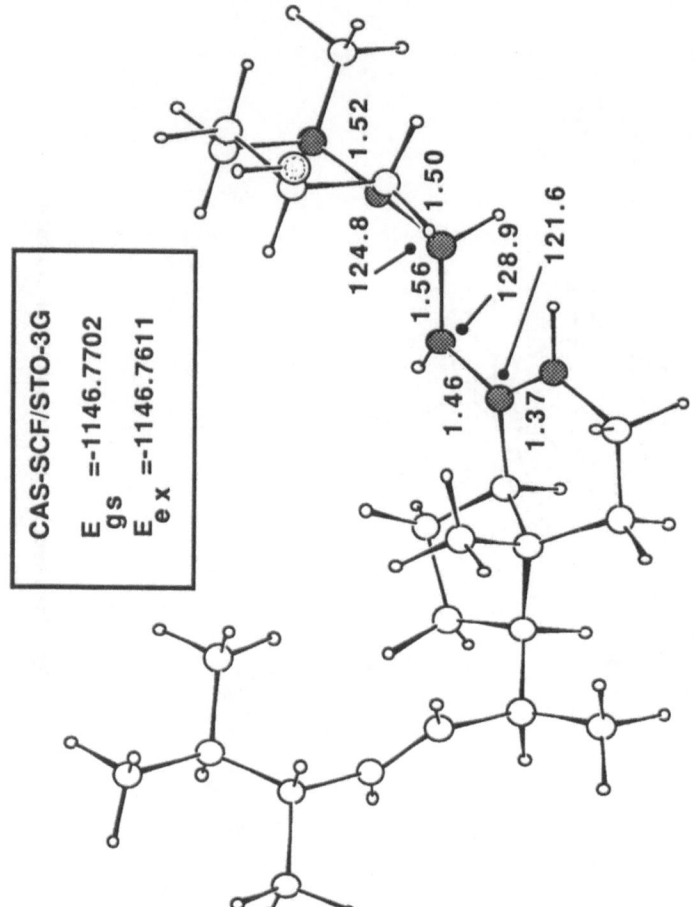

CAS-SCF/STO-3G

$E_{gs} = -1146.7702$

$E_{ex} = -1146.7611$

1.52
1.50
124.8
1.56
128.9
121.6
1.46
1.37

Figure 3   Second Conical Intersection on Precalciferol Surface

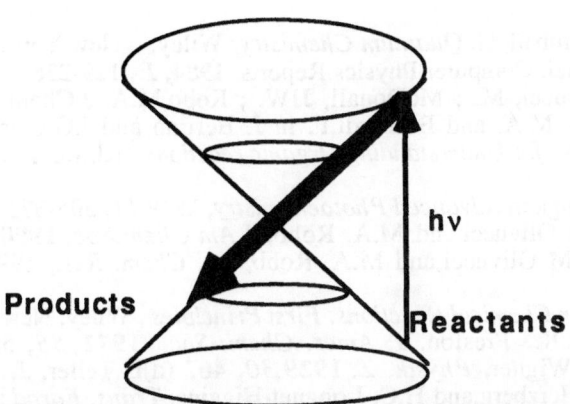

This mechanistic feature provides a "structural bottleneck" for the reaction ie the system must assume the specified geometry of the conical intersection or the decay probability is vanishingly small. Of course the conical intersection is really an n-2 dimensional surface and the molecule can undergo fully efficient decay to the ground state from any geometry that lies on this n-2 dimensional surface. In Figures 2 and 3 we show MM-VB geometries that lie on two different conical intersections and the ab-inito CAS-SCF energies at these geometries. The ab-initio CAS-SCF energies for the ground and excited state are virtually equal confirming the existence of this topological feature. Thus the combination of MM and VB methods permits the modelling of the delicate interplay between ground and excited state surfaces the is so relevant in photochemistry.

Experimentally, irradiation of precalciferol within 250-300 nm yields electrocyclic ring-closure products (Ergosterol and Lumisterol) and long-term irradiation products (Toxisterol C1,C2). The existence of these two conical intersections provides mechanistic pathways for both these possibilities and might explain the origin of the two classes of products. We are currently investigating.

Acknowledgements

This work was supported by the SERC (UK) under grant numbers GR/F 48029 and GR/G03335. The ab-initio computations were carried out using Gaussian 91 [15]

References

[1]J. Almlof and K. Faegri, in **1990** Ed. R. Carbo and M. Klobukwski, *Self-Consistent Field- Theory and Applications*, Elsevier Publishers, 278-311
[2] J.J.P. Stewart, *J. Comp. Chem* . **1989**, *10*, 209,201
[3]N.L. Allinger, *Adv.Phys.Org. Chem.* **1976**, *13*,1 , Allinger, N.L., Yuh, Y QCPE, **1980**, *12*, 395, Burkert U.; Allinger, N.L., *Molecular Mechanics* ACS Monograph 177; **1982** Amer. Chem.Soc.
[4]T. Clark,T *Handbook of Computational Chemistry;* Wiley, New York, 1985
[5]D.C. Spellmeyer,and K.N. Houk, *J. Org. Chem.* **1987**, *52*, 959
[6] Ph.Durand,and J.P. Malrieu, *Adv. Chem. Phys.* **1987**,*68*, 931
[7] D.L. Cooper,J. Geratt, M. Raimondi, Adv. Chem. Phys. **1987**, *67*, 319
[8]McWeeny, R.; Sutcliffe, B. *Methods of quantum mechanics*; Academic: New York,

248

1969.

[9]Eyring, H.; Walter, J.; Kimball, G. *Quantum Chemistry;* Wiley: New York, 1944.

[10] M. A. Robb and U. Niazi. Computer Physics Reports, **1984**, *1* , 129-236

[11] (a.)Bernardi, F. ; Olivucci, M. ; McDouall, JJW. ; Robb,M.A. J.Chem. Phys. **1988**, *89* ,6365 (b.) Robb, M.A. and Bernardi,F. in J. Bertran and I.G.Csizmadia (ed.) *New Theoretical concepts for Understanding Organic reactions* (Kluwer Academic Publishers) **1989**, 101-146

[12] J.C Jacobs and E. Havinga in *Advanced Photochemistry*, **1979** *11*, 305-373

[13] F. Bernardi, S. De, M Olivucci,and M.A. Robb, *J.Am.Chem.Soc.* **1990**, *112*, 1737-1743. , F. Bernardi, M Olivucci,and M.A. Robb,*Acc. Chem. Res.*, 1990, 23, 405-412.

[14] (a) L.Salem, *Electrons in Chemical Reactions: First Principles* , Wiley, New York, **1982** (b) J.C.Tully, and R.K. Preston, *J. Amer. Chem. Soc.* **1971**, *55*, 562 (c) J.Von Neumann,J. and E Wigner, *Physik. Z.* **1929**,*30*, 467 (d)E Teller, *J. Phys. Chem.* **1937**,*41*, 109 (e)G. Herzberg,and H.C. Longuet-Higgins, *Trans. Faraday Soc.* **1963**,*35*, 77 (f)G.Herzberg,*The Electronic Spectra of Polyatomic Molecules* Van Nostrand, Princeton **1966** pp442 (g)C.A. Mead,and D.G.Truhlar, *J.Chem.Phys.* **1979**, *70*, 2284 (h) C.A. Mead, *Chem.Phys.* **1980**, *49*, 23 (i)S.P Keating, and C.A Mead, *J.Chem.Phys.***1985**, *82,* 5102 (j) S.P Keating,and C.A. Mead *J.Chem.Phys.***1987**, *86*, 2152   (j) W.Gerhartz,R.D. Poshusta, and J. Michl,*J.Amer.Chem.Soc* ., **1977**, 99, 4263 (k)V. Bonacic-Koutecky, J. Koutecky, and J. Michl, *Agnew. Chem. Ed. Engl.* **1987**, 26, 170-189. (l) E.R.Davidson.,W.T.Borden, J. Smith, *J.Amer.Chem.Soc*, **1978**, 100, 3299-3302. (m)C.A. Mead, C. A., *The Born-Oppenheimer approximation in molecular quantum mechenics*, in Thrular, D. G., editor, *Mathematical frontiers in computational chemical physics*, chapter 1, pages 1-17, Springer, New York, **1987**. (n)N.C Blais, D.G. Thrular, and C.A. Mead, *J.Chem.Phys*, **1988**, 89, 6204-6208. (o)U. Manthe, and H Koppel, *J.Chem.Phys*, **1990**, 93, 1658-1669.

[15] **Gaussian 91**, (Revision C), M. J. Frisch, M. Head-Gordon, G. W. Trucks, J. B. Foresman, H. B. Schlegel, K. Raghavachari, M. Robb, M. W. Wong, E. S. Replogle,J. S. Binkley, C. Gonzalez, D. J. Defrees, D. J. Fox,J. Baker, R. L. Martin, J. J. P. Stewart, and J. A. Pople,Gaussian, Inc., Pittsburgh PA, 1991.

Discussion:

Professor Kollman: What is the size of the matrix one must use to study systems with MM-VB and 6 active orbitals.

One needs only the 5 canonical VB structures that are familiar from resonance theory of benzene. In general one needs only the canonical Rumer functions. This set grows much more slowly than the CAS-SCF space.

Professor Mezey: The molecular mechanics VB method you have proposed appears as a natural combination of two approaches, both based on local bond concepts, one of local force fields, the other an effective hamiltonian obtained from MC-SCF, high quality quantum chemical approach. I think that this development was never foreseen by the original developers of the VB method.

Obviously the early work on VB methods was in fact carried using semi-emprical potentials for Q and K (See for example G Evans, , M Polanyi, M. *Trans.Far.Soc.* **1938**, *34*, 11. and G. Evans, E Warhurst, *Trans Far.Soc.***1938**, *34*, 614). It seems that VB fell into disuse primarily because when AO basis was used in ab-initio there were so many contributions from ionic terms. The work of Geratt and his co-workers at Bristol has demonstrated that one merely needs to allow for orbital distortion to eliminate this problem.

Professor Warshell: What about the effect of environment on K.

If the effect of the environment can be localized into Q then one can clearly develop algorithms to deal with the problem. We do not really understand how important the orbital relaxation is as part of this problem. Clearly we need ab-initio studies to understand this problem. There are no technical difficulties associated with computing H_{eff} and this H_s within the SCRF approach.

Professor Rival: In the case of an enzymatic reaction, the reacting part undergoes a strong perturbation by a very heterogeneous potential. Did you check how your coulomb and exchange terms are modified by this perturbation?

We are still in the early days of our development of this model so we do not have am answer to this question. Provided the effect is localized in Q then there are no problems. Again we need to study model problems at the ab-initio level.

Professor Miller: Do the coulomb and exchange integrals in your VB method depend only on the atom-atom distance of the two relevant atoms or do they depend on the geometry of all the surrounding atoms? The simpler "diatomics-in-molecules" model assumes that they are a function only of the atom-atom distance.

The Q and K in our model depend on a) the distance between the atoms b) the "hybridisation status" of each of the atoms and the relative orientation of the orbitals (which is in turn coupled via MM to the geometry of the surrounding molecular structure) and c) a delocalization model that describes the distortion of the AO in the molecular environment. The coupling of the "hybridisation status" to the MM force field is vital to the success of the model. Without this effect we cannot obtain any quantitative results. Our first attempts (M. Bearpark,F. Bernardi, M. Olivucci,M. and M.A. Robb, *J.*

Am.Chem.Soc. **1990**, *112*, 1732-1736 or   F. Bernardi, M Olivucci,and M.A. Robb,*Acc. Chem. Res.*, 1990, 23, 405-412)   using only the atom-atom distances correctly reproduced the surface topology in a well defined subspace and could be used to obtain useful qualitative mechanistic information. However full geometry optimization is impossible without a much more refined model.

Challenges in Computer Modeling
Complex Molecular Systems

Peter Kollman

Professor of Pharmaceutical Chemistry
University of California, San Francisco

Abstract

We present a personal perspective on some of the challenges in computer modeling of complex molecular systems as well as reviews of some of the applications in these areas.

Introduction

Applications of computer modeling to biotechnology would ideally involve models which can predict the protein structure from amino acid sequence, predict the catalytic properties of this protein, as well as its stability and pathway to folding/unfolding. We are still a long way from this ideal, but considerable progress has occurred in recent years to bring us closer to it. What are some of the recent exciting developments that have occurred? In our opinion, there are four areas of research that have seen exciting developments in various labs including ours and they are: (1) the ability to calculate relative free energies for complex molecular systems using molecular dynamics and Monte Carlo methods; (2) advances in our ability to characterize the conformational space of medium sized molecules; (3) combined quantum/molecular mechanical-dynamical approaches to study chemical reactions in solutions and in enzymes and (4) improvements in molecular mechanical/dynamical force fields, including the incorporation of non-additive effects.

Free Energy Calculations

The basis for free energy calculations on complex molecular systems is in thermodynamic cycles, such as presented first by Tembe and McCammon. [1] A typical cycle is presented below, in which one studies the association of two ligands L and L´ to a protein P.

J. Bertrán (ed.), Molecular Aspects of Biotechnology: Computational Models and Theories, 251–262.
© 1992 *Kluwer Academic Publishers.*

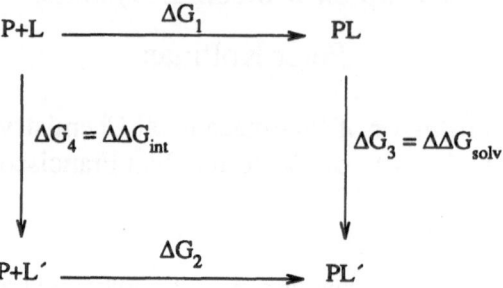

$$\Delta\Delta G_{bind} = \Delta G_2 - \Delta G_1 = \Delta G_4 - \Delta G_3$$

$$= \Delta\Delta G_{int} - \Delta\Delta G_{solv}$$

The free energy of association of ligand L to the protein is ΔG, and that of ligand L′ to the protein

is ΔG_2; these can be measured experimentally. Because free energy is a state function and the

sum of the free energies in a closed cycle is zero, the difference between these association (bind-

ing) free energies $\Delta\Delta G_{bind} = \Delta G_4 - \Delta G_3$, which are the free energies associated with mutating L to

L′ when bound to P ($\Delta\Delta G_{int}$ and when free in solution ($\Delta\Delta G_{solv}$). These latter two free energies

are easier to calculate, since they may involve much smaller structural changes than ΔG_1 and

ΔG_2.

The above cycle has been applied to the study of ligand association to proteins and iono-

phores, among others, [2] with the goal of predicting new ligands L′ that will bind more strongly

to P. The key point from a conceptual view is that the $\Delta\Delta G_{bind}$ is a compromise between the rela-

tive interaction of L and L′ with P and the relative solvation of L and L′. For example, Merz

et al. [3] showed that, surprisingly, a replacement of a hydrogen bonding NH group with a CH_2

in a thermolysin inhibitor, where the NH group formed an interaction with a protein C=O group,

led to a $\Delta\Delta G_{bind}$ of ~ 0. This was because the replacement of NA by CH_2 increased (led to a less

favorable) $\Delta\Delta G_{int}$ by ~ 2 kcal/mole, but since a CH_2 is easier to desolvate than NH, $\Delta\Delta G_{solv}$ also

increased by ~ 2 kcal/mole. Analogously, the enzyme ribonuclease T1 binds 2′GMP ~ 3

kcal/mole more tightly than 2′AMP, according to both experiments and calculations. The calculations suggest, however, that this $\Delta\Delta G_{bind}$ of ~ 3 kcal/mole results from a $\Delta\Delta G_{int}$ ~10 kcal/mole, compensated by a $\Delta\Delta G_{solv}$ ~7 kcal/mole. [4] Simply stated, guanine, having a much larger dipole moment than adenine, has a much more negative solvation free energy; in order to bind G better than A, the enzyme forms 7 H-bonds with the guanine base and is in a position to form only two H-bonds with A.

Other applications of free energy cycles include calculation of relative solvation free energies of heterocyclic tautomers, [5] which when combined with accurate quantum mechanical on the relative tautomeric energies in the gas phase, can be used to predict tautomeric preferences in solution. Variants of the above cycle have been used to study different site specific mutated proteins P and P′ binding to a ligand L, both in a Michaelis complex and in a model for the transition state in serine protease catalysis. A specific application to the Asn 155 → Ala mutation [6] in subtilisin has been most insightful and has led to further experimental and theoretical studies.

Free energy calculations have been applied to study relative protein stability, using a small peptide model of the denatured state of the protein. Such calculations have led to interesting insights on why Thr 157 is more stable than the mutant Val or Ala 157 in T4 lyzozyme. [7]

One can, in favorable cases, calculate the free energy profile directly for molecular association of two ligands, but this has so far been accomplished for rather small molecules such as 2 CH_4, [8] 18-crown-6/K^+, [9] and nucleic acid base pairs. [10] Nonetheless, the same technique can be used to evaluate conformational free energies of molecules in solution and such studies have considerable potential in leading to further understanding of molecular solvation of complex molecules.

All of the above calculations involve the use of molecular dynamics or Monte Carlo to create representative configurations of the system. For example, calculating the relative solvation free energy of methanol and ethane involves "mutating" the molecular mechanical parameters of the O-H of methanol into the CH_3 of ethane, evaluating the free energy from the ensemble average of the difference in Hamiltonian (ΔH) which is approximately the difference in portential energy of interaction of water with the O-H and the CH_3:

$$\Delta G = -\sum_{\lambda=0}^{\lambda=1} RT \ln < e^{-\Delta H/RT}>_\lambda \tag{2}$$

$\lambda=1$ corresponds to CH_3OH; $\lambda=1$ corresponds to ethane; in practice, one mutates one into the other via a number of artificial Hamiltonia (λ between 0 and 1), evaluating the free energy for small increments in λ and adding these up. It should be clear that the accuracy with which one can calculate ΔG via equation (2) (thermodynamic perturbation), or the related equation (3), (thermodynamic integration)

$$\Delta G = -\int_{\lambda=0}^{\lambda=1} < \frac{\partial H}{\partial \lambda} > d\lambda \tag{3}$$

depends on two factors: first, the accuracy of the molecular mechanical Hamiltonian H and, secondly, the adequacy of the sampling of configurational space (how converged the ensemble average $<>$ in equations (2) and (3) are. In practice, to calculate $\Delta\Delta G_{solv}$ for simple molecules via equations (2) and (3) requires $\sim$ 40 psec (electrostatic dominated perturbations) of molecular dynamics or $\sim$ 100-200 psec (van der Waals dominated perturbations), but even these times may not be adequate if many conformations of complex molecules with different intrinsic solvation free energies must be sampled. To accurately calculate $\Delta\Delta G_{int}$ in protein active sites might require much longer simulations, depending on the characteristic relaxation time of the active site.

Conformational Sampling

The single greatest difficulty facing computational studies of complex molecules is the number of degrees of conformational freedom to be searched, coupled with the difficulty of evaluating the free energy in solution of each conformation. In a protein with n residues, there are $\sim 3^n$ backbone conformations, which is already a large number even for small n. Professor Scheraga's group has been developing a variety of methods to efficiently search conformationl space and applying these to small peptides and proteins. [11]

As noted above, one of the difficulties in studying peptides in this way is, if one wishes to accurately describe their properties, one must include the effect of solvent. We thus chose to study a simpler system, 18-crown-6, whose energy has an important contribution from dipolar electrostatic effects. However, since it does not have any proton donor groups, its conformational properties are not dominated by solvation effects. We wished to assess how well "brute force" molecular dynamics calculations would perform on this system, so Yax Sun [12] carried out three 2 nanosecond and 10 200 picosecond molecular dynamics trajectories on the molecule and after each psec sampled the coordinates. These sampled coordinates were energy-minimized, leading to a set of conformations. The 30 lowest energy conformations were chosen for analysis. One result of these studies was that, in terms of the number of low energy conformations found, ten 200 psec simulations were similar in efficiency to one 2 nsec simulation. Has one begun to approach conformational equilbrium in these systems, i.e. is the frequency of sampling each conformation related to its Bolzmann factor $e^{-Ei/KT}$? The answer is, for ~ 6 nsec of total simulation at 600K (the three 2 nsec trajectoreis), a qualified yes. After correcting for conformational degeneracy (a highly symmetric conformation such as D_{3d} has 1/6 the statistical weight of a completely unsymmetrical conformation), and taking into account the conformational entropy of the various conformations with normal mode analysis, Sun found a correlation coefficient (r^2) of ~ 0.8 between the calculated probability of finding a conformation and the probability with which it was actually found. Thus, for a molecule like 18-crown-6, ~ 6 nsec at 600K seems enough for reasonably complete sampling of conformational space. However, the largest rotational barriers in 18-crown-6 are ~ 5 kcal/mole, and no H-bonds must be broken to undergo conformational transition, so it is difficult to extrapolate the 18-crown-6 results to other systems.

Quantum/Molecular Mechanical Studies on Enzyme Catalysis

Molecular mechanical methods are inherently unable to study chemical reactions and excited states, situations with significant electronic as well as nuclear structure changes. Because of the importance and interest in studying enzyme catalysis, a number of approaches to coupling quantum and molecular mechanical methods have been proposed, including the EVB approach of Warshel and co-workers, the *ab initio*/molecular mechanics coupling of Singh, and the semi-empirical molecular orbital/molecular mechanics coupling of Field *et al.*. Each of these has their strong points and weaknesses, but it is fair to say that no one has developed an accurate and general model, which, without calibration on each system, can predict the free energy profile of enzyme catalysis. We refer the reader to a series of studies on the enzyme triose phosphate isomerase, where we have employed molecualr mechanics, quantum mechanics, molecualr dynamics and free energy perturbation approaches to attempt to understand the catalytic properties of this "perfectly evolved" enzyme and its site-specific mutants. [13] Recently, we have returned to the enzyme trypsin and have studied its complete catalytic pathway with the semi-empirical method PM3. [14] Without including environmental effects, the results of the calculations can only be compared in a qualitative way with experiments, but one is able to connect them with experiments in useful ways. Further refinements are possible and desireable, coupling molecular mechanics more directly to the PM3 results.

One of the interesting issues that is often raised is: what makes an enzyme a good catalyst, i.e. how does it stabilize its transition state? The earliest studies focused on "strain" effects, in that the sofa conformation of a sugar fit the site of lyzozyme, but as first pointed out by Warshel and Levitt, [15] electrostatic effects can be much more efficient at stabilizing transition states in chemical reactions.

So the question remains: are "electrostatic" effects what makes enzymes catalyze reactions so much faster than the corresponding reaction in water? To some extent, the question is sematic, but we think the answer is no; what makes an enzyme a good catalyst is that the enzyme has paid a free energy price upon synthesis and folding to allign its dipoles/charges to stabilize the transition states for catalysis. The point we are making is not new, it has been made before by Warshel [16] and ourselves, [17] but the important structural point is as follows: in aqueous solution, water probably forms much "better" (more ideal) H-bond interactions with transition states found in enzyme active sites, but it must pay the free energy price of alligning its dipoles to do so; in the enzyme, no such free energy price need be paid. The analogy we made some time ago [17] is that 18-crown-6 binds K^+ despite the fact that the alignment of K^+ with water dipoles in water is far better than with the $-CH_2-O-CH_2-$ groups in the crown. The point is that for every kcal/mole of free energy gained by cation-water interaction, roughly 1/2 kcal/mole must be "paid" to break water-water H bonds and align the dipoles. Thus, an enzyme catalyzes its reaction compared to the solution reaction by providing a favorable binding site for which favorable interactions compensate for the entropy cost of binding and place the functional groups in an orientation so prealigned enzyme groups can effect their catalysis. Of course van der Waals and electrostatic effects must be favorable for this, but need not be as structurally "ideal" as the corresponding interactions achievable in solution.

Molecular Mechanical Force Fields

It is surprising how well molecular mechanical force fields can reproduce molecular interactions and conformations. Nonetheless, there is clearly room for improvement for these, both in improving the parameters within a given force field and in modifying the functional form of force fields. Among the most notable efforts in this direction are the development of the MM3 force field by Allinger and co-workers and the BIOSYM force field consortium inspired by the work of Hagler et al. [18]

Our own efforts in this area include the incorporation of non-additive effects in force fields and the calibration of these on water and water/ion systems. [19, 20] One expects that non-additive effects will be largest for systems with the largest electric fields and, thus, ionic solutions would be expected to have the largest contribution from non-additive effects. We have developed a set of parameters, including non-additive effects, that can reproduce the properties of gas phase ion-water clusters and ionic solutions. Eventually, we hope to have developed a general non-additive force field for studying organic and biochemical molecules.

Summary

Above we have reviewed some of our recent work in simulating the properties of complex molecules. Further technical developments such as new algorithms and increased computer power will likely lead to an ever-increasing ability of theoretical chemistry to simulate real chemistry.

References

1. B. Tembe and A. McCammon, *Comp in Chem* **8**, 281 (1984).

2. P. A. Kollman and K. M. Merz, Computer Modeling of the Interactions of Complex Molecules, *Acc Chem Res* **23**, 2460252 (1990).

3. K. Merz and P. Kollman, Free Energy Perturbation Simulations of the Inhibition of Thermolysin: Predictions of the Free Energy of Binding of a New Inhibitor, *J Amer Chem Soc* **111**, 5649 (1989).

4. S. Hirono and P. Kollman, Calculation of the Relative Binding Free Energy of 2GMP and 2AMP to Ribonuclease T1 Using Molecular Dynamics/Free Energy Perturbation Approaches, *J Mol Biol* **212**, 197-209 (1990).

5. P. Cieplak, P. Bash, U. Singh and P. Kollman, A Theoretical Study of Tautomerism in the Gas Phase and Solution: A Combined Use of *Ab Initio* Quantum Mechanics and Free Energy Perturbation Methods, *J Amer Chem Soc* **109**, 6283 (1987).

6. S. Rao, P. Bash, U. Singh and P. Kollman, Free Energy Perturbation Calculations on Binding and Catalysis: Mutating Asn155 in Subtilisin, *Nature* **328**, 551-554 (1987).

7.   L. Dang, K. Merz Jr and P. Kollman, Free Energy Calculations of Protein Stability: The Thr 157 → Val 157 Mutation to T4 Lyzozyme, *J Amer Chem Soc* **111**, 8505-8508 (1989).

8.   W. Jorgensen, J. Chandrasekhar, J. Madura, R. Impey and M. Klein, *J Chem Phys* **79**, 926-935 (1983).

9.   L. X. Dang, D. A. Pearlman and P. A. Kollman, Why Do AT Base Pairs Inhibit ZDNA Formation? *Proc Nat Acad Sci* **87**, 4630 (1990).

10. L. Dang and P. Kollman, Molecular Dynamics Simulations of the Free Energy of Association of 9-CH_3 Adenine and 1-CH_3 Thymine Bases in Water. *J Amer Chem Soc* **112**, 503 (1990).

11. H. Scheraga, *This Volume*.

12. Y. Sun and P. Kollman, Conformational Sampling and Ensemble Generation by Molecular Dynamics Simulations: 18-crown-6 as a Test Case, *J Comp Chem* (in press).

13. V. Daggett and P. A. Kollman, Molecular Dynamics Simulations of Active Site Mutants of Triosephosphate Isomerase, *Prot Engineering* **3**, 677--690 (1990).

14. V. Daggett, S. Schroeder and P. Kollman, The Catalytic Pathway of Serine Proteases. Classical and Quantum Mechanical Calculations, *J Amer Chem Soc* (in press).

15. M. Levitt and A. Warshel, *J Mol Biol* **103**, 227 (1976).

16. A. Warshel, *Proc Nat Acad Sci* **75**, 5250 (1978).

17. G. Wipff, P. Weiner and P. A. Kollman, A Molecular Mechanism Study of 18-Crown-6, its Open Chain Analog and Their Complexes with Alkali Cations, *J Amer Chem Soc* **104**, 3249-3258 (1982).

18. Force Field Consortium. Sponsored by Biosym Technologies, San Diego, CA..

19. J. Caldwell, L. X. Dang and P. A. Kollman, Implementation of Nonadditive Intermolecular Potentials by Use of Molecular Dynamics: Development of a Water-Water Potential and

Water-Ion Cluster Interactions. *J Amer Chem Soc* **112**, 9144-9147 (1990).

20. L. X. Dang, J. Caldwell and P. Kollman, *J Amer Chem Soc* **113**, 2481 (1991).

Discussion

(1) Prof. Scheraga: (a) Can you tell us the principle of your method to search conformational space for 18-crown-6. Is it brute force? (b) Also, how does your new model of water do on the radial distribution function? (c) When you put a charged molecule in water, how do you handle the reaction field of the polarized water on modifying the charge distribution of the solute?

(a)   The principle of conformational search of 18-crown-6 is indeed brute force, high temperature (600K) dynamics for 8 nanoseconds.

(b)   The radial distribution functions for POL1 water [19] is nearly identical to that of SPC/E water, from which it was developed. They are in rather good agreement with experiment.

(c)   We allow the charge distribution of solute and solvent to be determined self-consistently. Thus, both solute and solvent respond to altered charge distributions of each other by changing the magnitude and direction of their induced dipoles.

(2) Prof. Naray-Szabo: I will defend the electrostatic concept of enzymatic rate acceleration; e.g. in subtilisin the rate increase, as compared with water, is 8 orders of magnitude. If the buried aspartate of the enzyme is replaced by a central side chain, 4 orders of magnitude are lost, as proven both by theory and experiment. On the other end of the active site, in the oxyanion hole further 4 orders of magnitude are lost if hydrogen bonds - the negatively charged oxygen are destroyed. Both effects are purely electrostatic with no hydration involved. Even if they may not be additive the role of electrostatics is unquestionable.

I am not arguing that if one changes the correct electrostatic configuration (Asp 102 and the oxyanion hole) to the wrong one (Asn 102 and no oxyanion hole), one will not reduce the rate of catalysis. I emphasize again the comparison is the solution reaction vs the enzyme catalyzed reaction -- there is no reason to expect that the solution reaction is related to the enzyme reaction with the "wrong electrostatic orientations".

(3) Lluch: In order to minimize hysteresis in free energy perturbation calculations, which is the best choice of the windows?

> To minimize hysteresis in solution free energy calculations, the calculations should be run for as long and for as many windows as possible, but probably 200 psec in each direction will be more than adequate.

(4) Scheraga: (a) Do you have any information as to how the field of the enzyme, and the residues in the P_2, P_3 ... P_2', P_3' ... sites affect the kinetics of the serine proteases? (b) The experimental stacking enthalpy for 2 adenines is positive -- do your evaluations account for this apparent hydrophobic interaction?

> (a) No, but A. Fersht (Enzyme Structure and Mechanism, W.H. Freeman, 1985) has a discussion of this topic.

> (b) We calculate free energies, and calculations of ΔH (and ΔS) have approximately an order of magnitude greater error, so we have not attempted to calculate the ΔH for base stacking.

(5) Perez: How large are the induction contributions in the systems you are modeling? And how is the polarization included in the potential?

> The induction contribution is about ~ 20% of the electrostatic energy in the water and water-ion systems. The polarization energy is included as described in ref 19 and 20.

Theoretical Study of the Catalyzed Hydration of CO_2 by Carbonic Anhydrase: A Brief Overview.[1]

MIQUEL SOLA, AGUSTI LLEDOS,
MIQUEL DURAN[2], and JUAN BERTRAN⁻
Departament de Química
Universitat Autònoma de Barcelona,
08193 Bellaterra, Catalonia, Spain.

ABSTRACT. The catalytic cycle of the hydration of CO_2 by the Carbonic Anhydrase enzyme has been studied by means of ab initio calculations. Environmental effects have been taken into account by a continuum model. All the four steps of the catalytic mechanism have been studied. It is shown that the two different mechanisms proposed for the CO_2 hydration (step 3) can be competitive if a more detailed description of the active site is made. The inclusion of environmental effects turns out to be essential to understand the HCO_3^- release and the water binding (step 4). The problem of modelling enzymatic reactions is discussed.

1. Introduction

The high catalytic activity of enzymes is one of the fundamental factors of bio-chemical processes. In order to know why enzymatic reactions are so fast and so specific it is necessary to understand the details of catalytic mechanisms. One of the mechanisms studied most lately has been that of Carbonic Anhydrase (CA) enzyme. This enzyme is found both in plants and animals, its only known biological function being to catalyze the reversible hydration of CO_2 and dehydration of the bicarbonate anion with an extreme efficiency. [1-7] X-Ray data [8-11] show that the zinc atom placed at the active site of CA is bound to three imidazole groups coming from His94, His96, and His119, whereas a water molecule completes a nearly sym-metrical tetrahedral coordination geometry. A simplified scheme of the active site is displayed in Figure 1. The hydroxyl of Thr199 is hydrogen-bound *via* its oxygen to the zinc-bound water and *via* its hydroxyl hydrogen to Glu106. The enzyme site can be divided into hydrophilic and hydrophobic halves. The hydrophilic half contains the proton acceptor group His64 together with partially ordered water molecules of functional significance. The inner water environment is almost completely separated from the outer channel by a ring of residues including His64, His67, Phe91, Glu92, and His200. Therefore, the inner water molecules are not in contact with those which

---

[1] A contribution from the "Grup de Química Quàntica de l'Institut d'Estudis Catalans"

[2] Present address: Laboratori de Química Computacional, Estudi General de Girona; Plaça de l'Hospital, 6; 17071 Girona; Catalonia, Spain

J. Bertrán (ed.), Molecular Aspects of Biotechnology: Computational Models and Theories, 263–298.
© 1992 *Kluwer Academic Publishers.*

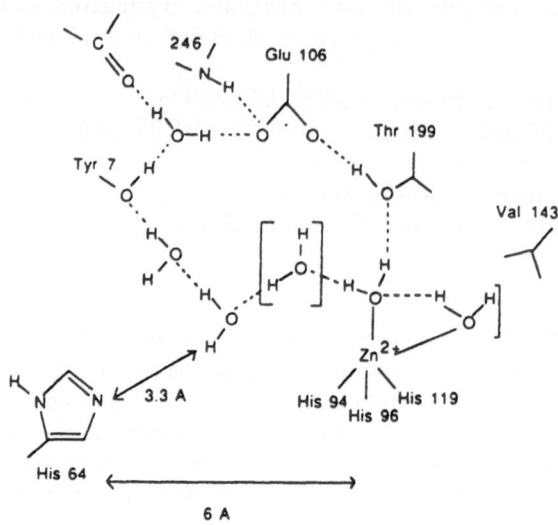

Figure 1: Scheme of the CA active site.

are outside the active site, His64 bridging these two solvent areas. In turn, the hydrophobic half provides a binding pocket for CO_2. In this region, a water molecule found about 3.2 Å away from the zinc ion is thought to be displaced by CO_2 during substrate binding.

The efficiency of CA to hydrate CO_2 depends on the nucleophilic power of its basic form generated by ionization of a water molecule with a pk_a around 7. [12-17] It is almost universally accepted that a zinc-bound hydroxide ion is the nucleophile in this catalytic reaction, [3-7,9] which was first suggested by Davis. [18] The overall chemical process can be simply sketched with the following separated reactions: [4]

$$EZn^{II}(H_2O) \rightleftharpoons EZn^{II}(OH^-) + H^+ \tag{1}$$

$$EZn^{II}(OH^-) + CO_2 \rightleftharpoons EZn^{II}(OH^-)CO_2 \tag{2}$$

$$EZn^{II}(OH^-)CO_2 \rightleftharpoons EZn^{II}(HCO_3^-) \tag{3}$$

$$EZn^{II}(HCO_3^-) + H_2O \rightleftharpoons EZn^{II}(H_2O) + HCO_3^- \tag{4}$$

which involve the following four steps: (1) proton transfer from $EZn^{II}(OH_2)$ in order to generate the catalytic species $EZn^{II}(OH^-)$; (2) binding of CO_2 near the active species $EZn^{II}(OH^-)$; (3) generation of the EZn^{II}-bound HCO_3^- species; and (4) binding of a water molecule and release of HCO_3^-.

Step (1) is the best known in the mechanism. Different experimental studies [13,19-22] have shown this first step to be rate-limiting at high buffer concentrations.

Given that the maximal turnover of Human Carbonic Anhydrase II (HCAII) is measured to be 10^6 s^{-1} at pH=9 and T=25 °C, the energy barrier for this intramolecular proton transfer must be about 10 kcal/mol. [23] It has also been clearly demonstrated that His64 acts as a proton shuttle in this process. The deprotonation is carried out first, through a two-water bridge [19,24] from $EZn(H_2O)$ to His64, and second from this group to the buffer. Using the PRDDO method, Liang and Lipscomb [25] reported an energy barrier of 34 kcal/mol for this deprotonation step whereas Merz, Hoffman, and Dewar, [23] using the AM1 method, found 18 kcal/mol for the same energy barrier.

As to step (2), the pioneering work performed by Davis [18] lead him to suggest that CO_2 is directly bound to Zn^{+2} during the initial interaction of CO_2 with the active site. Conversely, IR data [26] show CO_2 to be scarcely distorted in this initial step, thus leading to the conclusion that the substrate is not directly bound to the Zn^{+2} ion. Moreover, ^{13}C NMR studies [27-29] with a metal-substituted CA suggested a metal-carbon distance between 3.2 and 3.6 Å. The results furnished by theoretical studies are not coincident at all. Pullman et al. [30] reported a value of 3.4 Å for the Zn-C distance using an electrostatic approximation. With the PRDDO method the Zn-C distance obtained was 2.9 Å[31] which forced the authors to conclude that the nucleophilic attack of the $EZn(OH^-)$ group to CO_2 is already initiated in this early stage of the interaction. On the contrary, Merz et al. [23] with the AM1 method, Jacob et al. [32] and Krauss et al. [33] using an ab initio methodology, found that the nucleophilic attack has been not initiated yet. Finally, recent molecular dynamics and free-energy perturbation simulations results [34] have identified the presence of two CO_2 binding sites; one of them has a Zn-C distance of 3-4 Å; its calculated binding energy of -3.4 kcal/mol is also in reasonable agreement with the experimental value of -2.2 kcal/mol. [35]

For step (3), two mechanisms have been proposed, namely the Lipscomb and the Lindskog mechanisms (see Figure 2). In the first one, following the nucleophilic attack of the $EZn(OH^-)$ group to the CO_2 molecule, there is a proton transfer between two oxygens, [36] whereas in the second mechanism, there is no proton transfer but rather a change of the precise oxygen atom that becomes directly coordinated to Zn^{+2}. [37] As a support to the Lipscomb mechanism, experiment shows [38] that when the hydrogen of HCO_3^- is substituted by an alkyl group (R), the resulting alkyl carbonate anions ($ROCO_2^-$) exhibit no substrate activities in the enzyme-catalyzed reaction. If steric and pk_a factors are not a problem, this result may support the proton transfer mechanism. ^{13}C NMR [4,39] and isotope effect [22] experiments favor the Lindskog mechanism, showing that there is no rate-limiting proton transfer during the CO_2/HCO_3^- interconversion. From a theoretical point of view, four leading theoretical groups have studied recently the hydration of CO_2 catalyzed by CA. However, their results have not clarified the controversy yet. Liang and Lipscomb, [25,31,40] using the PRDDO method, showed that the Lipscomb mechanism is favored over the Lindskog mechanism because the energy barrier for the internal proton transfer is

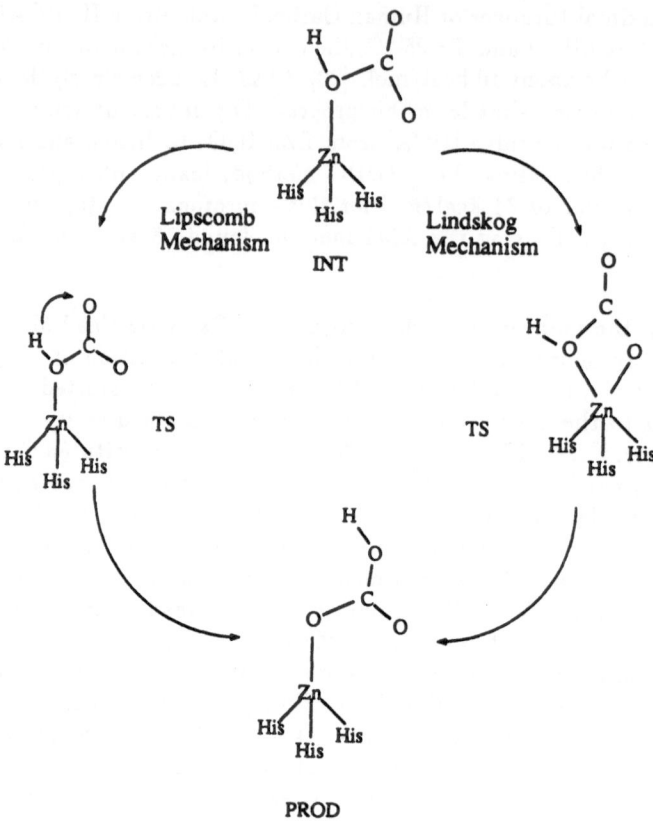

Figure 2: Lipscomb and Lindskog mechanisms for the formation of the $EZn^{II}(HCO_3^-)$ complex.

much lower when an appropriate proton relay is considered. Merz, Hoffmann, and Dewar [23] used the AM1 method to obtain the transition state corresponding to the Lindskog mechanism with a calculated energy barrier of 13.1 kcal/mol. Nonetheless, they were unable to locate a transition state for the Lipscomb mechanism, so the comparison between the two mechanisms was not performed. Jacob, Cardenas, and Tapia [32] carried out *ab initio* calculations with an extended basis set finding that, at the SCF level, the transition state for the Lindskog mechanism is 50 kcal/mol more stable than the transition state for the Lipscomb mechanism. This difference in energy barriers decreases by 20 kcal/mol if correlation energy is taken into account. Finally, in a very recent paper, Krauss and Garmer [33] carried out an ab initio pseudopotential calculation, where they found low barriers for both mechanisms.

Finally, for step (4), which corresponds to the substitution of a HCO_3^- ligand by a H_2O molecule in the coordination sphere of Zn^{+2}, to our knowledge only two experimental studies have appeared so far. In the first one, [41] from ESR spectra in a Co(II)-substituted CA it is found that a transient pentacoordinate intermediate exists, and that this complex might facilitate the ligand exchange between bicarbonate and water. In the second study, [7] analysis of anionic inhibition in CA showed that during the exchange a pentacoordinate metal transient intermediate is formed. A number of experimental work dealing with zinc pentacoordination reinforces the possibility of the existence of such an intermediate. [14,42-55] As to theoretical studies, this exchange has been studied only in three leading papers. In the first one, Liang and Lipscomb [40] showed that release of a HCO_3^- anion is facilitated by the previous binding of water to Zn^{+2}. In particular, their results indicated that the energy barrier for the unfacilitated displacement of EZn^{II}-bound HCO_3^- with a water molecule is 169 kcal/mol, and that this barrier is reduced to 109 kcal/mol when a fifth water ligand is considered. They also found that inclusion of some proteinic residues reduces further the energy barrier to 43 kcal/mol, and that deprotonation of this water ligand completely removes the energy barrier. In the second theoretical paper, Merz, Hoffmann, and Dewar, [23] with the AM1 semiempirical method, found that during the HCO_3^-/H_2O exchange, a pentacoordinate trigonal bipyramidal (tbp) complex is formed. In this complex, the water molecule occupies an equatorial position, whereas the bicarbonate anion is located at an axial site of the tbp. Furthermore, they showed that loss of a bicarbonate ion is aided by the presence of a number of water molecules. Finally, the authors proposed a new HCO_3^-/H_2O exchange mechanism, involving an intramolecular proton transfer from the water ligand to the coordinated bicarbonate, prior to release of carbonic acid. However, this mechanism does not follow the experimental ping-pong kinetics, [4] and furthermore, carbonic acid has never been detected, [56] either. Krauss and Garmer [33] found a pentacoordinate structure with water and bicarbonate simultaneously bound to Zinc. The removal of a proton from this water favours the bicarbonate release. This effect is reinforced if the deprotonation of a coordinated histidine is also considered.

In spite of a general agreement with the four steps enumerated above, some points of the CO_2 hydration by CA remain still unclear. For instance, the exact location and nature of the initial binding of the substrate CO_2 to the active site are unknown, and the possibility that the active site might accommodate the CO_2 as a fifth ligand to zinc is challenged. For step (3), as mentioned above, two mechanisms have been proposed (Figure 2), a great controversy between these two possible mechanisms existing both from an experimental and theoretical points of view. Finally, as to step (4), it is difficult to understand how a strongly bound anion can be released and substituted by a neutral ligand like the water molecule is. Further, it is still an open question whether a pentacoordinate complex is formed during the HCO_3^-/H_2O exchange, and even if it does exist, its structure is still unknown. Finally, the mechanism of this exchange process has not been definitely elucidated.

Therefore, our aim is to perform a theoretical study of all the four steps of the mechanism of the CO_2 catalyzed hydration by CA. Special attention will be focused on step (3) and on the comparison between the Lipscomb and the Lindskog mechanisms.

2. Methodology

Ab initio all-electron SCF calculations were performed for all the steps of the CA mechanism by means of the Hartree-Fock-Roothaan method. [57] AM1 [58-60] semiempirical calculations were also carried out, but solely for step (1) of the CA mechanism. In some cases, correlation energy was included at the MP2 level. [61] Owing to computational limitations, in *ab initio* calculations three ammonia groups were used to simulate the three imidazole ligands of the zinc ion, the $(NH_3)_3Zn^{II}(OH^-)$ complex becoming our model of the CA active site; the validity of this substitution which is necessary to perform *ab initio* calculations, is supported by previous studies of Pullman *et al.* [62,63] who demonstrated that ammonia and imidazole transfer a similar amount of charge to Zn^{+2}. The geometry of each NH_3 group was kept frozen, with the N-H distance being 1.05 Å and the $\widehat{HNH}$ angles being tetrahedral. C_{3v} symmetry was imposed on the $(NH_3)_3Zn^{II}$ fragment, except for the Lindskog transition state where this fragment had only C_s symmetry. Under the aforementioned geometric constraints, full geometry optimizations of minima and transition states (TS) were performed using Schlegel's method. [64] Minima and transition states were characterized by the correct number of negative eigenvalues of their Hessian matrices. In *ab initio* calculations, the 3-21G basis set [65,66] was used for all atoms, except for hydrogens of the NH_3 groups for which the STO-3G basis set [67] was used.

To introduce the environment effect, the SCRF model due to Tomasi *et al.* [68-70] was used. In this model the environment is represented by a continuous polarizable dielectric with permittivity ϵ, and the solute is placed inside a cavity accurately defined by its own geometry. [71] Dielectric polarization due to the solute is simulated by the creation of a system of virtual charges on the cavity surface, the charge distribution on the surface polarizing in turn the charge distribution in the solute. This process is iterated until obtaining the self-consistency in the solute electron density. In this way, the feedback effect of the environment on the solute is taken into account. The electrostatic contribution to the solvation free energy is found as the difference between the energies computed with the continuum model and without it minus one half of the interaction energy between the charge distribution of the solute and the reaction field potential. [69] Moreover, the cavitation free energy is calculated with Pierotti's equation. [72] The optimized geometries in the gas phase ($\epsilon=1.00$) were used throughout.

All calculations presented in this paper were carried out with the help of the GAUSSIAN 86 [73] and MONSTERGAUSS [74] programs. The surrounding medium effect was considered through a FORTRAN subroutine written by Tomasi and co-workers and incorporated in the MONSTERGAUSS program. Finally, semiempirical results were obtained by means of the AMPAC program. [75]

3. Modelling enzymatic reactions

The main problem found in theoretical biochemistry is how to model enzymatic reactions. [76] In the case of CA, the simplified description of the active site given above evidences its complexity and hence the difficulty of its modelling. This great complexity forces introduction of important simplifications in order to perform calculations. Different approaches have been suggested depending on the problem studied. For instance, molecular mechanics, [77-79] semiempirical, [23,25,31,40] *ab initio*, [30,32,80-83] free energy simulations, [34,84] and molecular dynamics [85] methods have been used at several levels of sophistication and applied to different models of the CA active site.

The general purpose of this communication is to analyze the CA catalytic cycle. Since our particular aim is to obtain reliable structures for intermediates and transition states and also meaningful energy barriers for the different steps of the mechanism of CA, we think that an *ab initio* methodology must be used. However, because of the big size of the system studied, we are forced to simplify the theoretical description of the active site by excluding, in a first step, all proteinic residues and all water molecules. The simplest model for the active site is the Zn^{II}-bound hydroxide species, although this is not indeed the best choice. Charge transfer from the ligands translates into an $EZn^{II}(OH^-)$ species which is a more nucleophilic than the $Zn^{II}(OH^-)$ complex. [81] For instance, our results indicate that the charge on the hydroxyl group increases by -0.11 au when the ammonia ligands are taken into account. Therefore, a less simplified model involving ligands coordinated to Zn^{+2} should be incorporated in the model of the active site. Furthermore, previous theoretical studies have stressed the importance of considering additional water molecules in some proton transfer processes. [86-98] These water molecules facilitate the proton relay by acting as bifunctional catalysts. Because in the Lipscomb mechanism there is a proton transfer between two oxygens, we should introduce additional water molecules in our model when studying this mechanism. Finally, there is one problem left: modelling the remaining groups still present at the active site, both the proteinic residues and the water molecules. This is the subject of the next section.

4. Modelling the proteinic environment

The theoretical modelling of environmental effects in enzymatic processes remains an open question. It is expected that the electric field created by the environment may modify the energy barriers, [99-102] so its effect should be included for a good description of the enzymatic reaction. There are two possible ways to represent the proteinic environment: a discrete representation of the environment, [103] or an average representation of it by means of a continuum model. [104-107] Presence of polar groups in the proteinic chains of CA creates an important electric field which surely will have an important influence on the reaction. Moreover, water molecules having a restricted motion due to the formation of hydrogen bonds will also contribute to the permanent electric field. The electronic distribution of the chemical system, which changes along the process, will in turn produce an electronic polarization of the environment and also some polarization due to partial reorientation. In addition, the electric field of the environment will electronically polarize the chemical system which will reorientate itself in front of the electric field created by the environment. Therefore, an electronic coupling between the chemical system and the environment, and a certain dynamic coupling through mutual reorientation is found simultaneously.

The simplest model in a discrete representation considers the environment by a number of fixed point charges. [103] A first drawback of this model is that electronic polarization of the environment is not included. This limitation has been overcome by some authors through inclusion of atomic polarizabilities. [108-111] Another drawback of this model is the freezing of the rigid structure of the initial charge distribution. Modern Molecular Dynamics studies in proteins [112] have shown the inadequacy of this hypothesis. For this reason, more recent models try to account for the coupling between the chemical system and the environment dynamics. The most simple way to study this coupling is the use of a continuum model, where the orientation and intensity of the reaction field depends on the charge distribution in the chemical system at each point of the reaction coordinate. Furthermore, this reaction field produces in turn a polarization of the chemical system. Nevertheless, a continuum model has also strong limitations; for instance, it represents inadequately the permanent electric field created by the environment. For this reason, we should discuss now how to correct this weakness of the continuum model.

One solution would be to introduce a permanent electric field besides that created by polarization of the dielectric. [113] However, we have adopted here another strategy: we have taken a large value for the dielectric constant of the environment. Given that the reorientation of proteinic residues and water molecules at the active site is partially blocked, it seems that a low dielectric constant to model the environment should be used. But as pointed out above, a proteinic environment differs from a free solvent, because in a protein there is a permanent electric field which has to be taken into account. Therefore, a continuum model developed for free solvents must be corrected in some way to model correctly an enzymatic environmental effect. The use of a large dielectric constant will create a reaction field which accounts much better for the real electric field acting on the chemical system. In this work, the study of surrounding medium effects has been performed using the Tomasi continuum model with two different values of the dielectric constant, one corresponding to an n-hexane solution with a low dielectric constant ($\epsilon=1.88$), and the other corresponding to a water solution with a large dielectric constant (ϵ=78.36). We think that this latter value represents better the environmental effects in CA.

5. Results

Presentation of the results obtained is split into three sections. First, the generation of the active species $EZn^{II}(OH^-)$ from the $EZn^{II}(H_2O)$ complex is studied; second, binding of CO_2 near the active species $EZn^{II}(OH^-)$ and generation of the $EZn^{II}(HCO_3^-)$ complex via the Lipscomb and the Lindskog mechanisms are analyzed; and third, the HCO_3^-/H_2O exchange is discussed. In all the cases the effect of the surrounding medium in the process is introduced.

5.1. GENERATION OF THE ACTIVE SPECIES $EZn^{II}(OH^-)$.

The process corresponding to step (1) of the CO_2 hydration by CA can be formulated as:

$$EZn^{II}(H_2O) + buffer \rightleftharpoons EZn^{II}(OH^-) + bufferH^+ \tag{5}$$

In this process, the His64 group plays the role of a proton shuttle, accepting a proton coming from the water directly coordinated to Zn^{+2}, and releasing another proton to the buffer. The proton transfer from the $EZn^{II}(H_2O)$ species to His64 seems to take place through a bridge of two water molecules. [19] Due to the large number of molecules intervening in this proton transfer, which makes the problem not affordable with ab initio methodology, we have studied the proton transfer from $(NH_3)_3Zn^{II}(H_2O)$ to a water molecule as a simplified model for this deprotonation process.

The intermediate and the transition state found in this study are depicted in Figure 3. In Table 1 the most important geometrical parameters, charges on the zinc atom and the H_3O group, and relative energies referred to reactants for the different intervening species are reported.

From the values of the the O_1-H_1 and O_2-H_1 bond lengths, and from the values of their Pauling bond orders in the $(NH_3)_3Zn^{II}(H_2O)_2$ intermediate ($B(O_1$-$H_1)$=0.774 and $B(O_2$-$H_1)$=0.228), it can be shown that H_1 is not transferred yet in $(NH_3)_3Zn(H_2O)_2^{+2}$. Further, the intermediate can be seen as a $(NH_3)_3Zn^{II}(H_2O)$ complex interacting with a water molecule. On the contrary, in the transition state the H_3O^+ group has been already formed and is moving away from the other fragment. This can be also seen from the values of the eigenvector associated to the only negative eigenvalue of the Hessian matrix. The main component of this vector, which corresponds to the $r(O_1$-$H_1)$ distance, has a value of 0.992, showing that the separation of the H_3O^+ and the $(NH_3)_3Zn^{II}(OH)^-$ fragments is the most important component in the transition state.

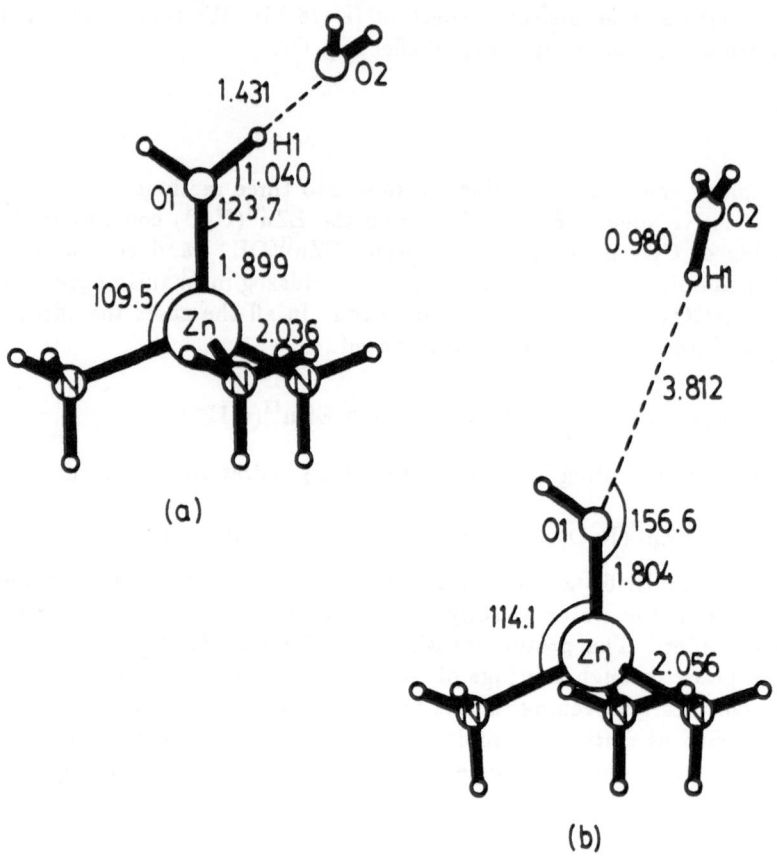

(a)

(b)

Figure 3: Optimized structures of (a) the hydrogen bond complex and (b) the transition state both computed with the 3-21G basis set for the $(NH_3)_3Zn^{+2}(H_2O)_2$ system.

Table 1: Distances from O_1 to the metal, distances from oxygens to the transferring hydrogen, charges on the metal (q_M) and H_3O fragment $(q_{(H_3O)})$, and relatives energies referred to reactants computed with the 3-21G basis set for the $(NH_3)_3Zn^{II}(H_2O)_2$ species. Distances are given in Å, charges in atomic units, and energies in kcal/mol.

Species	$r(ZnO_1)$	$r(O_1H_1)$	$r(O_2H_1)$	q_M	$q_{(H_3O)}$	ΔE
Reactants	1.927	0.949	∞	1.314	-	0.
Intermediate	1.899	1.040	1.431	1.293	0.674	-39.0
Transition state	1.804	3.812	0.980	1.167	0.999	17.6
Products	1.751	∞	0.974	1.193	1.000	-15.5

The charges on the metal (Table 1) indicate that the ligands transfer an important amount of charge to the central ion. For instance, in the $(NH_3)_3Zn^{II}(H_2O)$ reactant the charge transfer from the three ammonia ligands is 0.494 au. This charge transfer from the ligands to zinc reduces the positive charge about the central ion. Moreover, the s and p metal orbitals receiving the charge transfer from the water dimer are destabilized by the ligands. Both aspects contribute to the decrease of the electrostatic and charge transfer effects, difficulting the deprotonation process. Two reasons account for this fact: first, a larger positive charge polarizes the water molecule even further so the deprotonation process is facilitated; second, the charge transfer to zinc makes water more positive, so deprotonation is eased. In fact, the energy barrier in the $(NH_3)_3Zn^{II}(H_2O)_2$ system (58.6 kcal/mol) is 42.6 kcal/mol higher than in the $Zn^{II}(H_2O)_2$ species. Consideration of environmental effects in the $(NH_3)_3Zn^{II}(H_2O)_2$ system does not change very much the energy barrier, which becomes 60.0 kcal/mol in a medium with a dielectric constant of 78.36. The main difference found in the energy profile when environmental effects are included is the fact that the reaction becomes endothermic, in agreement with the experimental pk$_a$ value of the water directly coordinated to Zn^{+2} in CA, which is close to 7. [12-17]

The large value for the energy barrier obtained shows that the process studied cannot be directly related to the proton transfer taking place in CA. In this enzyme, the driving force necessary for this proton transfer is provided by the His64 group, which is a much more basic group than water. Since *ab initio* calculations with a more realistic model which includes an imidazole group (ImH) to represent His64 in CA are not practical, the AM1 semiempirical method has been used to study the proton transfer from $(ImH)_3Zn^{II}(H_2O)$ to an ImH group. In Figure 4, the intermediate and transition state obtained for this proton transfer are represented. In this case, the energy barrier obtained is 6.1 kcal/mol, which is close to the 10 kcal/mol expected from an experimental point of view. This notwithstanding, consideration of (a) the presence of a two-water bridge between water and the His64 group, and (b) the effect of the environment should increase the energy barrier by a few kcal/mol, as pointed out by different authors. [23]

5.2 FORMATION OF THE $EZn^{II}(HCO_3^-)$ SPECIES.

In this section the binding of CO_2 to the biochemically active species $EZn^{II}(OH^-)$ and the generation of the $EZn^{II}(HCO_3^-)$ species *via* the Lipscomb and the Lindskog mechanisms are studied. The results obtained are split into two subsections. In the first one, we analyze the CO_2 hydration through the $(NH_3)_3Zn^{II}(OH^-)$ model for the CA active site, whereas in the second section we discuss the effect of the surrounding medium in the reaction.

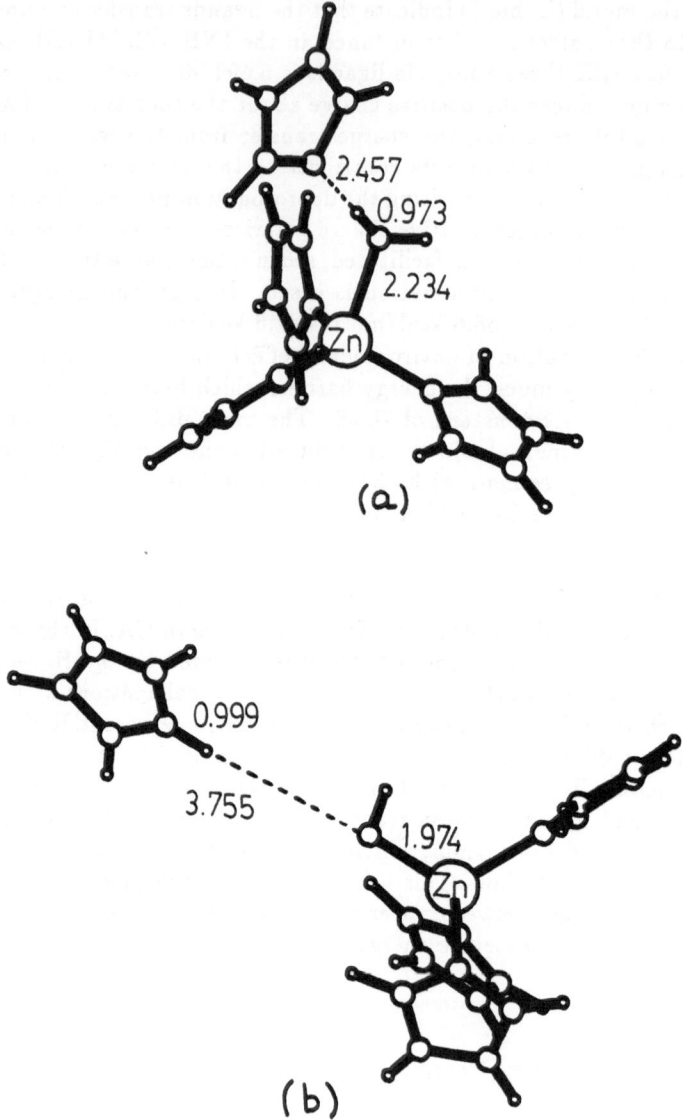

Figure 4: Optimized structures of (a) the hydrogen bond complex and (b) the transition state both obtained with the AM1 method for the $(ImH)_3Zn^{+2}(H_2O)(ImH)$ system.

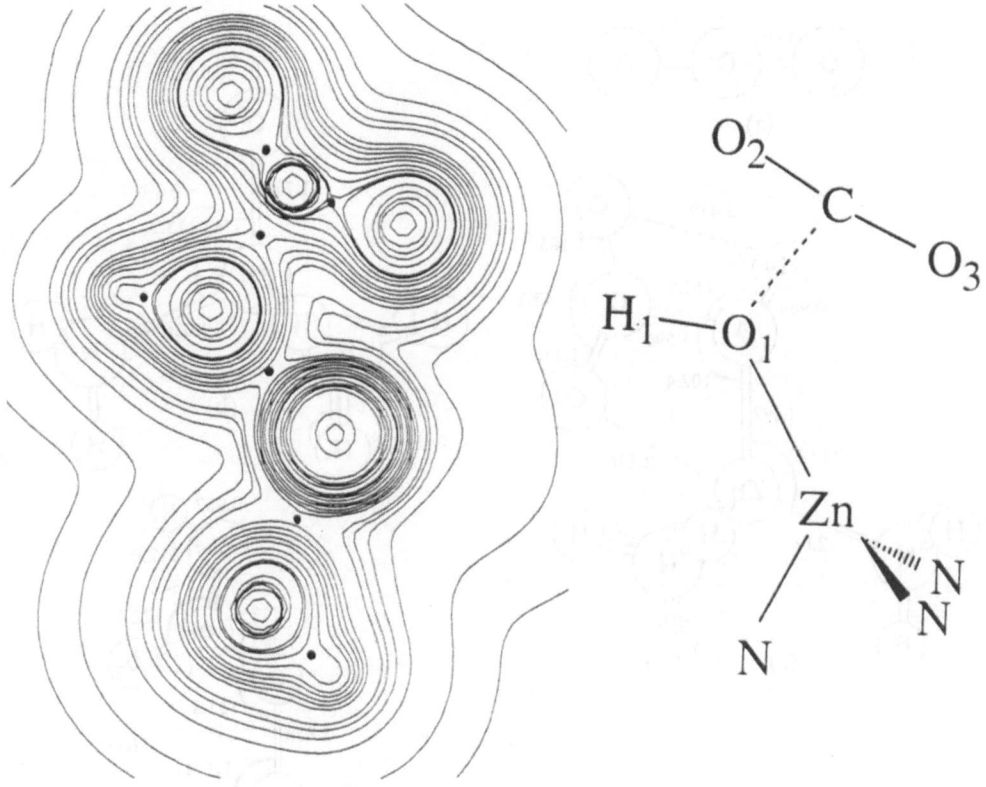

Figure 6: Electron density plot in the plane formed by the hydroxide ion and the CO_2 for the intermediate. • indicates a bond critical point.

5.2.1 Hydration of CO_2 by the $(NH_3)_3 Zn^{II}(OH^-)$ model. Figures 5a-d show the geometries of the separated reactants, intermediate, and final product obtained when the CO_2 interacts with the $(NH_3)_3Zn^{II}(OH^-)$ complex. In the intermediate, the deformation of the $Zn\widehat{O_1H_1}$ and $O_2\widehat{C}O_3$ angles, as compared to their values in the separated reactants, together with the value of the C-O_1 distance, indicate that the nucleophilic attack of the Zn^{II}-bound OH^- to CO_2 is quite advanced. The density map [114] for this intermediate (Figure 6) reinforces this point, showing the existence of a bond critical point [115] between C and O_1. The charge transfer from the Zn^{II}-bound OH^- to CO_2 is 0.246 au, thus showing again that the nucleophilic attack has already been produced to a great extent in this intermediate.

The Zn-O_1 distance has changed from 1.75 Å in the reactants to 1.89 Å in the intermediate. This distance increases in a similar way to how solvent molecules separate from the hydroxyl in a nucleophilic attack.

The electron density map of the intermediate reveals clearly that no bond critical point between Zn and O_3 has been formed, although given the positive charge on the Zn atom and the negative charge on O_3 there is an important electrostatic interaction between Zn and O_3. The polarization of the CO_2 fragment induced by the Zn^{II}-bound hydroxide species favors the nucleophilic attack. This CO_2 activation due to the

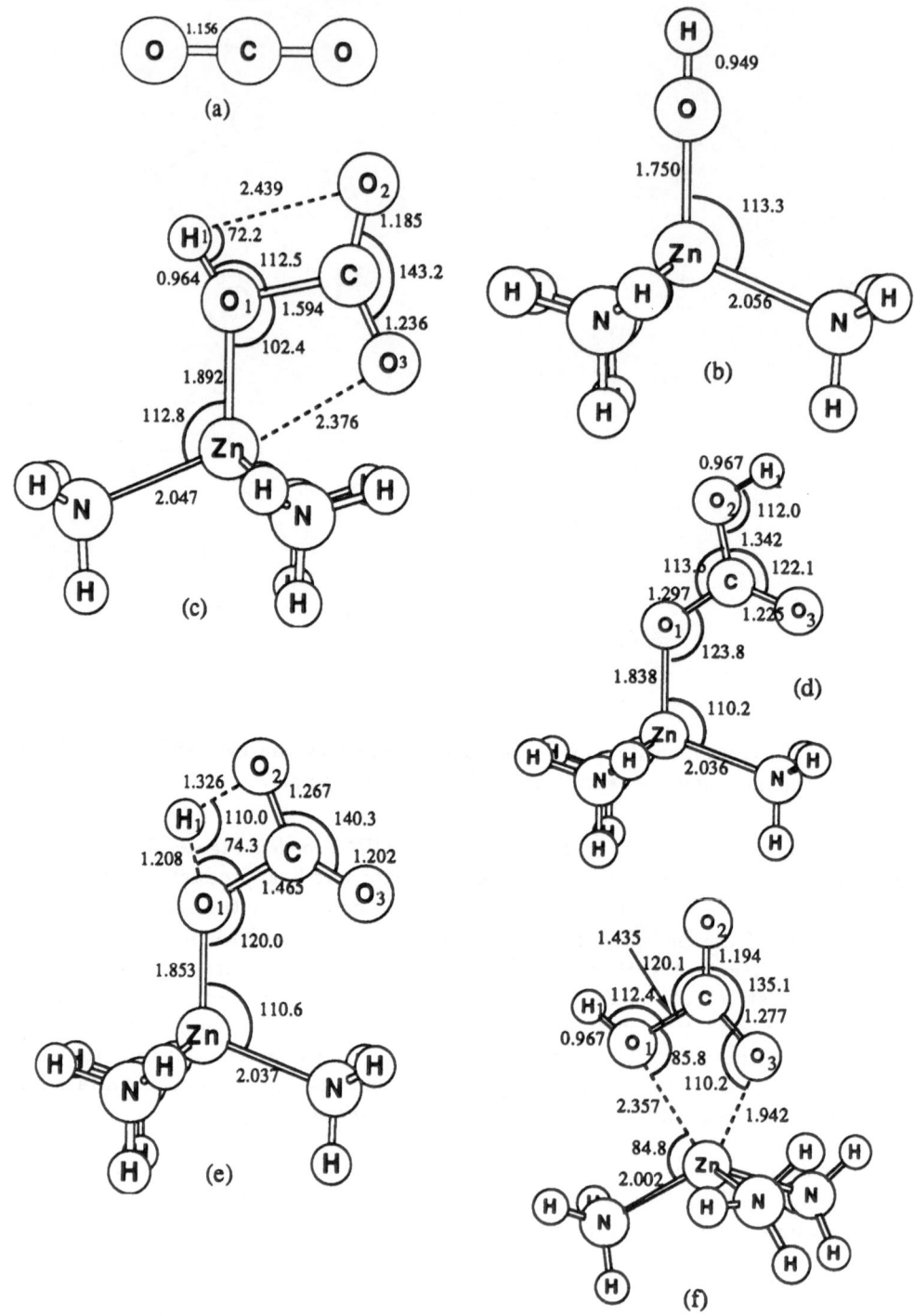

Figure 5: Optimized structures of (a) CO_2; (b) $(NH_3)_3Zn^{II}(OH^-)$; (c) $(NH_3)_3Zn^{II}(OH^-) \cdot CO_2$ intermediate; (d) $(NH_3)_3Zn^{II}(HCO_3^-)$ product *anti*, (e) $(NH_3)_3Zn^{II}(OH^-) \cdot CO_2$ Lispcomb TS; and (f) $(NH_3)_3Zn^{II}(OH^-) \cdot CO_2$ Lindskog TS.

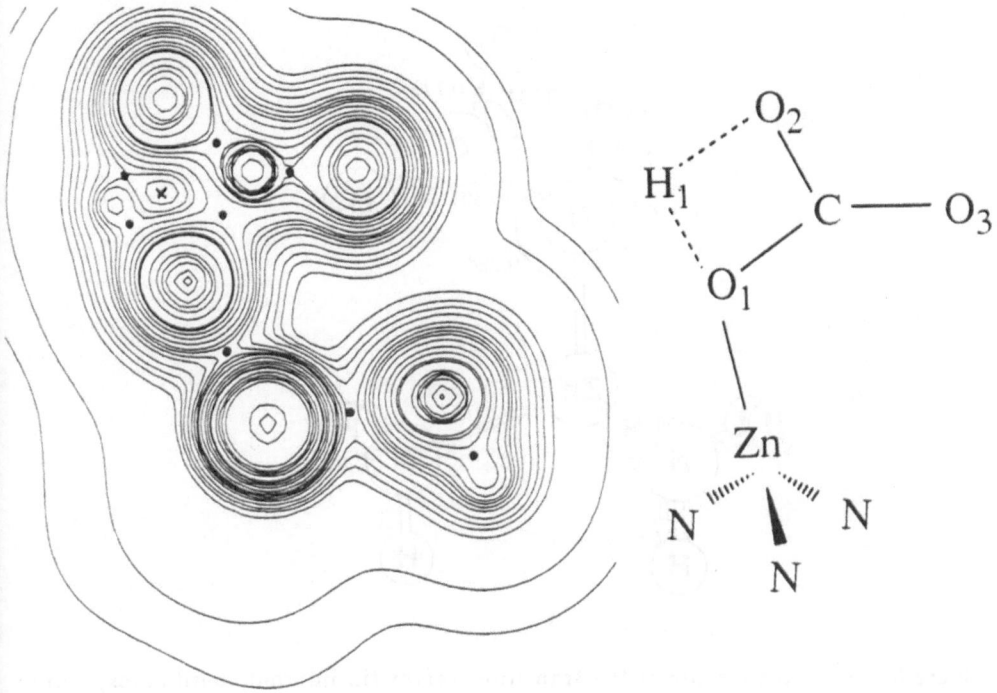

Figure 7: Electron density plot in the plane formed by the hydroxide ion and the CO_2 for the transition state which corresponds to the Lipscomb mechanism. • indicates a bond critical point, while x indicates a ring critical point.

presence of the Zn^{+2} dication has been previously emphasized [23,30,31,33] especially by Jacob *et al.* [32]

In Figure 5e the transition state found for the Lipscomb mechanism, which involves a proton transfer between two oxygens, is depicted. [116] The four atoms involved most in the process (O_1, H_1, O_2 and C) form a planar four-membered ring. The existence of a ring critical point in the electron density map of this transition state (Fig. 7) shows that a four-membered ring between H_1, O_1, O_2, and C has been indeed formed. To get a deeper insight into this mechanism, the main components of the transition vector (in internal coordinates) are drawn in Fig. 8. One can see that the nucleophilic attack and the proton transfer are concerted processes. However, these two processes are not advanced to the same extent; whereas the nucleophilic attack is almost completed in the TS, the proton is only half-way transferred. The largest components of the transition vector correspond to this proton transfer, even though there is an important component corresponding to the nucleophilic attack (-0.30).

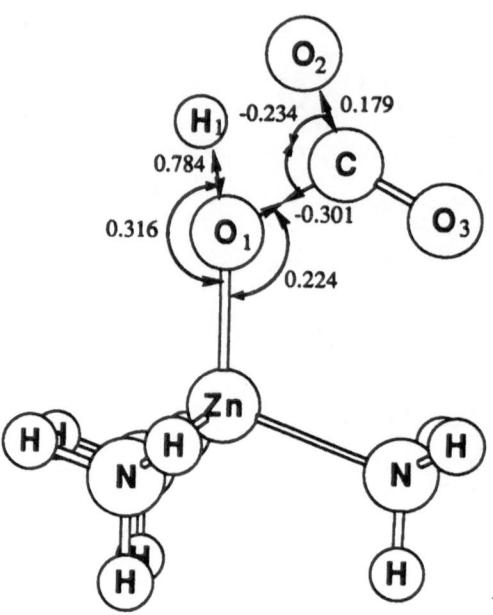

Figure 8: Main components of the transition vector (in internal coordinates) corresponding to the Lipscomb mechanism.

The transition state found for the Lindskog mechanism, which implies a change in the oxygen directly coordinated to Zn^{+2}, is depicted in Figure 5f. In this transition state the $Zn-O_1$ bond is almost broken and the $Zn-O_3$ bond is almost formed. The electron density map, drawn in Fig. 9, shows that no ring critical point between Zn, O_1, C and O_3 has been formed, and that the $Zn-O_1$ bond is already broken whereas the $Zn-O_3$ bond is almost completely formed. The main components of the transition vector in internal coordinates (Fig. 10) bring about the same conclusions: breaking of the $Zn-O_1$ bond together with the formation of a new $Zn-O_3$ bond. The O_1-H_1 distance remains unchanged, and participation of H_1 in the transition vector is due merely to spatial reorientation of the O_1-H_1 bond.

The final product of the reaction is shown in Figure 5d. The product exhibits two possible conformations, depending on whether H_1 is closer to O_1 than to O_3 (this product hereafter named syn-) or, in the contrary, H_1 is closer to O_3 than to O_1 (this product hereafter named anti-). The latter turns out to be the most stable, although given the slight energy differences between these two structures, a fast interconversion rate between them can be expected.

The energies of the different structures mentioned above referred to separated reactants, and the forward and backward energy barriers for the Lipscomb and the Lindskog mechanisms are collected in Table 2.

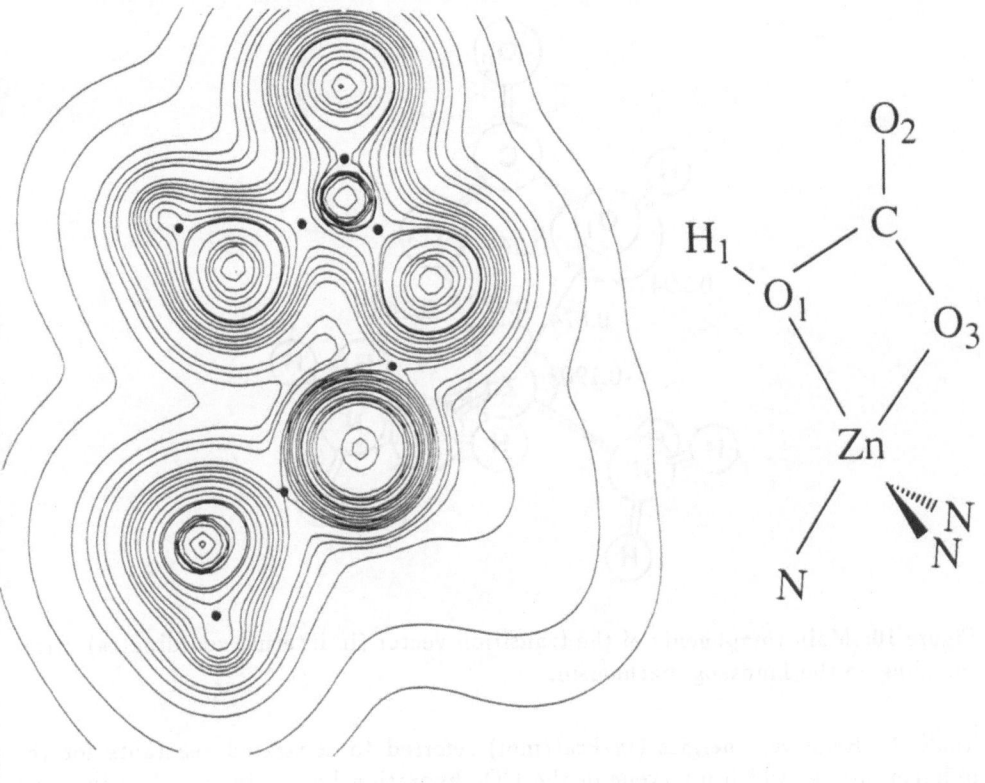

Figure 9: Electron density plot in the plane formed by the hydroxide ion and the CO_2 for the transition state which corresponds to the Lindskog mechanism. • indicates a bond critical point.

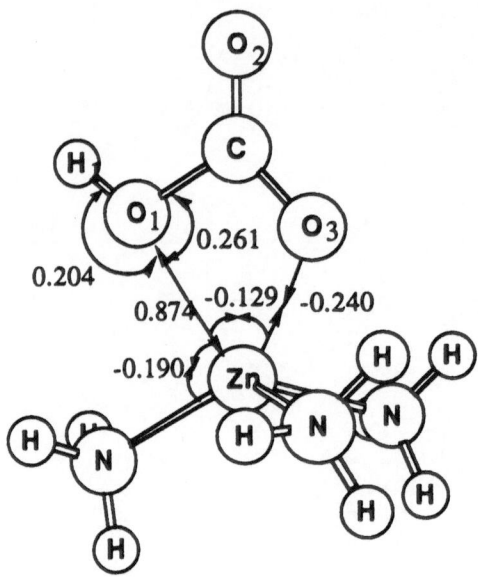

Figure 10: Main components of the transition vector (in internal coordinates) corresponding to the Lindskog mechanism.

Table 2: Relatives energies (in kcal/mol) referred to separated reactants for the different species which intervene in the CO_2 hydration by CA in the Lipscomb and Lindskog mechanisms, together with forward and backward energy barriers for these two mechanisms, obtained with the 3-21G basis set at the SCF and MP2 levels.

	SCF	MP2
$(NH_3)_3Zn^{II}(OH^-) + CO_2$	0.	0.
$(NH_3)_3Zn^{II}(OH^-) \cdot CO_2$ INT	-20.3	-12.4
$(NH_3)_3Zn^{II}(OH^-) \cdot CO_2$ Lipscomb TS	22.3	24.0
$(NH_3)_3Zn^{II}(OH^-) \cdot CO_2$ Lindskog TS	-18.1	-4.4
$(NH_3)_3Zn^{II}(HCO_3^-)$	-31.6	-15.1
Forward Lipscomb	42.6	36.4
Backward Lipscomb	53.9	39.1
Forward Lindskog	2.2	8.0
Backward Lindskog	13.5	10.7

The computed forward and backward barriers for the Lipscomb mechanism are 42.6 and 53.9 kcal/mol. For the Lindskog mechanism the forward and backward barriers turn out to be 2.2 and 13.5 kcal/mol. The final product of this CO_2 hydration is 11.3 kcal/mol more stable than the intermediate complex. This fact leads to sizeable differences between forward and backward energy barriers, unfavoring the necessary reversibility of all enzymatic processes. Introduction of correlation energy at the MP2 level (Table 2) improves this situation, the energy barriers for the forward and backward reactions being now more similar. Further, it diminishes the energy barriers for the Lipscomb mechanism, and increases the forward energy barrier for the Lindskog mechanism. Nevertheless, the Lindskog mechanism continues to be favored. As long as CO_2 hydration is not the rate-determining step in the CA catalytic cycle, the energy barrier for the CO_2 hydration must be lower than 10 kcal/mol. Therefore, from results of Table 2 the Lipscomb mechanism would not seem in agreement with the results obtained in the present study made with our simplified model.

5.2.2 Environmental Effects. So far, we have dealt with a quite simplified model of the CA active site. However, it is well known that in the active site there is a number of water molecules, some of which can participate directly in the proton transfer acting as bifunctional catalysts. [86-98]

Figs. 11a-c show the geometries of the intermediate, transition state, and product for the Lipscomb mechanism when an additional water molecule is included for the proton transfer. Comparing this intermediate with the intermediate which does not incorporate an additional water molecule, the $C-O_1$ distance has been reduced from 1.59 to 1.49 Å. This result shows that the nucleophilic attack is now more advanced as one can see also from the charge transfer of 0.324 au to the CO_2 molecule.

In the transition state the six atoms involved most in the process (O_1, H_1, O_4, H_4, O_2, and C) form a fairly regular six-membered ring, which is less strained than the four-membered ring found previously, and hence this transition state is stabilized with respect to the four-membered transition state. The O-H distances show that H_1 has been quite transferred to the water molecule, whereas this water has only began to transfer its H_2 to O_2. When going from the intermediate to the transition state, the $C-O_1$ distance has changed only from 1.49 to 1.44 Å. The component of the $C-O_1$ bond in the transition vector (Figure 12) has diminished when an additional water molecule is taken into account, being now only -0.18, thus showing that the nucleophilic attack is almost finished at this transition state. The largest components of the transition vector correspond to the motion of the H_1 and H_2 atoms during the proton transfer processes.

In Table 3 we collect the relative energies at the SCF and MP2 levels referred to separated reactants for the different species involved in the process, together with

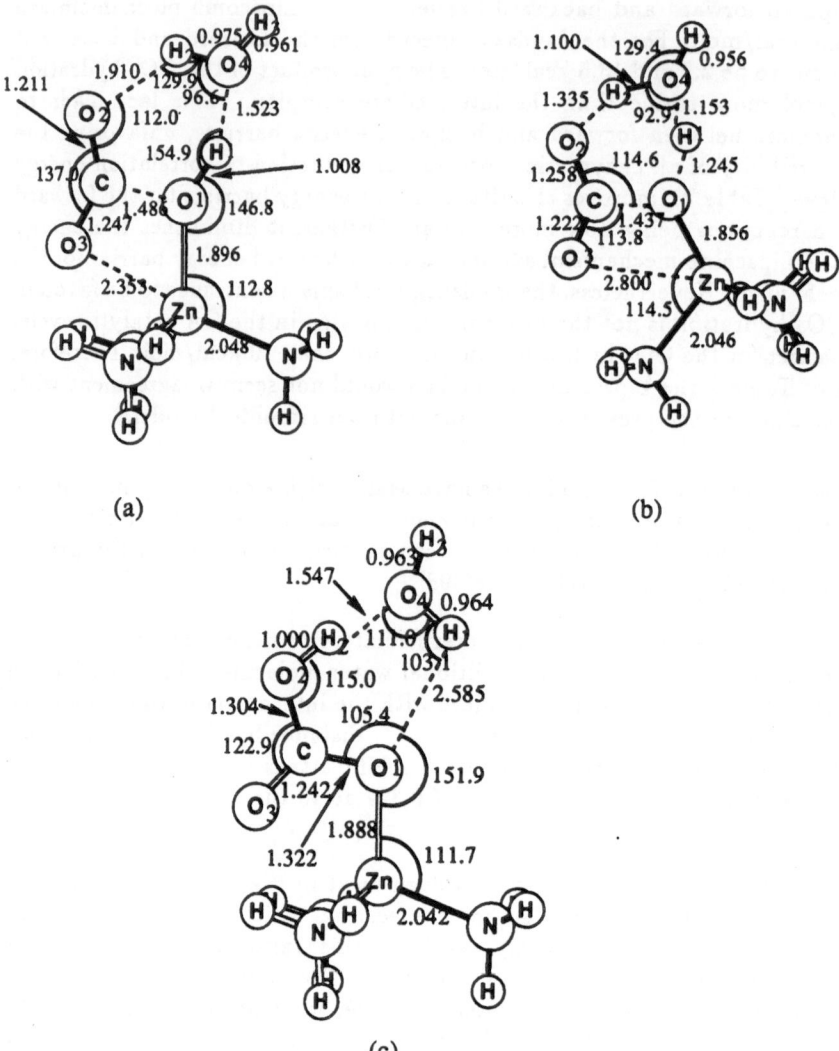

Figure 11: Optimized structures of (a) $(NH_3)_3Zn^{II}(OH^-)(CO_2) \cdot H_2O$ intermediate; (b) $(NH_3)_3Zn^{II}(OH^-)(CO_2) \cdot H_2O$ Lispcomb TS; (c) $(NH_3)_3Zn^{II}(HCO_3^-) \cdot H_2O$ product, when one additional water molecule has been considered for the proton release.

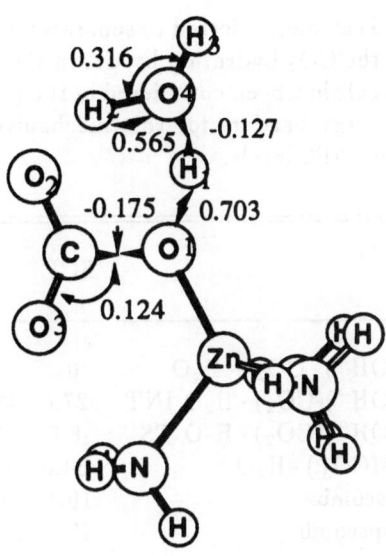

Figure 12: Main components of the transition vector (in internal coordinates) corresponding to the Lipscomb mechanism when one additional water molecule has been included in the description of the active site.

the values obtained for the forward and backward energy barriers for the Lipscomb mechanism when studied with an additional water molecule.

The most striking fact is the important reduction of the two energy barriers. In particular, the barrier for the forward reaction is now 19.1 kcal/mol, *i.e.*, it has decreased by 23.5 kcal/mol when an additional water has been considered for proton release. Likewise, the backward energy barrier is now 25.6 kcal/mol, being reduced by 28.3 kcal/mol. At the MP2 level, the forward and backward energy barriers for the Lipscomb mechanism with an additional water molecule become 17.0 and 17.7 kcal/mol, respectively. These values are closer to those presented above for the Lindskog mechanism at the same level of calculation.

So far, we have only considered one additional water molecule in the description of the active site for the Lipscomb mechanism. Actually, the active site contains a number of water molecules and proteinic residues which have not been taken into account yet. As explained in the methodology section, their effect has been introduced by means of a continuum model.

Relative energies referred to reactants, and forward and backward energy barriers of the CO_2 hydration according to the Lipscomb and the Lindskog mechanisms for

Table 3: Relatives energies (in kcal/mol) referred to separated reactants for the different species which intervene in the CO_2 hydration by CA in the Lispcomb mechanism when an additional water molecule has been considered in the proton release, together with forward and backward energy barriers for this mechanism, obtained with the 3-21G basis set at the SCF and MP2 levels.

	SCF	MP2
$(NH_3)_3Zn^{II}(OH^-) \cdot CO_2 + H_2O$	0.	0.
$(NH_3)_3Zn^{II}(OH^-)(CO_2) \cdot H_2O$ INT	-27.6	-27.5
$(NH_3)_3Zn^{II}(OH^-)(CO_2) \cdot H_2O$ TS	-8.5	-10.5
$(NH_3)_3Zn^{II}(HCO_3^-) \cdot H_2O$	-34.0	-28.2
Forward Lipscomb	19.1	17.0
Backward Lipscomb	25.6	17.7

three different values of the dielectric constant (ϵ=1.00, 1.88, and 78.36) are gathered in Tables 4 and 5, respectively. Upon examination of values in Table 4, it can be concluded that the increase of the dielectric constant diminishes the energy barriers for the Lipscomb mechanism and reduces the exothermicity of the reaction. In a dielectric medium of ϵ=78.36, the decrements of the forward and backward energy barriers are quite important: 6.3 and 14.1 kcal/mol, respectively, leading to barriers which are almost identical, and so favoring the reversibility of the process. Likewise, in the Lindskog mechanism the increase in the dielectric constant implies a decrease of the energy barrier and an increase of the exothermicity of the reaction, though now the effect is less important than in the Lipscomb mechanism. In a surrounding medium of ϵ=78.36, and referred to *in vacuo* results, the forward and backward energy barriers diminish by 2.0 and 3.8 kcal/mol, respectively.

Due to computational limitations we have been unable to introduce the effect of correlation energy and the effect of surrounding medium simultaneously. Nevertheless, we can make the approximation that the correction due to environmental effects and that owing to correlation effects at the MP2 level are additive. In this case, the final values for the energy barriers are collected in Table 6. These values indicate that both mechanisms are acceptable to explain experimental facts, because the values for their energy barriers lie near the maximum allowed value for this CO_2 hydration (10 kcal/mol).

Table 4: Relatives energies (in kcal/mol) referred to separated reactants for the different species which intervene in the CO_2 hydration by CA in the Lispcomb mechanism when an additional water molecule has been considered in the proton release, together with forward and backward energy barriers for this mechanism, obtained with the 3-21G basis set, in environments of three different dielectric constants: 1.00, 1.88 and 78.36.

	$\epsilon=1.00$	$\epsilon=1.88$	$\epsilon=78.36$
$(NH_3)_3Zn^{II}(OH^-) \cdot CO_2 + H_2O$	0.	0.	0.
$(NH_3)_3Zn^{II}(OH^-) \cdot CO_2 \cdot H_2O$ INT	-27.6	-26.1	-21.3
$(NH_3)_3Zn^{II}(OH^-) \cdot CO_2 \cdot H_2O$ TS	-8.5	-9.8	-8.6
$(NH_3)_3Zn^{II}(HCO_3^-) \cdot H_2O$	-34.0	-29.2	-20.0
Forward energy barrier	19.1	16.3	12.7
Backward energy barrier	25.6	19.4	11.4

Table 5: Relatives energies (in kcal/mol) referred to separated reactants for the different species which intervene in the CO_2 hydration by CA in the Lindskog mechanism, together with forward and backward energy barriers for this mechanism, obtained with the 3-21G basis set, in environments of three different dielectric constants: 1.00, 1.88 and 78.36.

	$\epsilon=1.00$	$\epsilon=1.88$	$\epsilon=78.36$
$(NH_3)_3Zn^{II}(OH^-) + CO_2$	0.	0.	0.
$(NH_3)_3Zn^{II}(OH^-) \cdot CO_2$ INT	-20.3	-25.5	-30.9
$(NH_3)_3Zn^{II}(OH^-) \cdot CO_2$ TS	-18.1	-24.2	-30.7
$(NH_3)_3Zn^{II}(HCO_3^-)$	-31.6	-36.1	-40.4
Forward energy barrier	2.2	1.3	0.2
Backward energy barrier	13.5	11.9	9.7

Table 6: Forward and backward energy barriers for the Lipscomb and Lindskog mechanisms, when it has been considered the approximation that environmental and MP2 corrections are additive.

	Lipscomb	Lindskog
Forward energy barrier	10.7	6.0
Backward energy barrier	3.6	6.9

5.3 THE HCO_3^-/H_2O EXCHANGE IN CARBONIC ANHYDRASE.

Once the $EZn^{II}(HCO_3^-)$ species has been formed, the next step in the CA catalytic cycle is the regeneration of the $EZn^{II}(H_2O)$ species. With our model for the CA active site this process can be formulated as:

$$(NH_3)_3Zn^{II}(HCO_3^-) + H_2O \rightleftharpoons (NH_3)_3Zn^{II}(H_2O) + HCO_3^- \tag{6}$$

For this reaction, the *in vacuo* energy gap between the products and the reactants at the SCF level is 205.4 kcal/mol. Inclusion of correlation energy at the MP2 level (MP2/3-21G//3-21G) [61] does not change the result obtained at the SCF level meaningfully, because it yields an energy gap of 205.3 kcal/mol. This reaction can be regarded as a charge separation process where the reactants with a global charge of +1 forms two products of global charges +2 and -1. Therefore, surrounding medium effects should be especially important, thus causing a larger stabilization in products than in reactants. Effectively, environmental effects calculated with $\epsilon=1.88$ reduce the endothermicity of reaction (3) to 112.9 kcal/mol. This value is further decreased to 11.3 kcal/mol when a dielectric constant of 78.36 is used.

As mentioned in the Introduction, experimental studies [7,41] show that during the HCO_3^-/H_2O exchange a pentacoordinate metal transient intermediate is formed, which might facilitate the exchange between bicarbonate and water ligands. All possible pentacoordinate complexes which can be obtained when water binds the $(NH_3)_3Zn^{II}(HCO_3^-)$ complex have been explored, three different $(NH_3)_3Zn^{II}(HCO_3^-)(H_2O)$ pentacoordinate complexes having been found as stable species: [117] (a) $(NH_3)_3Zn^{II}(HCO_3^-)_{ax}(H_2O)_{eq}$ (hereafter named **weba** because of the location of the water and bicarbonate ligands, and depicted in Fig. 13a), which was reported already using the AM1 method by Merz *et al.*; [23] (b), $(NH_3)_3Zn^{II}(HCO_3^-)_{eq}(H_2O)_{ax}$ (hereafter named **wabe**, and depicted in Fig. 13b);

Table 7: Dissociation energies of HCO_3^- (in kcal/mol) for the optimized pentacoordinate $(NH_3)_3Zn^{II}(HCO_3^-)(H_2O)$ complexes, and for three different values of the dielectric constant.

	$\epsilon=1.00$	$\epsilon=1.88$	$\epsilon=78.36$
weba	231.9	137.7	30.5
wabe	231.8	137.6	30.5
webe	227.7	134.3	28.7

and (c) $(NH_3)_3Zn^{II}(HCO_3^-)_{eq}(H_2O)_{eq}$ (hereafter named **webe**, and plotted in Fig. 13c).

All these pentacoordinate Zn^{II} complexes found in our study feature a trigonal bipyramidal rearrangement. Previous theoretical calculations reported the same result in analogous Zn^{II} complexes. [23,33,117,118] Interestingly, complexes **weba** and **wabe** have an intramolecular hydrogen bond which accounts partially for the distortion from the ideal tbp geometry.

Table 7 assembles the dissociation energy of HCO_3^- in the different pentacoordinate $(NH_3)_3Zn^{II}(HCO_3^-)(H_2O)$ species for three different dielectric constants of the surrounding medium, corresponding to the reaction:

$$(NH_3)_3Zn^{II}(HCO_3^-)(H_2O) \rightleftharpoons (NH_3)_3Zn^{II}(H_2O) + HCO_3^- \qquad (7)$$

In the gas phase ($\epsilon=1.00$) the dissociation energy of HCO_3^-, which is larger than 225 kcal/mol, shows the intrinsic difficulty of the HCO_3^- release. The most appealing aspect is the important reduction of this HCO_3^- dissociation energy with the increase on the value of the dielectric constant. This point can be easily interpreted taking into account that the final product of the release of HCO_3^-, which are $(NH_3)_3Zn^{II}(H_2O)$ and HCO_3^-, have the largest free energy of solvation, $i.e.$, the products in reaction (4) are far more stabilized by the surrounding medium than the reactants. Thus, as suggested earlier, [23,32,40] medium effects become essential to understand the HCO_3^- release in the CA mechanism. Nevertheless, the dissociation energies of HCO_3^- in a medium of $\epsilon=78.36$ are still too large to account for experimental data.

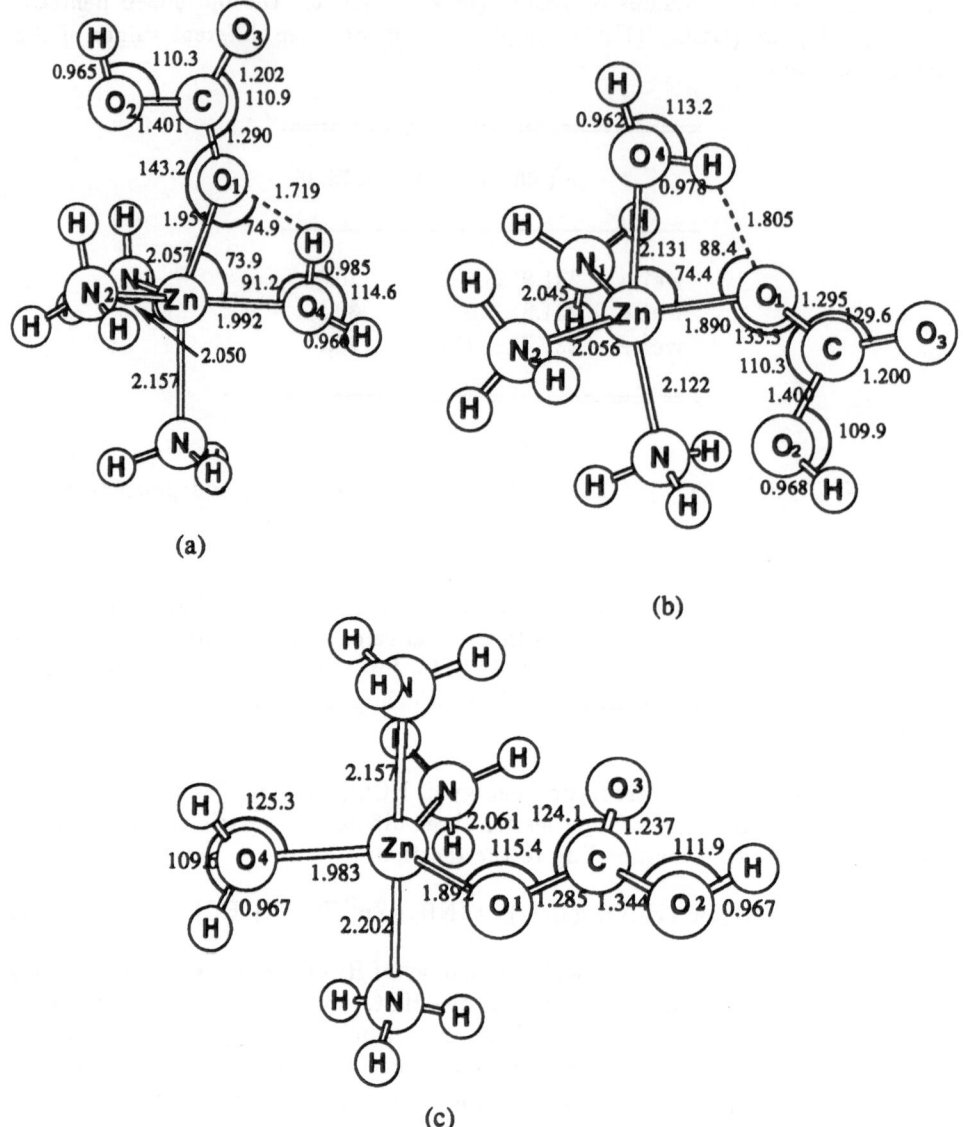

Figure 13: Optimized structures of the $(NH_3)_3Zn^{II}(HCO_3^-)_{ax}(H_2O)_{eq}$ (a), $(NH_3)_3Zn^{II}(HCO_3^-)_{eq}(H_2O)_{ax}$ (b), and $(NH_3)_3Zn^{II}(HCO_3^-)_{eq}(H_2O)_{eq}$ (c) complexes.

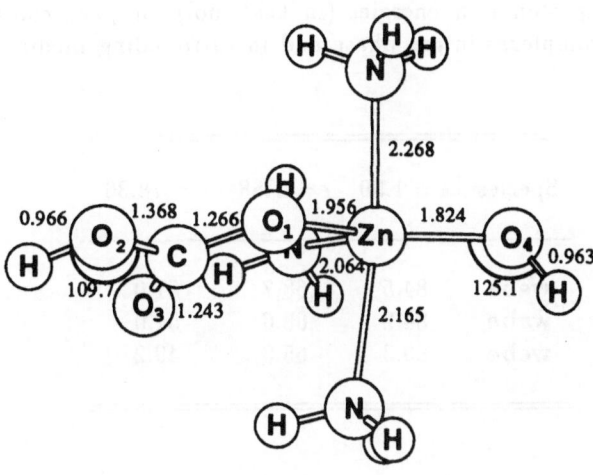

Figure 14: Optimized structure of the $(NH_3)_3Zn^{II}(HCO_3^-)_{eq}(OH^-)_{eq}$ complex.

To date, two other alternatives have been proposed to explain the release of HCO_3^-. The first one, suggested earlier by Pullman [63] and studied more recently by Merz *et al.* [23] with the AM1 method, concerns the intramolecular proton transfer from water to HCO_3^- to yield the $(NH_3)_3Zn^{II}(H_2CO_3)(OH^-)$ species, where release of H_2CO_3 would be very easy. Any attempt we made to optimize this complex has given negative results, complexes **weba** or **wabe** having been always obtained. Therefore, it has been concluded that complex $(NH_3)_3Zn^{II}(H_2CO_3)(OH^-)$ is not a true minimum in the potential energy hypersurface, and thus, it is not a stable species within our level of calculation. Besides, from an experimental point of view, H_2CO_3 has been never detected [56] during the CO_2/HCO_3^- interconversion. The second possibility, suggested first by Liang and Lipscomb, [40] deals with a previous deprotonation of water to give the $(NH_3)_3Zn^{II}(HCO_3^-)(OH^-)$ species before the release of HCO_3^-. In this complex the posterior loss of a bicarbonate anion will be easier because Zn^{+2} complexes having two anions in the coordination sphere are quite unstable. [117,119] To analyze this possibility, the $(NH_3)_3Zn^{II}(HCO_3^-)(OH^-)$ complex (hereafter named **hebe**) has been optimized, its geometry being depicted in Fig. 14.

From an energetic point of view there are two features worth being remarked. First, as previously emphasized, [117,119] Zn^{+2} pentacoordinate complexes with two anions in the coordination sphere are quite unstable, so formation of

Table 8: Water deprotonation energies (in kcal/mol) for pentacoordinate **weba**, **wabe**, and **webe** complexes in gas phase and in surrounding media of $\epsilon=1.88$ and $\epsilon=78.36$.

Species	$\epsilon = 1.00$	$\epsilon = 1.88$	$\epsilon = 78.36$
weba	84.5	68.7	51.0
wabe	84.2	68.6	51.0
webe	80.3	65.3	49.2

$(NH_3)_3Zn^{II}(HCO_3^-)(OH^-)$ complex will facilitate the release of HCO_3^-. Second, it is well known that binding of an anion originates an important increase in the deprotonation energies of the ligands in the complex, [14,54,55,81,117] and therefore, deprotonation of water in the $(NH_3)_3Zn^{II}(HCO_3^-)(H_2O)$ complexes, necessary to obtain the $(NH_3)_3Zn^{II}(HCO_3^-)(OH^-)$ complex, is more difficult than in the $(NH_3)_3Zn^{II}(H_2O)$ species. From now on, these two energetic aspects are being analyzed.

The dissociation energies of HCO_3^- in complex **hebe** are 131.8, 76.6 and 16.0 kcal/mol, in surrounding media of $\epsilon=1.00$, $\epsilon=1.88$, and $\epsilon=78.36$, respectively. These values can be compared with those given in Table 7 for the $(NH_3)_3Zn^{II}(HCO_3^-)(H_2O)$ complexes. One can see that in the gas phase the energy required to extract HCO_3^- from complex **hebe** is 100.1 kcal/mol lower than from complex **weba**. Therefore, deprotonation of water facilitates the HCO_3^- release. This difference decreases with the increase in the dielectric constant. In particular, in a surrounding medium of $\epsilon=78.36$ this difference is only 14.5 kcal/mol. In this case, loss of bicarbonate anion in complex **hebe** costs only 16.0 kcal/mol, which is closer to that expected from experimental data.

Table 8 collects the deprotonation energies (DPE) for the reaction

$$(NH_3)_3Zn^{II}(HCO_3^-)(H_2O) + H_2O \rightarrow (NH_3)_3Zn^{II}(HCO_3^-)(OH^-) + H_3O^+ \quad (8)$$

It is interesting to compare these values with those obtained for the tetracoordinate complex $(NH_3)_3Zn^{II}(H_2O)$. In particular, for the reaction

$$(NH_3)_3Zn^{II}(H_2O) + H_2O \rightarrow (NH_3)_3Zn^{II}(OH^-) + H_3O^+ \quad (9)$$

the values of the DPE for $\epsilon=1.00$ and $\epsilon=78.36$ are -16.7 and 40.3 kcal/mol, respectively. In the gas phase, the difference between the DPE of the tetracoordinate complex and, for instance, complex **weba**, is greater than 100 kcal/mol. However, in a surrounding medium of $\epsilon=78.36$, the difference in the DPEs has decreased to merely 10.7 kcal/mol. Therefore, it can be concluded that in a surrounding medium having a large dielectric constant, presence of a bicarbonate anion in the coordination sphere does not change very much the deprotonation energy of water. It is worth noting that, as commented in section 5.1, the change of H_2O by a more realistic imidazole group as a proton acceptor, would decrease remarkably the value of the DPE required to obtain the $(NH_3)_3Zn^{II}(HCO_3^-)(OH^-)$ species, which should be then close to 10 kcal/mol.

In the formation of $(NH_3)_3Zn^{II}(HCO_3^-)(OH^-)$ complex prior to release of HCO_3^- there are two factors which counteract. On one hand, deprotonation of water facilitates the HCO_3^- release; on the other hand, presence of HCO_3^- in the coordination sphere difficults the deprotonation of water. The importance of both factors is reduced with the increase in the dielectric constant. This notwithstanding, our results suggest that prior to bicarbonate release, there is a deprotonation of the coordinated water in order to generate the $(NH_3)_3Zn^{II}(HCO_3^-)(OH^-)$ species.

6. Concluding remarks

In this work theoretical calculations to study the CO_2 hydration by CA have been performed. All four steps of the catalytic cycle of carbonic anhydrase have been analyzed.

For the deprotonation of the water directly coordinated to Zn^{+2} to generate the active species, it has been found that the proton transfer to another water molecule in the active site it is not possible because of the appearance of a large energy barrier. Thus, the presence of a more basic group, such as His64, is required; this group provides the driving force required for this proton transfer. It has been also shown that the minimum model required to study the CO_2 hydration in CA, must incorporate ligands coordinated to the zinc ion, and consider the effect of the environment.

As to the initial interaction between CO_2 and the active site, our model shows that the nucleophilic attack has been already initiated, and that CO_2 has been activated for the nucleophilic attack due to the presence of the Zn^{+2} ion.

Referring to $EZn^{II}(HCO_3^-)$ formation, the Lipscomb and the Lindskog mechanisms have been studied, finding that both mechanisms are acceptable to explain experimental data, because the values of their energy barrier are close to the maximum allowed value for this CO_2 hydration.

For the HCO_3^-/H_2O exchange, it has been suggested that there is first a coordination of water to $(NH_3)_3Zn^{II}(HCO_3^-)$ complex, followed by a deprotonation

292

of this water molecule, and finally loss of a bicarbonate and regeneration of the $(NH_3)_3Zn^{II}(OH^-)$ catalytic active species, which reiniciates the catalytic cycle.

The model used in the present work, due to computational constrains imposed by the size of the system studied, exhibits obvious limitations such as the level of calculation, the representation of the environment, and the lack of reoptimized geometries when including correlation energy or environmental effects. In any event, from our results it emerges clearly the great influence of considering the environmental effects in the study of the CO_2 hydration in CA. In fact, to get a definitive answer of the mechanism of CO_2 hydration catalyzed by the CA enzyme, it would be necessary to follow these steps: first, to make a complete description of the active site; and second, to carry out a dynamical study where the coupling between the dynamics of the environment and that of the chemical system is well introduced, which requires that mutual electronic coupling is also well considered. Although this is not possible nowadays, it may be attained in the future.

Acknowledgment. This work has been supported by the Spanish "Dirección General de Investigación Científica y Técnica" under Project No. PB86-0529, and by the Commission of the European Communities (CEE) under contract SC1.0037.C. Partial support has been received from the Universitat Autònoma de Barcelona through the Program fro Precompetitive Research Groups, 1990 and 1991.

7. References

[1] R.H. Prince and P.R. Woolley, J. Chem. Soc. **1972**, *Dalton Trans.*, 1548-1554.

[2] T.J. Williams and R.W. Henkens, *Biochemistry* **1985**, *24*, 2459-2462.

[3] A.C. Sen, C.K. Tu, H. Thomas, G.C. Wynns, and D.N. Silverman, in "Zinc Enzymes" p **329-339**, *I. Bertini, C. Luchinat, W. Maret and M. Zeppezauer, Eds.; Birkhäuser Boston*, 1986; Inc., Vol. I.

[4] D.N. Silverman and S. Lindskog, *Acc. Chem. Res.* **1988**, *21*, 30-36.

[5] P. Woolley, *Nature* **1975**, *258*, 677-682.

[6] I. Bertini and C. Luchinat, *Acc. Chem. Res.* **1983**, *16*, 272-279.

[7] Y. Pocker and T.L. Deits, *J. Am. Chem. Soc.* **1982**, *104*, 2424-2434.

[8] K.K. Kannan, M. Petef, K. Fridborg, S. Lövgren, A. Ohlsson and M. Petef, Proc. Natl. Acad. Sci. USA **1975**, *72*, 51.

[9] K.K. Kannan, M. Petef, K. Fridborg, H. Cid-Dresdner and S. Lovgren, FEBS Lett. **1977**, *73*, 115.

[10] A. Liljas, K.K. Kannan, P.-C. Bergsten, I. Waara, K. Fridborg, B. Strandberg, U. Carlbom, L. Järup, S. Lövgren and M. Petef, Nature New Biol. **1972**, *235*, 131.

[11] E.A. Eriksson, T.A. Jones and A. Liljas, in "Zinc Enzymes" p **317-328**, *I. Bertini, C. Luchinat, W. Maret and M. Zeppezauer, Eds.; Birkhäuser Boston*, 1986; Inc., Vol. I.

[12] Y. Pocker and J.T. Stone, *J. Am. Chem. Soc.* **1965**, *87*, 5497-5498.

[13] J.H. Coates, G.J. Gentle and S.F. Lincoln, *Nature* **1974**, *249*, 773-775.

[14] E. Kimura, T. Koike and K. Toriumi, *Inorg. Chem.* **1988**, *27*, 3687-3688.

[15] D.D. Perrin, *J. Chem. Soc.* **1962**, 4500-4502.

[16] J. Chin and X. Zou, *J. Am. Chem. Soc.* **1984**, *106*, 3687-3688.

[17] E.T. Kaiser and K.-W Lo, *J. Am. Chem. Soc.* **1969**, *91*, 4912-4918.

[18] R.P. Davis, *J. Am. Chem. Soc.* **1959**, *81*, 5674-5678.

[19] Y. Pocker and N. Janjić, *J. Am. Chem. Soc.* **1989**, *111*, 731-733.

[20] D.N. Silverman and C.K. Tu, *J. Am. Chem. Soc.* **1975**, *97*, 2263-2269.

[21] Y. Pocker and C.H. Miao, *Biochemistry* **1987**, *26*, 8481-8486.

[22] H. Steiner, B.-H. Jonsson and S. Lindskog, *Eur. J. Biochem.* **1975**, *59*, 253-259.

[23] K.M. Merz, Jr., R. Hoffmann and M.J.S. Dewar, *J. Am. Chem. Soc.* **1989**, *111*, 5636-5649.

[24] K.S. Venkatasubban and D.N. Silverman, *Biochemistry* **1980**, *19*, 4984-4989.

[25] J.-Y. Liang and W.N. Lipscomb, *Biochemistry* **1988**, *27*, 8676-8682.

[26] M.E. Riepe and J.H. Wang, *J. Biol. Chem.* **1968**, *243*, 2779-2787.

[27] I. Bertini, E. Borghi and C. Luchinat, *J. Am. Chem. Soc.* **1979**, *101*, 7069-7071.

[28] R.H. Prince and P.R. Woolley, *Angew. Chem. Int. Ed. Engl.* **1972**, *11*, 408-417.

[29] Henkens, R.W.; Merrill, S.P.; Williams, T.J., Ann. New York Acad. Sci. **1984**, *143*, 429.

[30] A. Pullman and D. Demoulin, *Int. J. Quantum Chem.* **1979**, *16*, 641-653.

[31] J.-Y. Liang and W.N. Lipscomb, *Biochemistry* **1987**, *26*, 5293-5301.

[32] O. Jacob, R. Cardenas and O. Tapia, *J. Am. Chem. Soc.* **1990**, *112*, 8692-8705.

[33] Krauss, M. and Garmer, D.R., *J. Am. Chem. Soc.* **1991**, *113*, 6426.

[34] K.M. Merz, Jr., *J. Am. Chem. Soc.* **1991**, *113*, 406-411.

[35] J.J. Led and E. Neesgaard, *Biochemistry* **1987**, *26*, 183.

[36] W.N. Lipscomb, *Ann. Rev. Biochem.*. **1983**, *52*, 17.

[37] S. Lindskog, in "Zinc Enzymes" p. 77, T.G. Spiro, Ed.; Wiley, New York, 1983.

[38] Y. Pocker and T.L. Deits, *J. Am. Chem. Soc.* **1983**, *105*, 980-986.

[39] I. Simonsson, B.-H. Jonsson and S. Lindskog, *Eur. J. Biochem.*. **1979**, *93*, 409-417.

[40] J.-Y. Liang and W.N. Lipscomb, *Int. J. Quantum Chem.* **1989**, *36*, 299-312.

[41] P.H. Haffner and J.E. Coleman, *J. Biol. Chem.* **1975**, *250*, 996-1005.

[42] Y. Nakacho, T. Misawa, T. Fujiwara, A. Wakawars and K. Tomita, *Bull. Soc. Chem. Jpn.* **1976**, *49*, 595-599.

[43] H. Grewe, M.R. Udupa and B. Krebs, *Inorg. Chim. Acta* **1982**, *63*, 119-124.

[44] Y. Kai, M. Morita, N. Yasuoka and N. Kasai, *Bull. Soc. Chem. Jpn.* **1985**, *58*, 1631-1635.

[45] K. Takahashi, Y. Nishida and S. Kida, *Bull. Soc. Chem. Jpn.* **1984**, *57*, 2628-2633.

[46] A. Bencini, A. Bianchi, E. Garcia-España, S. Mangani, M. Micheloni, P. Orioli and P. Paoletti, *Inorg. Chem.* **1988**, *27*, 1104-1107.

[47] C. Kirchner and B. Krebs, *Inorg. Chem.* **1987**, *26*, 3569-3576.

[48] A.F. Monzingo and B.W. Matthews, *Biochemistry* **1984**, *23*, 5724-5729.

[49] M. Kato and T. Ito, *Inorg. Chem.* **1985**, *24*, 509-514.

294

[50] P.G. Harrison, M.J. Begley, T. Kikabhai and F. Killer, *J. Chem. Soc. Dalton Trans.*, **1986**, 929-938.

[51] L. Lebioda and B. Stec, *J. Am. Chem. Soc.* **1989**, *111*, 8511-8513.

[52] M.A. Holmes and B.W. Matthews, *Biochemistry* **1981**, *20*, 6912-6920.

[53] L.C. Kuo and M.W. Makinen, *J. Biol. Chem.* **1982**, *257*, 24-27.

[54] E. Kimura, T. Shiota, T. Koike, M. Shiro and M. Kodama, *J. Am. Chem. Soc.* **1990**, *112*, 5805-5811.

[55] E. Kimura and T. Koike, *Comments Inorg. Chem.* **1991**, *11*, 285.

[56] R.G. Khalifah, in "Biophysics and Physiology of Carbon Dioxide" p. 206, C. Bauer, G. Gros, and H. Bartels, Eds.; Springer-Verlag New York, 1980.

[57] C.C.J. Roothaan, *Rev. Mod. Phys.* **1951**, *23*, 69.

[58] M.J.S. Dewar, E.G. Zoebisch, E.F. Healy and J.J.P. Stewart, *J. Am. Chem. Soc.* **1985**, *107*, 3902-3909.

[59] M.J.S. Dewar and K.M. Merz, Jr., *Organometallics* **1988**, *7*, 522-524.

[60] J.Y. Choi, E.R. Davidson and I. Lee, *J. Comp. Chem.* **1988**, *10*, 163-175.

[61] C. Møller and M.S. Plesset, *Phys. Rev.* **1934**, *46*, 618.

[62] D. Demoulin and A. Pullman, *Theoret. Chim. Acta* **1978**, *49*, 161-181.

[63] A. Pullman, *Ann. N. Y. Acad. Sci.* **1981**, *367*, 340-355.

[64] H.B. Schlegel, *J. Comp. Chem.* **1982**, *3*, 214-218.

[65] J.S. Binkley, J.A. Pople and W.J. Hehre, *J. Am. Chem. Soc.* **1980**, *102*, 939-947.

[66] K.D. Dobbs and W.J. Hehre, *J. Comp. Chem.* **1987**, *8*, 861-879.

[67] W.J. Hehre, R.F. Stewart and J.A. Pople, *J. Chem. Phys.* **1969**, *51*, 2657.

[68] S. Miertuš, E. Scrocco, and J. Tomasi, *Chem. Phys.* **1981**, *55*, 117-129.

[69] J.L. Pascual-Ahuir, E. Silla, J. Tomasi, and R. Bonaccorsi, *J. Comp. Chem.* **1987**, *8*, 778-787.

[70] F. Floris and J. Tomasi, *J. Comp. Chem.* **1989**, *10*, 616-627.

[71] The sphere radii used for atoms were 20% larger than the van der Waals (or ionic) radii (H, 1.44 Å; C, 1.94 Å; N, 1.80 Å; O, 1.68 Å; Zn, 0.84 Å). T=298.15 K.

[72] R.A. Pierotti, *Chem. Rev.* **1976**, *76*, 717.

[73] M.J. Frisch, J.S. Binkley, H.B. Schlegel, K. Raghavachari, C.F. Melius, R.L. Martin, J.J.P. Stewart, F.W. Bobrowicz, C.M. Rohlfing, L.R. Kahn, D.F. Defrees, R. Seeger, R.A. Whiteside, D.J. Fox, E.M. Fleider and J.A. Pople., GAUSSIAN 86, Carniege Mellon University, Pittsburgh, PA, 1984.

[74] M.R. Peterson and R.A. Poirier, Program MONSTERGAUSS, Department of Chemistry, University of Toronto, **1981**, *Ontario*, Canada.

[75] M.J.S. Dewar, J.J.P. Stewart and M. Eggar, Program n. 506, QCPE, Department of Chemistry, Indiana University, Bloomington, Indiana, USA.

[76] G. Alagona and C. Ghio, in "The enzyme catalysis process" pp. 345-355, A. Cooper, J.L. Houben and L.C. Chien, Eds.; Plenum Publishing Corporation, 1989.

[77] A. Vedani, D.W. Huhta and S.P. Jacober, *J. Am. Chem. Soc.* **1989**, *111*, 4075-4081.

[78] A. Vedani and D.W. Huhta, *J. Am. Chem. Soc.* **1990**, *112*, 4759-4767.

[79] A. Vedani, *J. Comp. Chem.* **1988**, *9*, 269-280.

[80] J.-Y. Liang and W.N. Lipscomb, *Biochemistry* **1989**, *28*, 9724-9733.

[81] I. Bertini, C. Luchinat, M. Rosi, A. Sgamellotti and F. Tarantelli, *Inorg. Chem.* **1990**, *29*, 1460-1463.

[82] A. Vedani, M. Dobler and J.D. Dunitz, *J. Comp. Chem.* **1986**, *7*, 701-710.

[83] P.G. De Benedetti, M.C. Menziani, M. Cocchi and G. Frassineti, *J. Mol. Struct. (THEOCHEM)* **1989**, *183*, 393-401.

[84] M.C. Menziani, C.A. Reynolds and W.G. Richards, *J. Chem. Soc. Chem. Commun.*, **1989**, 853-855.

[85] J.C.L. Reynolds, K.F. Cooke and S.H. Northrup, *J. Phys. Chem.* **1990**, *94*, 985-991.

[86] A. Lledós and J. Bertrán, *J. Mol. Struct. (THEOCHEM)* **1985**, *120*, 73-78.

[87] A. Lledós and J. Bertrán, *J. Mol. Struct. (THEOCHEM)* **1984**, *107*, 233-238.

[88] M.T. Nguyen and T.-K. Ha, *J. Am. Chem. Soc.* **1984**, *106*, 599-602.

[89] A. Lledós and J. Bertrán, *Tetrahedron Lett.* **1981**, *22*, 75.

[90] A. Lledós and J. Bertrán, *J. Mol. Struct. (THEOCHEM)* **1984**, *107*, 233.

[91] O.N. Ventura, A. Lledós, R. Bonaccorsi, J. Bertrán and J. Tomasi, *Theoret. Chim. Acta* **1987**, *72*, 175.

[92] P. Ruelle, U.W. Kesselring and H. Nam-Tram, *J. Mol. Struct. (THEOCHEM)* **1985**, *124*, 41.

[93] P. Ruelle, U.W. Kesselring and H. Nam-Tram, *J. Am. Chem. Soc.* **1986**, *108*, 371.

[94] P. Ruelle, *Chem. Phys.* **1986**, *110*, 263.

[95] P. Ruelle, *J. Comp. Chem.* **1987**, *8*, 158.

[96] P. Ruelle, *J. Am. Chem. Soc.* **1987**, *109*, 1722.

[97] M.T. Nguyen and P. Ruelle, *Chem. Phys. Lett.* **1987**, *138*, 486.

[98] T.Oie, G.H. Loew, S.K. Burt and R.D. MacElroy, *J. Am. Chem. Soc.* **1983**, *105*, 2221.

[99] J. Bertrán, in "New Theoretical Concepts for Understanding Organic Reactions". p. 231, J. Bertrán and I.G. Csizmadia, Eds., Kluwer Academic Press, New York, 1989.

[100] J.L. Andrés, A. Lledós, M. Duran and J. Bertrán, *Chem. Phys. Lett.* **1988**, *153*, 82-86.

[101] M. Solà, A. Lledós, M. Duran and J. Bertrán, *Int. J. Quantum Chem.* , in press.

[102] M. Solà, A. Lledós, M. Duran, J. Bertrán and J.L.M. Abboud, *J. Am. Chem. Soc.* **1991**, *113*, 2873-2879.

[103] J. Tomasi, G. Alagona, R. Bonaccorsi and C. Ghio, in "Modelling of structures and properties of molecules" pp. 330-355, E. Horwood, Ed.; Ellis Horwood Ltd., Chichester, England, 1987.

[104] L.I. Krishtalik and V.V. Topolev, *Mol. Biol. (Mos.)* **1984**, *18*, 721.

[105] L.I. Krishtalik, *Mol. Biol. (Mos.)* **1974**, *8*, 75.

[106] L.I. Krishtalik, *J. Theor. Biol* **1985**, *112*, 251.

[107] L.I. Krishtalik, *J. Theor. Biol* **1980**, *86*, 757.

[108] G.A. Melcier, Jr., J.P. Dijkman, R. Osman and H. Weinstein, in "Quantum Chemistry: Basic Aspects, Actual trends". Amsterdam, R. Carbó Ed., Elsevier Scientific Publ, 1989.

[109] O. Tapia and G. Johannin, *J. Chem. Phys.* **1981**, *75*, 3624-3635.

[110] B.T. Thole and P.T. van Duijnen, *Theoret. Chim. Acta* **1983**, *63*, 209-221.

[111] A. Warshel, *Biochemistry* **1981**, *20*, 3167-3177.

[112] J.A. McCammon and S.C. Harvey, "Dynamics of Proteins and Nucleic Acids" 1987, Cambridge University Press, Cambridge.

[113] A. Warshel, S.T. Rusell and F. Sussman, *Isr. J. Chem.* **1987**, *27*, 217.

[114] The lines plotted in all the electron isodensity maps presented in this work, correspond to the values of 0.0001, 0.001, 0.01, 0.02, 0.03, 0.04, 0.08, 0.1, 0.12, 0.15, 0.18, 0.24, 0.30, 0.37, 0.40, 0.60, 0.80, 1.0, 1.5, 10., and 100. au.

[115] R.F.W. Bader, *Acc. Chem. Res.* **1985**, *18*, 9-15.

[116] M. Solà, A. Lledós, M. Duran and J. Bertrán, *J. Am. Chem. Soc.* **1991**, in press.

[117] M. Solà, A. Lledós, M. Duran and J. Bertrán, *Inorg. Chem.* **1991**, *30*, 2523.

[118] C. Giessner-Prettre and O. Jacob, *J. Comput.-Aided Mol. Design* **1989**, *3*, 23-37.

[119] Y. Pocker and T.L. Deits, *J. Am. Chem. Soc.* **1981**, *103*, 3949-3951.

8. Discussion

PROF. WARSHEL: The calculation with additional water gave a concerted mechanism with a 20 kcal/mol barrier. The reason for this is clear from your nice continuum calculations where adding the environment encourages the stepwise mechanism relative to the concerted mechanism.

PROF. BERTRAN: You have addressed a very interesting point, because obviously introduction of the solvent effect by means of a continuum model should stabilize the stepwise mechanism due to formation of ions. Although we have not yet carried out such a calculation in this particular case, we have a similar experience on the ketoenolic tautomerism (Theoret. Chim. Acta 72 (1987) 175). In the hydrated-cluster process, there is a concerted mechanism in the gas phase that is very favored with respect to the stepwise mechanism. Continuum model calculations show that the stepwise mechanism is dramatically stabilized in solution, although the concerted mechanism is still preferred.

PROF. KOLLMAN: What kind of continuum model have you used and how does it behave in protein calculations ?

PROF. BERTRAN: We use Tomasi's model (J. Chem. Phys. 55 (1981) 177; J. Comput. Chem. 4 (1983) 567) to include the solvent effect. It is worth noting than such a model has been developed to deal with a free solvent, and that a proteinic environment differs from that of a free solvent by having only a partial reorientation capability and by possessing a permanent electric field. As can be read in the text of my lecture, although the low reorientation should imply use of a low dielectric constant, presence of an electric field compels use of a large dielectric constant in order to simulate the real electric field present inside proteins.

PROF. SMITH: What is the current status of the crystallography of this enzyme? Have there been any high-resolution structures recently determined which can give clues useful to the ab initio studies? This of course being despite the fact that the transition states and intermediates of interest are inaccessible using standard crystallographic methods.

PROF. BERTRAN: Really, crystallographic data on this enzyme are not scarce. If charges on the different proteinic residues be calculated, one might determine the electric field created by the proteinic environment at the beginning of the process. However, as you mention in your question, the geometrical parameters of the protein will change along the reaction, so the electric field will also change. Moreover, charges of proteinic residues will not only modify their position, but modify themselves since evolution of the chemical system will cause an evolutive electronic polarization of the proteinic residues, so charges will also change along the process. This is a fundamental and more difficult aspect to be taken into account in dynamical studies, where the electronic coupling between the chemical system and the proteinic environment must be considered. This point is well accounted for in continuum models, although they exhibit other limitations.

PROF. MAGGIORA: In Industry we use theory primarily as a guide to facilitate the design of enzyme inhibitors. We also seek to check or "calibrate" our results with experimental tests of building potency or evaluation of inhibitory constants.

PROF. BERTRAN: I think that in the present standard of Industry, understanding of chemical mechanisms can be not very useful. However, as happens always to basic science, the deep understanding of chemical mechanisms may emerge as an important tool in industrial progress in the future.

AB INITIO STUDIES AND QUANTUM-CLASSICAL MOLECULAR DYNAMICS SIMULATIONS FOR PROTON TRANSFER PROCESSES IN MODEL SYSTEMS AND IN ENZYMES

P. BALA[a,b], B. LESYNG[a,c], T. N. TRUONG[a] and J. A. McCAMMON[a]

a University of Houston
   Institute for Molecular Design
   Department of Chemistry
   University of Houston
   Houston, TX 77204, USA

b Nicolaus Copernicus University
   Institute of Physics
   Grudziadzka 5
   87-100 Torun, Poland

c University of Warsaw
   Department of Biophysics
   Zwirki i Wigury 93
   02-089 Warsaw, Poland

ABSTRACT: Present supercomputer technology allows for exploration of quantum dynamical phenomena in biomolecular structure and dynamics - the limiting factors are mostly theoretical models and efficient software. We undertook systematic studies in both areas, concentrating our attention on exploring the applicability of a combined quantum-classical or quantum-stochastic dynamics method for studying proton transfer processes, which are very important for many biomolecular phenomena. We present the new dynamical approach and its applications to the proton-bound amonia dimer, proton-bound amonia-water complex, and a model of the active site in phospholipases. Possible extensions of the models to enzyme catalysis with proton or electron transfer reactions are also discussed.

J. Bertrán (ed.), Molecular Aspects of Biotechnology: Computational Models and Theories, 299–326.
© 1992 Kluwer Academic Publishers.

1. Introduction

Proton and electron transfer plays an important role in biological processes, especially in enzyme catalysis, see [1-10,32-33,44,48]. From the point of view of physics, enzymes are mixed quantum-classical many body systems which act as "molecular machines". Most theoretical studies dealing with these machines have applied the Ehrenfest model. With this approach, the atomic motions evolve classically on an effective Born-Oppenheimer potential surface $V(\{R_k\})$. Quantum semi-empirical or *ab initio* techniques have been combined with classical dynamics in order to describe proton transfer events in lysosyme and proteolytic enzymes [7,8]. Although quantum mechanical methods were used to determine the proton potential energy, the protons were treated as nontunneling classical particles. Experimental data and theoretical models both indicate that due to the small transferring mass, quantum dynamical effects that have been neglected in previous inter- and intramolecular proton and electron transfer processes play an important role in many chemical and biological phenomena including acid-base equilibria, nucleophilic addition reactions, photosynthesis, enzyme catalysis, ionic transport across membranes, etc. For this reason, these reactions have received considerable experimental and theoretical attention. Recent theoretical studies provide information about potential energy surfaces for proton transfer processes in hydrogen-bonded systems [10-19,22-23,25-28,34].

In the past few years, effective techniques have been developed to integrate directly the time-dependent Schroedinger equation [21,24,30-31,35,53-55,63-65]. In particular, new simulation techniques were successfully applied by Carr, Parinello and coworkers to mixed systems of quantum and classical particles [35, 69]. The use of real-time dynamics allowed them to investigate hopping phenomena, diffusion and tunneling. In particular, it was possible to simulate the mobility of an excess electron in disordered materials or submicron devices; for review see [52].

These techniques open the way for a more detailed analysis of phenomena such as proton transfer. So far, according to our knowledge, these techniques have never been applied to bound systems, and in particular to hydrogen-bonded (more precisely proton-bonded) molecules/complexes. It is important to test the validity of the approximations for such quantum dynamics models and explore possible practical applications for bigger systems, like enzymes with proton transfer.

The purpose of the present study is to explore possible applications of the combined Quantum-Classical and Quantum-Stochastic Molecular Dynamics models (QCMD and QSMD models, respectively) to simple molecular bound systems, in which a time-dependent quantum-mechanical method is used to describe a quantum particle which binds the classical particles/atoms, and the classical molecular dynamics is used to describe the dynamics of the classical particles/atoms. Coupling between the quantum particle (proton/deuteron) and classical atoms is accomplished *via* Hellmann-Feynman forces* as well as

---

* The Hellmann-Feynman forces are known in molecular quantum mechanics literature, as the forces acting on nuclei, generated by the electron charge distribution, which vanish at equilibrium (see e.g. electrostatic theorem in: Ira N.Levine, "Quantum Chemistry", Allyn and Bacon, Inc., Boston, 1983). In this study, we use this notion in its more general meaning. This is an effective force generated by any quantum particle, e.g., a proton, which interacts with classical nuclei and binds them.

a time dependence of the potential energy function in the Schroedinger equation. Interaction of the system with the environment is described by stochastic dynamics of the classical atoms.

The QCMD/QSMD algorithms are tested using the following molecular systems: the proton-bound ammonia-ammonia dimer ($[H_3N\text{-}H\text{-}NH_3]^+$, 1), the proton-bound ammonia-water dimer ($[H_3N\text{-}H\text{-}OH_2]^+$, 2), Fig.1, and malonaldehyde (3), Fig.2. The imidazole-water-Ca^{++} complex ($[Im\text{-}H\text{-}OH]Ca^{++}$, 4), Fig.3, contains the active site region elements of phospholipases. This system is studied systematically using *ab initio* methods, and is being prepared for integration with the molecular structure of phospholipase A_2 and the combined quantum-classical simulation.

Fig.1.   Model proton-bound dimers : $[H_3N\text{-}H\text{-}NH_3]^+$ and $[H_3N\text{-}H\text{-}OH_2]^+$.

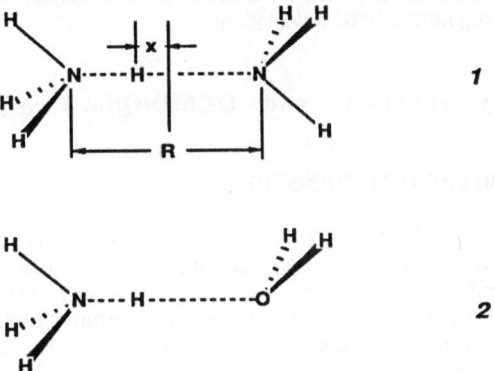

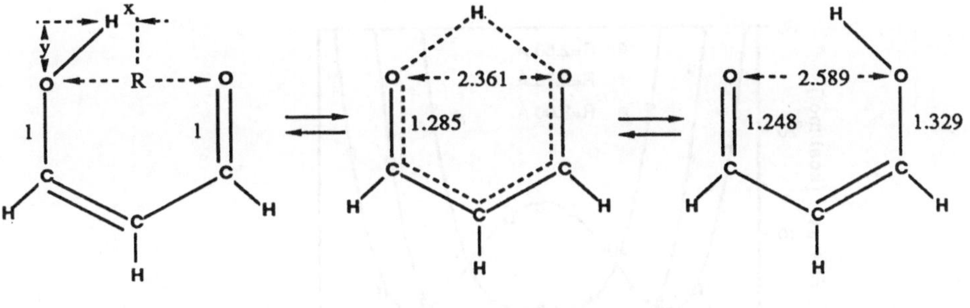

Fig.2.   Malonaldehyde, **3**. E1 and E2 denote equilibrium, tautomeric forms. TS is the transition state. R and $\Delta I = l_1\text{-}l_2$ are the most important degrees of freedom, and x and y are the proton quantum degrees of freedom.

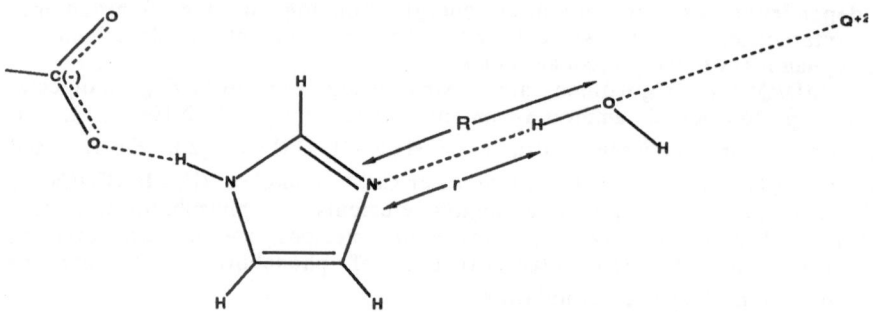

Fig.3. Our model of the active site of phospholipase A2, **4**. The calcium cation was modeled with the formal charge (+2) in the MP2/6-31G* calculations.

2. Potential Energy Functions and QCMD/QSMD Algorithms

2.1 BARRIERS FOR THE PROTON TRANSFER

Each molecular system in this study consists of two types of particles: the central quantum proton, and the classical atoms. In the case of the proton bound dimers, **1** and **2**, the donor and the acceptor are the united atoms consisting of the nitrogen or oxygen atoms with their hydrogens (Fig.1), and the proton transfers along the x-axis. In **3** the quantum proton is localized between the oxygen atoms (Fig.2). In equilibrium the molecule has C_s symmetry and in the transition state it adopts the C_{2v} symmetry. In **4** the water molecule is the proton donor and the imidazole ring is the acceptor (Fig.3). The calcium cation generates the electrostatic field which modifies the potential energy surface for the proton transfer.

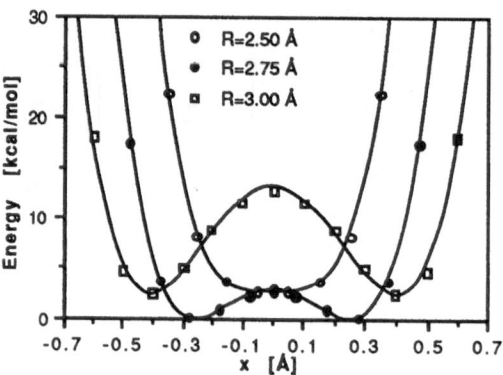

Fig.4. One-dimensional MP2/6-31G* energy barriers [34] for **1**.

Since the potential energy function in **1** and **2** is separable with respect to the x, y and z coordinates, and in addition, the potential is approximately harmonic in the y and z direction, the dynamics of the systems can be reduced to 1D in their quantum and classical domains (see discussion below). Thus, the potential energy function $V=V(x;R)$ depends on two variables, the proton position x, and the intermolecular distance R. The distance R strongly influences the barrier height for the proton transfer, which has been demonstrated in the *ab initio* study [34]. The squeezing of the donor acceptor distance in **1** can cancel the barrier totally, whereas the stretching of the distance increases it. This is shown in Fig.4. A similar effect is observed in **2**. The two dimensional potential energy surfaces for these systems, determined by fitting polynomials to the MP2/6-31G* *ab initio* energies at selected points [34,45], are shown in Fig. 5 and Fig. 6.

3 is a much more complicated system and the development of its full potential energy function, based for example on precise *ab initio* calculations, is not a simple task. Several theoretical models have been developed to describe the proton tunneling in this system. Makri and Miller [47] developed a semiclassical, "instanton-type" model. Shida et al., [57], determined an effective few-dimensional barrier which governs the proton tunneling. In turn, one-dimensional and bidimensional tunneling splitting have been calculated by Bosch et al. [58], and used to determine the tunneling frequency. All these models predict correctly the experimental tunneling frequency using microwave and far infrared spectroscopy [59-61]. A general disadvantage of these models is that they cannot be easily generalized to

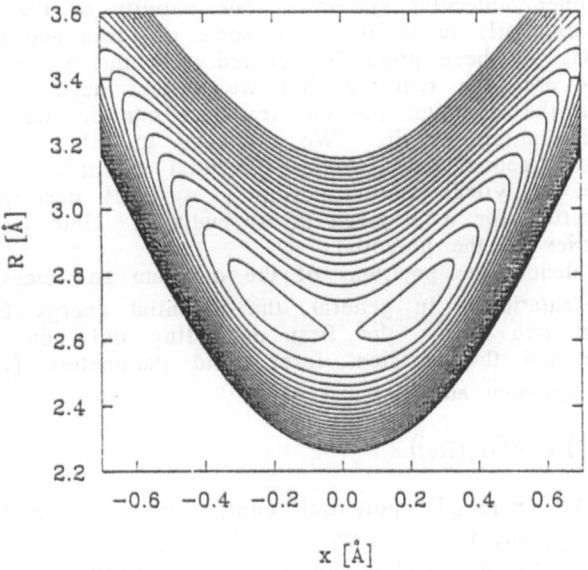

Fig. 5. Two-dimensional potential energy contour map for **1**. Equipotential contours are separated by an energy difference of 1 kcal/mol. The lowest energy contours, which represent the energy minima, are at the 0.8 kcal/mol.

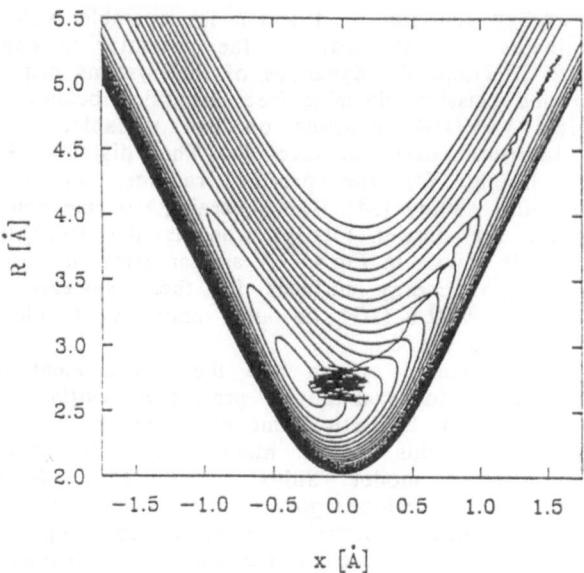

Fig. 6. Two-dimensional potential energy contour map for **2**. Equipotential contours are separated by an energy difference of 5 kcal/mol. A QSMD trajectory is plotted (see text).

proton transfer in other molecular systems. The validity of the mean barrier approximation, used explicitly or implicitly in some of these and other studies, to our knowledge has not been precisely verified. In the model we propose, some approximations are also required, but we neither need to assume any effective barrier for the quantum particle transfer, nor do we need any *a priori* determined tunneling path. We do have to select "quantum" and "classical" degrees of freedom, and express the total potential energy function in the form of an analytical approximation. A brief description of the potential energy function for **3** is given in Appendix 1. Here we will discuss only selected properties of the potential.

Let **r** and $\{\mathbf{R}_k\}$ denote the positions of the quantum and classical particles in the vector representation. In general, the potential energy functions can be decomposed into two terms, the first depending only on the classical degrees of freedom and the classical force field parameters $\{p_l\}$, and the second one on the quantum and classical ones:

$$V(\mathbf{r},\{\mathbf{R}_k\}) = V^c(\{\mathbf{R}_k\},\{p_l\}) + V^{qc}(\mathbf{r},\{\mathbf{R}_k\}) \tag{1}$$

Let us discuss the $V^{qc}(\mathbf{r},\{\mathbf{R}_k\})$ potential energy term for malonoaldehyde, first. Similar to the systems **1** and **2** , V^{qc} is determined by fitting a polynomial to MP2/6-31G** energies at selected points. The resulting two dimensional energy maps for its equilibrium state (form (E_1) in Fig.2) and the transition state (form TS in Fig.2), are presented in Fig.7a and Fig.7b, respectively.

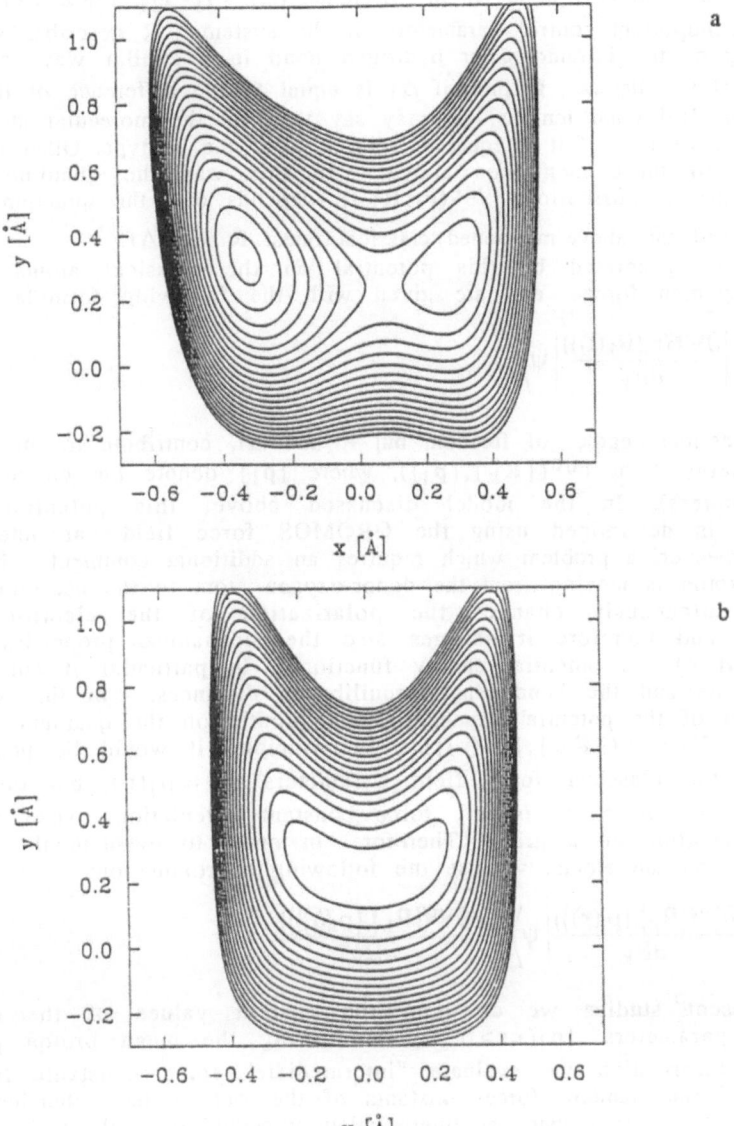

Fig. 7. Two-dimensional potential energy contour map for malonaldehyde, **3**. Equipotential contours are separated by an energy difference of 2 kcal/mol.
(a) for the equilibrium state, E₁ (see Fig. 2). The map for E₂ can be obtained by applying the x -> -x symmetry operation. (b) for the transition state, TS (see Fig. 2).

Note that R and the C-O bond length difference ($\Delta l = l_1 - l_2$), Fig.2, can be used as the most important control parameters of the system. R describes squeezing or stretching of the intramolecular hydrogen bond in a similar way as it does in the case of the dimers. In turn, if Δl is equal to the difference of the C-OH and C=O classical bond lengths, we may say that the the molecular structure is of the E-type, whereas, if it is equal to zero, it is of the TS-type. Other forms are intermediate to those mentioned above. In this way the quantum-classical potential energy contribution, $V^{qc}(r,\{R_k\})$, depends on the quantum degrees of freedom and the above-mentioned classical ones: R and Δl.

The forces generated by this potential on the classical atoms are the Hellmann-Feynman forces, and are given with the following formula:

$$F_k^{HF}(t) = -\left\langle \psi \left| \frac{\partial V^{qc}(r,\{R_k(t)\})}{\partial R_k} \right| \psi \right\rangle \tag{2}$$

All other classical degrees of freedom but R and Δl, contribute to the classical potential energy term ($V^c(\{R_k\},\{p_l\})$, where $\{p_l\}$ denote the classical force field parameters). In the model discussed above, this potential energy contribution is determined using the GROMOS force field parameters [62]. There is, however, a problem which requires an additional comment. Note that when the proton is moving from the donor-oxygen atom to the acceptor one, it almost instantaneously changes the polarization of the electron charge distribution, and therefore it changes also the mechanical properties of the classical part of the potential energy function. In particular it changes the force constants and the bond length equilibrium distances. In this way, the classical part of the potential depends also implicitly on the quantum degree(s) of freedom $V^c = V^c(\{R_k\},\{p_l(r)\})$. In principle, it would be possible to parametrize the classical force field parameters $p_l = p_l(r)$, but this would require very extensive *ab initio* force constant calculations when changing the proton position on a grid. Therefore, in order to evaluate the classical forces acting on the atoms, we use the following approximation:

$$F_k^c(t) = -\left\langle \psi \left| \frac{\partial V^c(\{R_k\},\{p_l(r)\})}{\partial R_k} \right| \psi \right\rangle \cong -\frac{\partial V^c(\{R_k\},\{p_l(\langle r \rangle)\})}{\partial R_k} \tag{3}$$

In the present studies we calculate the current values of the classical force-field parameters $\{p_l(\langle r \rangle)\}$ by monitoring the mean proton position. Using this information we evaluate "intermediate" force constants from the knowledge of the standard force constants of the well defined chemical bonds (C-OH, C=O, C-C, C=C), using an interpolation procedure (e.g., the hydroxy group changes into the keto one after the proton hopping, and *vice versa*). We use also another procedure in which the current values of the force-field parameters are calculated as the average values with the weights equal to the probabilities of the quantum proton being in the left and the right well, respectively. The initial results show that the final QD results depend very little on the two procedures applied here.

For one isolated molecule, like malonaldehyde, decomposition of the total potential function into the classical and quantum-classical contributions is partly a problem of notation. Regardless of where the potential is coming from (*ab initio* , empirical parametrization, guessed, etc.), the general rule is that if some parameters depend on the quantum degrees of freedom, the forces generated by these potential energy contributions should be averaged quantum-mechanically over the quantum degrees of freedom.

Everything said above regarding the determination of the classical force-field parameters should apply to other systems that display proton transfer. An extension of the model to macromolecular structures, like enzymes, with long-range electrostatic and/or non-bonded interactions, as well as with solvent effects, requires additional comments. This will be discussed in sections 4 and 5 below on the model system for phospholipases, **4**, and on enzyme mechanisms with proton transfer in general. First the QCMD and QSMD models are described, and selected simulation results for **1, 2** and **3** are presented.

2.2 QCMD AND QSMD ALGORITHMS

The wave function $\psi(r,t)$ is obtained from the time-dependent Schroedinger equation:

$$i\hbar\frac{\partial\psi(r,t)}{\partial t} = \left(-\frac{\hbar^2}{2m}\Delta + V^{qc}(r,\{R_k(t)\})\right)\psi(r,t) \tag{4}$$

The dynamics of the classical atoms is described by the Newtonian equations of motion with the classical forces, and the updated Hellmann-Feynman forces :

$$M_k\ddot{R}_k = F_k^c(\{R_k(t)\},\{p_l((r)(t))\}) - \int \psi^*(r,t)\frac{\partial V^{qc}(r,\{R_k\}(t))}{\partial R_k}\psi(r,t)\,dr \tag{5}$$

In this way the classical atoms/particles can move along unique, well-defined trajectories, which depend on the classical forces as well as on the quantum proton charge distribution. In turn, the dynamics of the quantum particle depends, through the potential, on the motion of all the classical atoms (eq.4). The propagation of the classical atoms has been performed using the well known "leap-frog" algorithm; see [1,2]. Description of the time dependent quantum mechanical wavefunction evolution methods is given in the next subsection. The model described above will be referred as the QCMD model.

In a case where the motion of the atoms is influenced by random forces from a thermal bath, the classical equations of motion (5) can be replaced by stochastic equations:

$$M_k\ddot{R}_k = F_k^c(\{R_k(t)\},\{p_l((r)(t))\}) - \int \psi^*(r,t)\frac{\partial V^{qc}(r,\{R_k\}(t))}{\partial R_k}\psi(r,t)\,dr + T_k - M_k\gamma_kv_k \tag{6}$$

where v_k is the velocity of the classical atom k, γ_k is the atomic friction coefficient, and $T_k(t)$ is the random force acted on the classical atom k satisfying the fluctuation-dissipation condition:

$$\langle T_i(0)\, T_j(t)\rangle = 2M_i\gamma_i k_b T\delta_{ij}\, \delta(t) \qquad (7)$$

Here k_B and T are the Boltzmann constant and temperature of the system, respectively. In the limit $\gamma_i \to 0$, Eq. (6) becomes identical to Eq. (5).

2.3 INTEGRATION OF THE TIME DEPENDENT SCHROEDINGER EQUATION

The formal solution of the time-dependent Schroedinger equation

$$i\hbar\frac{\partial\psi(r,t)}{\partial t} = H(r,\{R_k(t)\})\, \psi(r,t) \qquad (8)$$

can be written in the form:

$$\psi(r,t+\tau) = U_t(\tau)\, \psi(r,t) = T\, \exp\left\{-\frac{i}{\hbar}\int_t^{t+\tau} Hdt'\right\}\psi(r,t') \qquad (9)$$

where T is a time-ordering operator, and τ is a discrete time interval. There are several techniques which are being applied to integrate the time-dependent Schroedinger equation. For review of the methods see [52-54]. We will discuss the Chebychev polynomial technique [34] and the modified Cayley method [63-65] since, when studying proton transfer processes, these methods appear to be numerically precise and computationaly effective.

With the Chebychev technique, one expands the time evolution operator

$$U = \sum_{n=0}^{N} a_n\Phi_n(-iH\tau) \qquad (10)$$

where, a_n are the expansion coefficients, and Φ_n are complex Chebychev polynomials. U in this representation is the global propagator, but it can also be applied as a short time propagator for the time dependent potentials. With the time step close to $\tau = 10^{-16}$ s , and N varied in the range 25 - 50, the method works reasonably well, conserving the norm of the wave function and the energy of the system. The discretization of the problem introduces some limits on the momentum which can be represented on the grid, namely

$$p_{max}=\frac{\pi}{\Delta x} \qquad (11)$$

which implies that the energy range of the discretized Hamiltonian is given by

$$\Delta E = E_{max} - E_{min}$$
and $\qquad (12)$

$$E_{max} = V_{max} + \frac{p_{max}^2}{2m}$$
$$E_{min} = V_{min}$$

(13)

When using the normalized Hamiltonian

$$H_{norm} = 2\frac{H - I(\Delta E/2 + V_{min})}{\Delta E}$$

(14)

the evolution operator can be expressed as

$$U = e^{-i(\Delta E/2 + V_{min})\tau} \sum_{n=0}^{N} a_n(\zeta) \Phi_n(-iH_{norm})$$

(15)

with the coefficients a_n

$$a_0 = J_0(\zeta)$$
$$a_n = 2J_n(\zeta) \qquad \text{for } n = 1,N$$

(16)

and $\zeta = \Delta E\tau/2$

where J_n are the cylinder Bessel functions. The operators Φ_n are evaluated using the recursion procedure

$$\Phi_{n+1} = -2iH_{norm}\Phi_n + \Phi_{n-1}$$

(17)

Since the effect of the operator H on Ψ is calculated using the FFT technique twice, this method requires 2N FFT's per time step. The grid and the time step data used in simulations for **1, 2** and **3** are collected in Table 1.

TABLE I. Grid and time step data (in Å, s).

	1	2	3
Grid points	128	256	128x128
Δx	0.029	0.029	0.029
Δy	-	-	0.033
Δt	0.25×10^{-16}	0.25×10^{-16}	0.2×10^{-15}
Number of steps	4×10^4	4×10^4	5×10^3

In some QD simulations, and particularly in preparation of the initial wave function for the QD simulations, the kinetic referenced modified Cayley method was used. In this method the wave function is propagated using the following formula

$$\psi(r,t+\tau) = \left(1 + \frac{i\tau}{2}V\right)^{-1} e^{-iK\tau} \left(1 - \frac{i\tau}{2}V\right) \psi(r,t) \qquad (18)$$

where K and V are the kinetic and the potential energy operators, respectively. The modified Cayley method requires only two FFT's per step, however, the grid size (Δx) for the applications discussed here should be smaller than the one in the Chebychev scheme, e. g. the number of grid points is 512 in 1D.

In order to minimize the energy of the initial wave function the following two procedures were used. Either the Gaussian wave packet was optimized by changing its position and its standard deviation or an imaginary time dynamics (ITD, see e.g. [54]) was applied, to optimize the wave packet on the grid. In practical applications of ITD the wave function was propagated using the modified Cayley method with the imaginary time step $\tau' = i\tau$.

3. Results and Discussion

3.1 SMALL VIBRATIONAL MOTIONS AT THE ZEROTH STATIONARY VIBRATIONAL LEVEL IN THE PROTON BOUND DIMERS

As the check of the QCMD method we made an attempt to reproduce small vibrational motions of the dimers at the zeroth oscillation level. In each case the proton wave function was minimized using the ITD technique, which allowed generation of the proton eigenfunctions for the given initial R values. Starting with the amplitude ΔR, estimated from the curvature of the potential energy function for the heavy atom motion, the QCMD/Chebychev propagation scheme was applied. Table II shows the simulation results and compares them with *ab initio* calculations including harmonic oscillation frequency analysis at the Hartree-Fock level. QCMD reproduces fairly well the equilibrium positions of the heavy atoms (comp. $<R>_{min}$, $<R>_t$ and $R_d(t=\infty)$ with R_{min}) and

the classical oscillation frequencies (ν). Note that the equilibrium positions are shortened with respect to the potential energy minimum (R_{min}) by about 0.04-0.06 Å, which is known in literature as the vibronic or Ubbelohde effect (see e.g. [66]). The method also predicts correctly the partition between the kinetic and potential energy. In the case of the harmonic motion $<K> = <V> = 1/2E_{vib}$, which results from the virial theorem. In the QCMD simulations, due to anharmonicity, the time averaged kinetic energy $<K>_t$ is slightly smaller than $1/2 E_{vib}$.

TABLE II. The simulation results and reference data for the model systems (Å, kcal/mol, Hz). R_{min} - the minimum of the potential energy function in the Hamiltonian. $<R>_{min}$ - the distance which corresponds to the minimum of the total energy averaged over the wave function on the grid. $E_{vib}(\upsilon=0)$ - vibrational zero-point energy contribution in the harmonic approximation. $<R>_t$ - the mean distance between the classical atoms averaged over their trajectories. $R_d(t=\infty)$ - damped case, the limit distance between the classical atoms in the QSMD simulations with T=0, and the high atomic friction coefficient. ΔR - the mean vibrational amplitude. ν - vibrational frequency. $<T>_t$ - the mean kinetic energy of the classical particles averaged over their trajectories.

	1	2
R_{min}	2.723 [a] 2.731 [b,e] 2.729 [c]	2.805 [a] 2.736 [e]
$<R>_{min}$	2.69 2.660 [c] 2.67 [d]	2.74
$E_{vib}(\upsilon=0)$	0.64 0.46 [e]	0.52 0.45 [e]
$<R>_t$	2.677	2.733
$R_d(t=\infty)$	2.665	2.724
ΔR	0.102	0.105
ν	13.62×10^{12} 9.57×10^{12} [e]	10.14×10^{12} 9.36×10^{12} [e]
$<T>_t$	0.29	0.24

[a] Refs. [34,45], [b] Geometry optimized at the MP2/6-31G* level, ref. [17], [c] Ref. [66], [d] Value which corresponds to the minimal energy of the time independent Hamiltonian. The hamiltonian was diagonalized and its lowest eigenvalues plotted against R, L.Jaroszewski, Master Degree Thesis, Warsaw University,1991. [e] This study - MP2/6-31G* optimization and harmonic vibrational frequency analysis.

3.2 SMALL NONSTATIONARY VIBRATIONAL MOTIONS IN 1 - QCMD SIMULATIONS

The optimized Gaussian wave packet, localized in its lowest energy level, characterized by the potential energy function at R=2.85 Å (Fig.8) was propagated with QCMD.

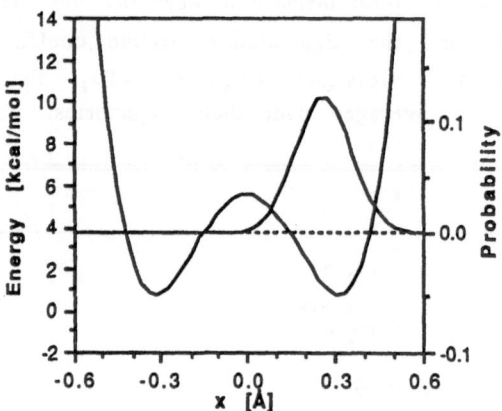

Fig.8 The one-dimensional energy barrier for 1 at R=2.85 Å, and the optimized Gaussian wave packet in its lowest energy level.

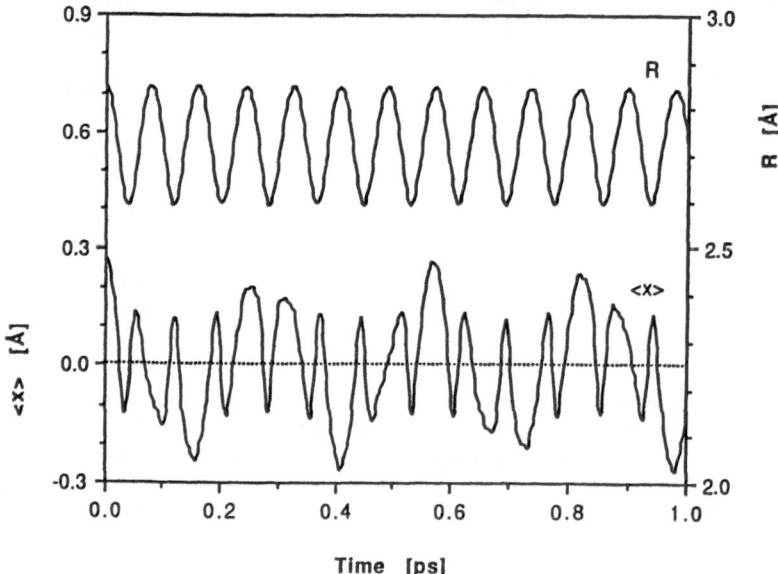

Fig. 9. The QCMD simulation results for 1. R and <x> refer to the positions of classical and quantum particles, respectively.

Note that the Gaussian wavepacket is not an eigenstate of the bistable potential. It can be, however, approximated with a linear combination of the symmetric and the antisymmetric lowest eigenstates, and its tunneling frequency, ν_{tun}, can be estimated based on the ground level energy splitting ($\nu_{tun} = (E_{asym}-E_{sym})/2h =$ 6.25×10^{12} 1/s [45], where h is the Planck constant). This can be compared with the direct QCMD simulations.

The relative position of the heavy atoms R=R(t), and the mean proton position as a function of time, <x>(t), obtained from the QCMD simulations, are shown in Fig.9. One observes two tunneling frequencies, 16.13×10^{12} 1/s and 3.70×10^{12} 1/s. Since during each oscillation period the barrier drops down, the proton can easily tunnel and the high tunneling frequency and the heavy atom oscillatory frequency are similar. In turn, the low tunneling frequency can be almost precisely reproduced (not shown here) with the simulation for the time independent barrier at $R_{min}+\Delta R$ distance. This is the frequency which is close to the theoretical estimate based on the ground energy level splitting. Note that at the $R+\Delta R$ distance, the classical kinetic energy as well as the velocities of the classical atoms are close to zero, and the proton is subjected to such potential for a relatively large fraction of the oscillation period of the heavy atoms.

3.3 LARGE NONSTATIONARY VIBRATIONAL MOTIONS IN 2 - QSMD SIMULATIONS

The simulation which is presented below imitates to some extent the quantum proton hopping in enzyme active sites. Due to the energy fluctuations in the macromolecules (of the order of 40-50 kcal/mol in a system consisting of 200 residues [41,42]), the displacement amplitudes of heavy atoms are fairly large, and squeezing or stretching distortions of the hydrogen

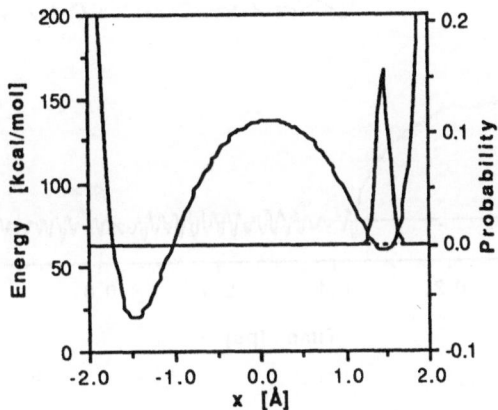

Fig.10 The one-dimensional energy barrier for **2** at R=2.85 Å, and the optimized Gaussian wave packet in its lowest energy level.

bonds are observed in many classical molecular dynamics simulations, as well as in spectroscopic experimental studies. The model system was coupled to the thermal bath which provides energy fluctuations and removes the excess kinetic energy in the system after the proton hopping. The QSMD simulation technique (eq.6), at T= 300 K has been applied. The coupling to the thermal bath was relatively strong with $\gamma =100$ ps^{-1}. For t=0, the ammonia and water molecules were separated with R=5.0 Å, and the optimized Gaussian wave packet was located in the metastable state close to water (Fig.10).

The QSMD simulation results, R(t) and <x>(t), are presented in Fig.11. The (R, <x>) trajectory on the 2D potential energy surface is shown in Fig.6. One can see that the proton moves, within the period of 0.4 ps, from the water molecule to ammonia. The (R,<x>) trajectory doesn't follow precisely the potential energy reaction path, making a kind of a "short cut" which penetrates the barrier. This is a typical behavior of the quantum particle during the transfer process. The proton localizes in the region of the potential energy minimum where it loses its kinetic energy, however, its mean position doesn't match precisely the minimum. It is shifted towards the top of the barrier (Fig.6), which is due to the tunneling penetration of the barrier by the quantum proton and repulsion of its wave packet by the steep slope of the potential energy boundary.

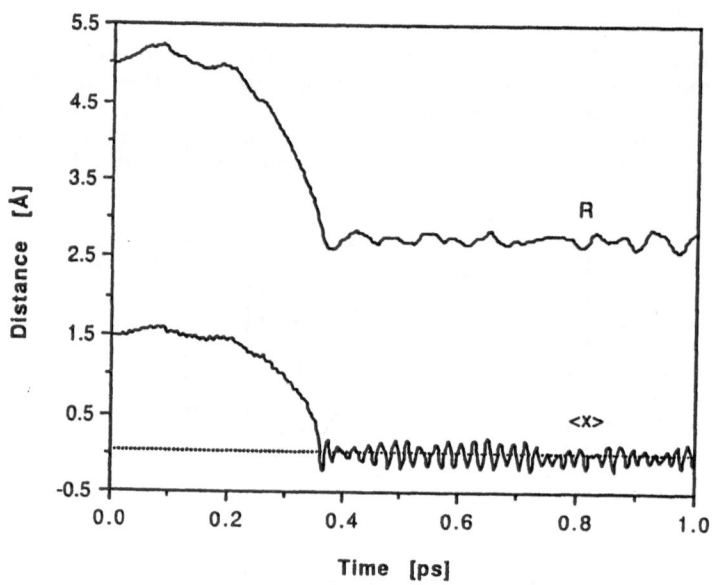

Fig. 11. The QSMD simulation results for **2**.

3.4 QSMD SIMULATIONS OF THE MULTIDIMENSIONAL PROTON TUNNELING IN MALONALDEHYDE

The proton transfer in malonaldehyde is the elementary tautomeric process. The simulation of this system required implementation of the QD module into a classical molecular dynamics library and this has been done using the GROMOS library. The combined QD/GROMOS package has the following structure. The classical forces are updated with the Hellmann-Feynman ones. In turn, the information of the classical atom positions is transferred to the QD module and is used to generate the current values of the potential energy function for the proton transfer on the grid. In addition, since the change of the proton position influences the classical force field parameters, the current values of these parameters are modified in the same way as is done when applying the thermodynamic perturbation method (see [1,2,50,51]). The mean proton position <x> or probability of localization of the proton in one well, play the role of the perturbational parameter l. As noted in the § 2.2, the present simulation results for malonaldehyde do not depend much on the selected approximation, however, this problem requires further, more detailed studies.

In the present simulations an analytical potential energy function based on MP2/6-31G** energies for the proton transfer was used. The analytical representation of the barrier (Appendix 1) gives the barrier height equal to 3.8 kcal/mol (the energy difference between the E and TS states). Far infrared experimental and MCPF *ab initio* data show that the barrier is probably higher, close to 6 kcal/mol [57-60]. Although we could rescale the barrier to get this barrier height, we decided rather at this stage of these studies to use

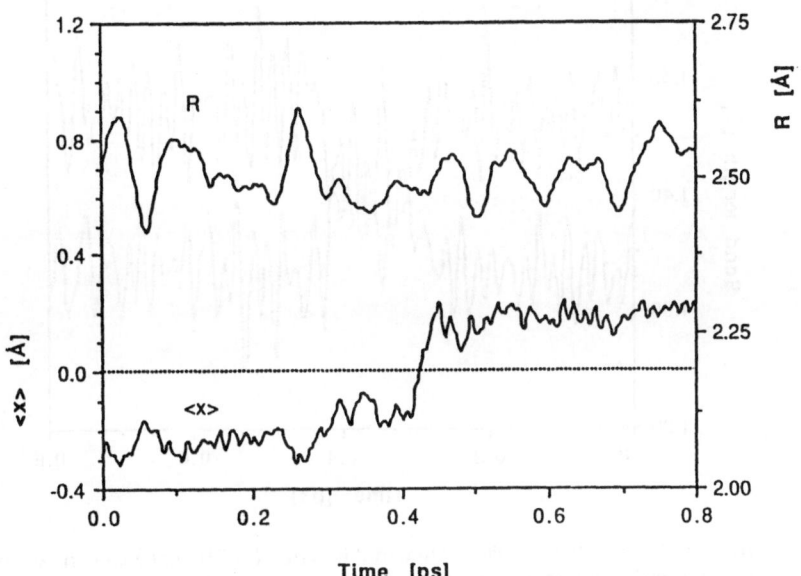

Time [ps]

Fig. 12. The QSMD simulation results for malonaldehyde. The sharp change in the mean proton position shows the tunneling transition from E_1 to E_2.

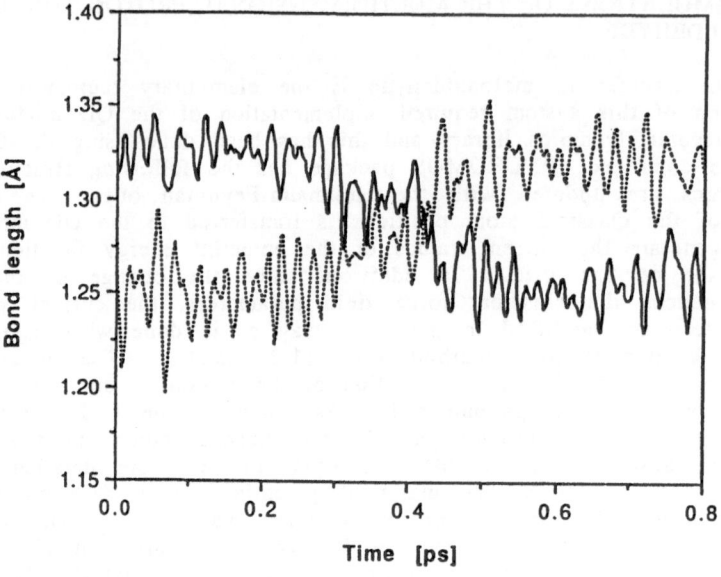

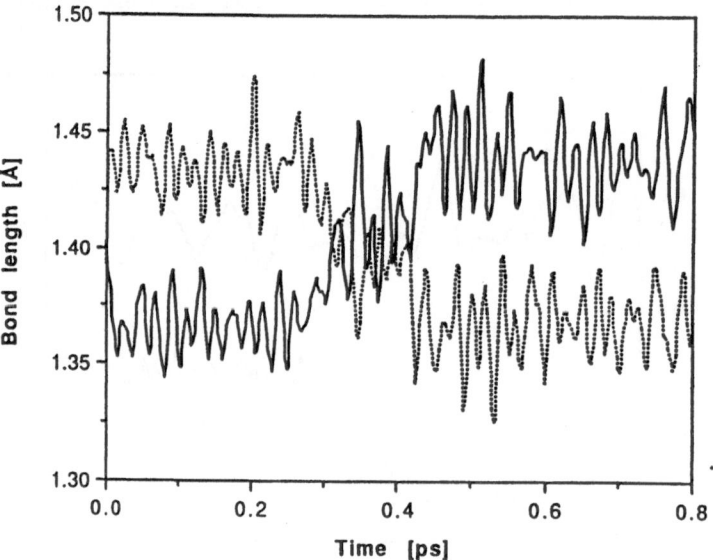

Fig. 13. The QSMD simulation results for malonaldehyde. (a) The solid line shows the C-O bond length of the left hand side hydroxyl group, and the dotted line shows the C=C bond length of the right hand site keto group in E_1 (see Fig. 2). (b) The solid line shows the left hand side C=C bond length, and the dotted line shows the right hand side C-O bond length in E_2.

the original MP2/6-31G** potential energy function representation, which should give an upper limit of the tunneling frequency. Simulations with the rescaled barrier are in progress.

The Gaussian proton wave packet was localized in the global energy minimum and optimized. The structure of malonaldehyde was optimized using the QSMD model with coupling to the thermal bath at T=0 and with $\gamma = 10$ ps^{-1} (cold, viscous isotropic medium). Then the kinetic energy was provided to the system in seven short, 0.2 ps QSMD runs (T=300 K, $\gamma= 10$ ps^{-1}). At this stage the system was thermalized. Next seven 0.2ps runs with the decreased γ up to 0.1 ps^{-1} moved the system into a vapour-like phase.

The 1 ps real time QSMD simulation followed this thermalization and equilibration procedure. During the QSMD simulation the proton tunnelled from the left to the right hand well after 0.4 ps. Fig.12 shows how the mean proton position changes in time. The tunneling transfer process itself is rapid and it takes approximately 0.02 ps to hop from the left to the right well. The microwave and far infrared spectroscopy data [59,60] yield a ~21 cm^{-1} tunneling splitting which gives the estimate of the tunneling period of 1.6 ps. Carrington and Miller [56] calculated the tunneling splitting of 60 cm^{-1} for the barrier of 4.3 kcal/mol, which is very similar to our present data. Shida et al. [57] based on the MCPF energy barrier (6.3 kcal/mol), got a tunneling splitting which was ~50% smaller than the experimental one. This comparison suggests that the QSMD model provides a pretty good estimate of the tunneling frequency.

The proton hopping is highly coupled to the geometry changes of the molecule. Fig. 13a and 13b show the C-O and C-C bond length changes, respectively. The left hand side hydroxyl group releases the proton and becomes the keto one and the right hand side keto group becomes the hydroxy one. At the same time the C-C and C=C bonds switch their positions.

The present calculations provide some dynamical insight to elementary processes where the quantum proton hopping is correlated with geometrical changes of the molecular system, which is quite common for biological systems. Tautomerism of nucleic acid bases or changes in the geometry of the imidazole ring during the protonation/deprotonation events in enzyme active sites are typical examples.

4. Ab Initio Study of the Potential Energy Surface for the Proton Transfer in Phospholipase A$_2$. Comments on Serine Proteases.

Phospholipases A$_2$ are calcium binding enzymes which catalyze the hydrolysis of the 2-acyl bond of phospholipids and perform a variety of biological functions. High-resolution crystal structures of pancreatic and snake venom phospholipases, crystallized amongst others with phosphonate phospholipid analogs, provided information about transition state geometry and mechanism of esterolysis [44].

The active site region of phospholipase A$_2$ contains amongst others histidine, aspartic acid, calcium and several water molecules [43,44,49]. Fig. 3 presents a simple model for the active site of this enzyme. The first step of the

318

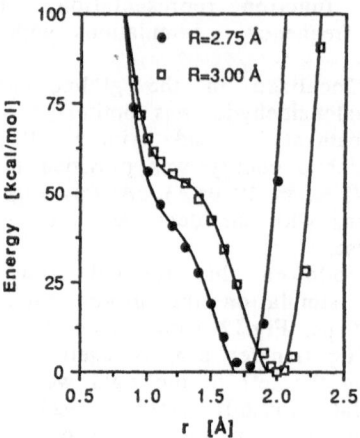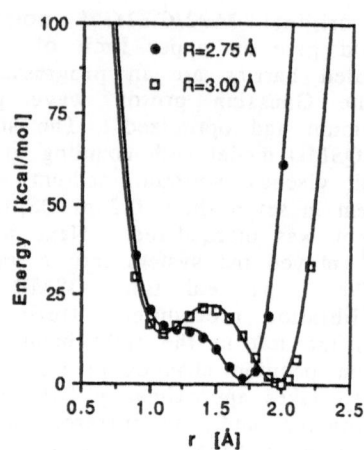

Fig.14 The MP2/6-31G* energy profiles for the proton transfer in the model active site of phospholipase A2 (comp. Fig.3):
a) in absence of the formal charge which models the Ca++,
b) with the formal charge (+2).

enzymatic action is the proton dissociation from water. In this way the hydroxyl ion (OH⁻) is generated, which acts as the nucleophile attacking the carbonyl carbon atom of the substrate. This results in tetragonalization of this carbon, and esterolysis of the chain C-O bond by another water molecule (for details see [38,44]). The first step is crucial for the whole reaction, and it should be possible to describe it using the theoretical models and the software tools developed during the former studies of the model proton bound systems.

The potential energy function for the proton transfer process between the water molecule and the imidazole ring was studied at the MP2/6-31G* level. The carboxyl group of the aspartic acid was modeled with the formal GROMOS charges (-0.635, +0.270, -0.635). Calculations were performed without and with the formal charge of the calcium cation (+2). The isolated elements of the system, the imidazole ring as well as the water molecule in their protonated and deprotonated forms, were optimized at the same MP2/6-31G* level.

Fig.14a shows the energy profiles in the absence of the formal calcium charge. The energy profiles do not contain any local energy minimum in the vicinity of the imidazole ring. The water molecule (more precisely the hydroxyl ion) is the better proton acceptor then imidazole. However, if the formal charge is located in the place of calcium, the situation changes markedly. A metastable state corresponding to the protonated form of the imidazole ring appears (Fig.14b), and the energy profile is similar to what we observed in the case of the proton bound water-ammonia dimer. Based on our experience with the model systems, we can think of the first enzyme reaction step as the quantum proton hopping process, with the water molecule as the donor, the imidazole ring as the acceptor and calcium as the source of the driving electrostatic field. The metastable state can catch the proton which, in turn, can result in the further OH⁻ nucleophilic substitution. Since the shape and the height of the barrier depends on R and on the electrostatic field, E, generated by the cation and other classical atoms, these two quantities may play the role of the control parameters of the system.

The local electrostatic field in enzyme active site regions is widely recognized in rhe literature as a driving force of enzymatic reactions. Much less is known about structural fluctuations and in particular squeezing/stretching deformations of the proton transfer regions which influences the barrier.

The SCF/6-31G *ab initio* results for the active site of a serine protease [67] show that the geometrical squeezing effects observed in the classical MD simulations can stabilize the metastable state with respect to the global energy minimum by about 15 kcal/mol. This is the same order as the polarization energy contribution induced by the presence of aspartic acid interacting with imidazole. The theoretical models and the software developed in the present studies can be used to provide more definite answers regarding these and other questions. In order to undertake such projects, the quantum-classical macromolecular dynamics requires, in addition to what was presented until now, description of the long-range interactions between remote enzyme or solvent classical atoms, and the quantum site region.

5. Interactions of Classical Remote Atoms with the Active Site Region in Enzymes.

In order to describe the dynamics of enzymes using the QCMD/QSMD models, the interaction of the remote atoms and solvent molecules with the quantum site region has to be incorporated in the multidimensional potential energy function. Several models have been developed until now, and most of them show that long range electrostatic forces dominate the total interaction energy. The forces are mostly generated by partial atomic charges. There are, however, no obvious indications how this coupling could be done when the quantum particles in the macromolecular structure are described by the time dependent Schroedinger equation.

One promising model was proposed several years ago by Zundel et al. [70] for the $[H_2O\text{-}H\text{-}OH_2]^+$ molecular ion located in an external electrostatic field. $H_5O_2^+$ has one global energy minimum at the symmetry center of the molecule, and is characterized by strong proton hyperpolarizability. Below we present an extension of this model, which can be used in the QCMD/QSMD approach when studying enzyme reactions.

Let V^{qc}_r denote the quantum-classical interaction potential function of the remote atoms with the quantum region. Systems with delocalized protons can be described by the dipole term

$$V^{qc}_r = -\mu_R(r, \{R_k\}) \cdot E_R(\{R_\alpha\}) \tag{19}$$

where $E_R(\{R_a\})$ is the electrostatic field generated by the remote atoms localized in the positions $\{R_a\}$, and the $\mu_R(r,\{R_k\})$ is the dipole moment of the quantum region (in the case of phospholipase A_2 the complex imidazole-water), which depends on the positions of the classical atoms $\{R_k\}$ and the proton, r. R is the position vector of a point between the proton donor and the proton acceptor (e.g. $(R_d + R_a)/2$).

In macromolecules, the electrostatic field is generated mostly by the atomic formal charges, q_a, and is equal to:

$$E_R(\{R_\alpha\}) = \sum_\alpha \frac{q_\alpha (R-R_\alpha)}{|R-R_\alpha|^3} \qquad (20)$$

The average of the potential interaction energy over the proton wave function is

$$\langle V_I^{qc}(r,\{R_k\},\{R_\alpha\})\rangle = -\sum_\alpha \frac{q_\alpha (R-R_\alpha)}{|R-R_\alpha|^3} \cdot \langle\mu(r,\{R_k\})\rangle \qquad (21)$$

which can be approximated, as

$$\cong -\sum_\alpha \frac{q_\alpha (R-R_\alpha)}{|R-R_\alpha|^3} \cdot \mu(\langle r\rangle,\{R_k\}) \qquad (22)$$

The dipole moment, as the function of the proton position and the molecular geometry of the quantum site region is calculated routinely using the *ab initio* methods. The practical difference between the equations (21) and (22) is that in the first case the dipole moment should be averaged on the grid, and in the second one its value is calculated as the function of the proton position and the molecular geometry. In both cases, the dipole moment function should be parametrized based on the quantum mechanical calculations prior to the QCMD/QSMD simulations.

Based on the 6-31G* *ab initio* calculations for the proton bound dimers we can say that the approximation given by eq. 22 fairly exact. It reproduces the *ab initio* energy difference with and without partial charges surrounding the systems with accuracy of 10%. The model for phospholipase is under development, and preliminary results also support validity of this approximation. It should be noted, that calcium ion can be treated in the same way as the other remote atoms with formal charges. This implies also, that the energy difference between the energy profiles presented in Fig.14a and Fig.14b can be approximated with the dipole model described above.

Atomic electron polarizability can in principle be incorporated in this model too. This results in a three body polarization energy contribution between the classical atoms, as well as the modification of the effective electrostatic field. There is, however, an additional complication due to the coupling between the classical and quantum regions which, as shown by Timoneda and Hynes [68], results in a nonlinear term in the Hamiltonian. This term depends on the magnitude of the electrostatic field in the quantum region. This problem can be the subject of future studies.

6. Comments on Rate Calculations for Proton Transfer Processes.

The activity of enzymes is measured experimentally using well established spectroscopic techniques. When the proton transfer process in the active site region is the rate limiting step of an enzyme reaction, it should be possible to compare directly the experimental and theoretical rate constants. The developments of theoretical methods for calculating quantum rates of proton transfer reactions are still limited to model systems with a few degrees of freedom [9,46,47,71]. Following the same idea of combined quantum-classical dynamis developed here, we are in progress of developing an effective method for calculating quantum thermal rate constants for such processes. This method is based on a flux-flux correlation function formalism which proposed ealier by Wahnstrom, Carmeli and Metiu [72]. Preliminary results for a model proton transfer reaction in aqueous solution are very encouraging [73]. Further testing of the new method and its applications to more realistic systems such as those considered here are now in progress.

7. Conclusions.

Combined QCMD/QSMD quantum-classical models were used to describe describe proton tunneling effects and classical/stochastic motions of the heavy atoms in $[H_3N-H-NH_3]^+$, $[H_3N-H-OH_2]^+$, and malonaldehyde.

These algorithms can be applied to other molecular systems, and in particular to macromolecules with biological activity, like enzymes, e.g. phospholipases, proteases, anhydrases, dehydrogenases, and lysosyme. The most important limitation is the development of the potential energy functions $V=V(r,(\{R_k\},\{R_a\}))$, which should be fitted prior to the basic simulations, using if possible, sufficiently reliable *ab initio* methods. This can be done for those molecular fragments which are involved in the quantum particle transfer processes. *Ab initio* results were presented for phospholipase A_2.

Since the barriers for the proton transfer in enzymes depend on the donor-acceptor distance and local electric fields, the dynamics of the quantum particle(s) depend also on these parameters. These parameters are controlled by the classical surroundings of the enzyme active site. It would be of interest to study these mechanisms as well as the feedback reaction of the quantum particle(s) on the classical atoms (Hellmann-Feynman forces), in order to find out a complete description of the enzymatic events.

The QCMD/QSMD models can be easily adapted for electron transfer processes in biomolecular systems, too, assuming that electron - atom/ion pseudo-potentials in the protein are known (see [36,40]).

It should be noted, however, that the precise determination of the proton wave function is required in order to conserve the energy of the system. If this is not satisfied the quantum subsystem behaves like a heat source or sink, gradually adding or draining energy from the classical part. The similar effects were described in QD electronic-structure calculations, see [69]. Systematic studies of the QCMD/QSMD algorithms and possible improvements of the models are in progress.

322

Acknowledgment

The authors are grateful to Dr. D. Kouri and Dr. J. Tanner at University of Houston, and Dr. J. S. Kwiatkowski at Nicolaus Copernicus Univerity in Torun (Poland), for very valuable discussions and reviewing the manuscript prior to publication. This work has been supported in part by the National Science Foundation, the Robert A. Welch Foundation, the Texas Advanced Research Program National Center for Supercomputing and the Ministry of National Education in Poland (grant MEN 106/90). T.N.T. is the recipent of a National Science Foundation Postdoctoral Fellowship. J.A.M. is the recipent of the G. H. Hitchings Award from the Burroughs Wellcome Fund.

Appendix 1. **Potential Surface for Malonaldehyde**

The potential energy surface for proton transfer depends on quantum, r, and classical degrees of freedom $\{R_k\}$. Based on our former studies for the proton bound dimers, [34,45], it can be expressed as:

$$V^{qc}(r,\{R_k\}) \approx V_1(x,y,\{R_k\}) + V_2(z,\{R_k\})$$ (1A)

where z is the direction perpendicular to the plane of the molecule. Since V_2 is approximately harmonic in the z direction, and depends weakly on $\{R_k\}$, the proton dynamics can be treated as 2 dimensional in the (x,y) plane. The most important classical degrees of freedom are R and Dl (see Fig.2). R is responsible for the squeezing/stretching deformations of the hydrogen bond, and Δl switches the structure between the E_1 and the E_2 forms. Other classical degrees of freedom are coupled indirectly to the quantum degrees *via* the V^c potential energy term (see the text).

The potential energy in the x direction can be described as a bistable, double well potential. In the transition state (TS) the potential is symmetric and has one or two wells, which depends on R. In the equilibrium states (E_1 and E_2) the potential has two asymmetric wells.

Following the procedure elaborated for the proton-bound dimers, [34], the MP2/6-31G* *ab initio* data were fitted with polynomials of the 4th order with respect to x and R. In the TS state the odd coefficients were set to zero due to the symmetry of the potential.

In the y direction a harmonic approximation was assumed. The position of potential energy minimum y_0 was also fitted with a polynomial of the 4th order with respect to x, both for the E and the TS states.

The results for the canonical E and TS forms, were then interpolated with a sine type function, which satisfies continuity requirements for the potential energy function and for its derivative. Finally, the potential energy function was normalized (F_0 term) to reproduce the *ab initio* data at the bottom of the potential. The formula is:

$$V^{qc}(x,y;\{R\}) = Ax^4+Bx^3+Cx^2+Dx+E + F(y-y_0)^2-F_0 \qquad (2A)$$

where:

$A = a_0+a_1R+a_2R^2+a_3R^3+a_4R^4$

$B = b_0\sin(\dfrac{\pi}{2}\dfrac{\Delta l}{\Delta l_0})$

$C = c_0+c_1R+c_2R^2+c_3R^3+c_4R^4$

$D = d_0\sin(\dfrac{\pi}{2}\dfrac{\Delta l}{\Delta l_0})$

$E = e_0+e_1R+e_2R^2+e_3R^3+e_4R^4$

$y_0 = u_0+u_1x+u_2x^2+u_3x^3+u_4x^4y_0$

$F_0=F(0.3 -y_0)^2$

The parameters are shown below:

index	0	1	2	3	4
a	8.0568×10^5	-1.2234×10^6	6.9915×10^5	-1.7783×10^5	1.6965×10^4
b	28.845				
c	3.9054×10^4	-6.0228×10^4	3.5591×10^4	-9537.4	971.87
d	-21.327				
e	5629.6	-7943.0	4195.6	-988.90	88.667
F	145.0				
u_{TS}	0.21299	0.0	2.1408	0.0	-1.9111
u_E	0.24037	0.032401	0.42739	-0.29880	2.3133
Δl	0.0803				

References:

[1] J.A.McCammon and S.Harvey, "Dynamics of Proteins and Nucleic Acids",
     Cambridge University Press, Cambridge, 1989
[2] Ch.L.Brooks II, M.Karplus and B.M.Pettit, in Adv. in Chem.Phys. LXXI,
     JohnWiley & Sons, pp. 1-259, 1988
[3] D.N.Silverman, S.Lindskog. Acc.Chem.Res., 21, 30-36 (1988)

[4]   D.Hadzi, J.Molec.Structure, **177**, 1-21(1988)
[5]   M.D.Harmony in Chem.Soc.Rev., **1**, pp. 211-228, 1972
[6]   S.J.Weiner, U.Ch.Singh and P.A.Kollman, J.Am.Chem.Soc., **107**, 2219-2229(1985)
[7]   A.Warshel, Proc.Natl.Acad.Sci., USA, **81**, 444- (1984)
[8]   A.Warshel and S.Russell, J.Am.Chem.Soc., **108**, 6569-6579(1986)
[9]   D.C.Borgis, S.Lee and T.Hynes, Chem.Phys.Letters, 162, 19-26(1989)
[10] S.Scheiner, in "Structure and Properties of Cell Membranes", ed.G.Banga, **3**, pp.1-17, 1985
[11] J.E.H.Koehler, W.Saenger and B.Lesyng, J.Comput.Chem., **8**, 1090-1098 (1987)
[12] B.Lesyng and W.Saenger, Biochim.Biophys.Acta, **678**, 478-413(1981)
[13] M.M.Szczesniak and S.Scheiner, Coll.Czechoslovak Chem. Commun., **53**, 2214-2229(1988)
[14] E.A.Hillenbrand and S.Scheiner, J.Am.Chem.Soc., **108**, 7178-7186(1986)
[15] E.A.Hillenbrand and S.Scheiner, J.Am.Chem.Soc., **107**, 7690-7696(1985)
[16] S.Scheiner and E.A.Hillenbrand, J.Phys.Chem., **89**, 3053-3060(1985)
[17] S.Ikuta, J.Chem.Phys., **87**, 1900-1901(1987)
[18] S.Scheiner, P.Redfern and E.A.Hillenbrand, Int.J.Quant.Chem., **XXIX**, 817-827(1986)
[19] M.M.Szczesniak and S.Scheiner, J.Chem.Phys., **84**, 6328-6335 (1986)
[20] J.S.Weiner, P.A.Kollman, D.T.Nguen and D.A.Case, J.Comput.Chem., **7**, 230-252 (1986)
[21] S.E.Koonin, "Computational Physics", The Benjamin/Cummings Publ., Inc., Menlo Park, 1986
[22] M.M.Szczesniak and S.Scheiner, J.Phys.Chem., **89**, 1835-1840(1985)
[23] J.Kucar and H.-D.Meyer, J.Chem.Phys., **90**, 5566-5577(1989)
[24] M.D.Feit, J.A.Fleck,Jr., and J.A.Steiger, J.Comput.Phys., **47**, 412-433(1982),
[25] H.Konwent, Phys.Letters A, **118**, 467-470(1986)
[26] S.-I.Sawada and H.Metiu, J.Chem.Phys., **84**, 6293-6311(1986)
[27] J.H.Bush and J.R. de la Vega, **99**, 2397-2406(1977)
[28] R.Janoschek, E.G.Weidemann, H.Pfeiffer and G.Zundel, J.Am.Chem.Soc., **94**,2387-2396(1972)
[29] M.-M.Zheng, "Double Well Potentials", PhD Dissertation, Univ.Libre de Bruxelles, Faculte des Sciences, 1984
[30] D.Kosloff and R.Kosloff, J.Comput.Phys., **52**, 35-53(1983)
[31] R.Kosloff, J.Phys.Chem., **92**, 2087-2100(1988)
[32] Y.Cha, Ch.J.Murray and J.P.Klinman, Science, **243**,1325-1330(1989)
[33] Ch.Zheng, C.F.Wong, J.A.McCammon and P.G.Wolynes, Chim. Scripta 29A, 171-179(1989)
[34] L.Jaroszewski, B.Lesyng, J.J.Tanner and J.A.McCammon, Chem.Phys. Letters, **175**, 282-288(1990)
[35] A.Selloni, P.Carnevali, R.Car and M.Parrinello, Phys.Rev.Letters, **59**, 823-826 (1987)
[36] A.Kuki and P.G.Wolynes, Science, **236**, 1647-1652(1987)
      D.Chandler and P.G.Wolynes, J.Chem.Phys., **74**, 4078-4089(1981)
[37] M.J.Field, P.A.Bash and M.Karplus, J.Comput.Chem., **11**, 700-733(1990)
[38] L.Streyer, *Biochemistry*, W.H. Freeman and Company, New York, 1989
[39] A.J.Sadlej, Coll.Czech.Chem.Commun., **53**,1995-2001 (1988)
[40] Ch.Zheng, J.Andrew McCammon and P.G.Wolynes, Proc.Natl.Acad. Sci. USA,

86, 6441-6444(1989),

R.A.Wheeler and J.Andrew McCammon, Chem.Phys.Letters, in press

[41]   B.Lesyng and E.F.Meyer, J.Computer-Aided Mol. Design, **1**, 211-217(1987)
       B.Lesyng and E.F.Meyer, Biopolymers, **30**, 773-780(1990)
       S.M.Swanson, T.Wesolowski, M.Geller and E.F.Meyer, J.Mol.Graphics, **7**,
       240-242(1989),

[42]   M.Geller, G.Carlson-Golab, B.Lesyng, S.M.Swanson and E.F.Meyer,
       Biopolymers, **30**, 781-796(1990)

[43]   H.M.Verheij, J.J.Volwerk, E.H.J.M.Jansen, W.C.Puyk, B.W.Dijkstra,
       J.Drenth and G.H.de Haas, Biochemistry, **19**, 743-750(1980)

[44]   D.L.Scott, S.P.White, Z.Otwinowski, W.Yuan, M.H.Gelb and P.B.Siegler,
       Science, **250**, 1541-1546(1990),
       S.P.White, D.L.Scott, Z.Otwinowski, M.H.Gelb and P.B.Sigler, Science, **250**,
       1560-1563(1990),
       D.L.Scott, Z.Otwinowski, M.H.Gelb and P.B.Siegler, **250**, 1563-1566(1990)

[45]   L.Jaroszewski, B.Lesyng and J.A.McCammon, submitted to J.Molec.Struct.

[46]   D.Chandler, J.Stat.Phys., **42**, 49-67(1986)

[47]   N.Makri and W.H.Miller, J.Chem.Phys., **91**, 4026-4036(1989)

[48]   D.N.Silverman and S.Lindskog, Acc.Chem.Res., **21**, 30-36(1988)

[49]   C.M.Dupureur, T.Deng, J.-G. Kwak, J.P. Noel and M.-D. Tsai,
       J.Am.Chem.Soc.,**112**,7074-7076(1990)

[50]   T.Lybrand, J.A.McCammon and G.Wipff, Proc.Natl.Acad.Sci. USA, **83**,
       833-835(1986)

[51] R.Wade. M.Mazor, J.A.McCammon and F.A.Quiocho, J.Amer.Chem.Soc., **112**,
       7057-7059 (1990)

[52]   R.K.Kalia, P.Vashishta, L.H.Yang, F.W.Dech and J.Rowlan,
       Int.J.Supercomputer Applications, **4**, 22-33(1990)

[53]   C.Leforestier, R.H.Bisseling, C.Cerjan, M.D.Feit, R.Friesner, A.Guldberg,
       A.Hammerich, G.Jolicard, W.Karrlein, H.-D.Meyer, N.Lipkin, O.Roncero
       and R.Kosloff, J.Comput.Phys., **94**, 59-80(1991)

[54] T.N.Truong, J.J.Tanner, P.Bala, J.A.McCammon, D.J.Kouri, B.Lesyng and
       D.K.Hoffman, J.Chem.Phys., in press

[55]   J.J.Tanner, J.Chem.Educ., **67**, 917-921(1990)

[56]   T. Carrington and W.H.Miller, J.Chem.Phys., **84**, 4364-4370(1986)

[57]   N.Shida, P.F.Barbara and J.E.Almlof, J.Chem.Phys., **91**, 4061-4072(1989)

[58]   E.Bosch, M.Moreno, J.M.Lluch and J.Bertran, J.Chem.Phys., **93**,
       5685-5692(1990)

[59]   S.L.Baughcum, Z.Smith, E.B.Wilson and R.W.Duerst, J.Am.Chem.Soc., **106**,
       2260-2265(1984)

[60]   S.L.Baughcum, R.W.Duerst, W.F.Rowe, Z.Smith and E.B.Wilson,
       J.Am.Chem.Soc., **103**, 6296-6303(1981)

[61]   D.W.Firth, P.F.Barbara and H.P.Trommsdorff, Chem.Phys., **136**,
       349-360(1989)

[62]   W.v.Gunsteren, GROMOS (Groningen Molecular Simulation Computer
       Program Package), Biomos, Laboratory of Physical Chemistry, University
       of Groningen

[63]   D.K.Hoffman, O.Sharafeddin, R.S.Judson and D.J.Kouri, J.Chem.Phys., **92**,
       4167-4177(1990)

[64]   R.S.Judson, D.B.McGarrah, O.A.Sharafeddin, D.J.Kouri and D.K.Hoffman,
       J.Chem.Phys., **94**, 3577-3585(1991)

[65]   D.K.Hoffman, N.Naresh, O.A.Sharafeddin and D.J.Kouri, J.Phys.Chem., in

326

press

[66] A.Tachibana, T.Inoue, M.Nagaoka and T.Yamabe, J.Phys.Chem, **93**, 220-225(1989)

[67] M.Geller, B.Lesyng and E.F.Meyer, submitted to Theoret.Chim.Acta

[68] H.J.Kim and J.T.Hynes, J.Chem.Phys., **93**, 5194-5210(1990)

[69] D.K.Remler and P.A.Madden, Mol.Phys., **70**, 921-966(1990)

[70] R.Janoschek, E.G.Weidemann, H.Pfeiffer and G.Zundel, J.Am.Chem.Soc., **94**, 2387-2396(1971)

[71] N. Rom, N. Moiseyev and R. Lefebvre, J. Chem. Phys. 95, 3562-3569(1991)

[72] G. Wahnstrom, B. Carmeli and H. Metiu, J. Chem. Phys. 88, 2478-2491(1988)

[73] T. N. Truong, J. J. Tanner, P. Bala, B. Lesyng and J. A. McCammon, in preparation

DISCUSSION

Comments of Prof. Miller:

I would like to point out a limitation of the "Ehrenfest model", i. e., treating some degrees of freedom quantum mechanically (via the time dependent Schrodinger equation) and others classically (with the classical forces given by the Ehrenfest force). Unphysical results are produced when the quantum wavefunction bifurcates and the forces depend on which quantum "state" (or arrangement) the system is in. The force seen by the classical degrees of freedom is given by the average over the quantum states, whereas it should be the conditioned average depending on which quantum state is considered.

Answers of Prof. Lesyng:

One could have some doubts if the mixed quantum-classical models work correctly in the case when the wavefunction is strongly delocalized between both energy minima, and the Hellmann-Feynman forces (also called the Ehrenfest forces) acting on the classical degrees of freedom are calculated as the average over the whole quantum probability distribution. I would like to point out, however, for the systems considered in this study and also for H_2^+ (with electron as the quantum particle) and $[D\mu T]^+$ (with muon as the quantum particle) in our unpublished results, we did not find any abnormal behavior in the dynamical results. In particular, there is a vast experimental data for $[D\mu T]^+$ due to its important in cold fusion, and our results agree reasonably well. What may cause some problem in future applications to enzymatic reactions is the small time step which required for accurate determinations of the wavefunction propagating through time and the Hellmann-Feynman forces. Our test simulations show that the time step in the mixed dynamics must be smaller than those normally used in the unmixed dynamics simulations. The mixed quantum-classical dynamics method is still under development, and we believe that it still requires more testing.